Diasporas in the New Media Age

Diasporas in the New Media Age

Identity, Politics, and Community

EDITED BY

ANDONI ALONSO AND
PEDRO J. OIARZABAL

UNIVERSITY OF NEVADA PRESS   RENO & LAS VEGAS

University of Nevada Press, Reno, Nevada 89557 USA
Copyright © 2010 by University of Nevada Press
All rights reserved
Manufactured in the United States of America
Library of Congress Cataloging-in-Publication Data
Diasporas in the new media age : identity, politics, and community / edited by Andoni Alonso and Pedro J. Oiarzabal.
p. cm.
Includes bibliographical references and index.
ISBN 978-0-87417-815-9 (pbk. : alk. paper)
1. Emigration and immigration—Social aspects.
2. Information technology—Social aspects.
3. Nationalism and technology.   I. Alonso, Andoni, 1966–   II. Oiarzabal, Pedro J.
JV6225.D53   2010
304.8—dc22   2009039101

The paper used in this book is a 100 percent recycled stock made from 30 percent post-consumer waste materials, certified by FSC, and meets the requirements of American National Standard for Information Sciences—Permanence of Paper for Printed Library Materials, ANSI/NISO z39.48-1992 (R2002). Binding materials were selected for strength and durability.

First Printing

19 18 17 16 15 14 13 12 11 10
5   4   3   2   1

Contents

List of Tables and Figures vii
Preface ix
The Immigrant Worlds' Digital Harbors: An Introduction 1
ANDONI ALONSO AND PEDRO J. OIARZABAL

PART I
INSIDE-OUT THE SCREEN:
DIASPORAS AT THE MARGINS OF CYBERSPACE

1 Interconnected Immigrants in the Information Society 19
   ADELA ROS

2 Migration, Information Technology, and International Policy 39
   JENNIFER M. BRINKERHOFF

3 Digital Diaspora: Definition and Models 49
   MICHEL S. LAGUERRE

4 An Activist Commons for People Without States
   by Cybergolem 65
   ANDONI ALONSO AND IÑAKI ARZOZ

PART II
DIALOGUES ACROSS CYBERSPACE

5 Oprah, 419, and DNA: Warning! Identity Under Construction 85
   TOLU ODUMOSU AND RON EGLASH

6 Cyber CVs: Online Conversations on Cape Verdean
   Diaspora Identities 110
   GINA SÁNCHEZ GIBAU

7 Nationalist Networks: The Eritrean Diaspora Online 122
   VICTORIA BERNAL

8 Keeping the Link: ICTs and Jamaican Migration 136
   HEATHER A. HORST

9	Maintaining Transnational Identity: A Content Analysis of Web Pages Constructed by Second-Generation Caribbeans	151
	DWAINE PLAZA	
10	Tidelike Diasporas in Brazil: From Slavery to Orkut	170
	JAVIER BUSTAMANTE	
11	Salvadoran Diaspora: Communication and Digital Divide	190
	JOSÉ LUIS BENÍTEZ	
12	3D Indian (Digital) Diasporas	209
	RADHIKA GAJJALA	
13	The Internet and New Chinese Migrants	225
	BRENDA CHAN	
14	The Migration of Chinese Professionals and the Development of the Chinese ICT Industry	242
	YU ZHOU	
15	"Cybernaut" Diaspora: Arab Diaspora in Germany	265
	KHALIL RINNAWI	
16	Net Nationalism: The Digitalization of the Uyghur Diaspora	291
	YITZHAK SHICHOR	
17	Migrate Like a Galician: The Graphic Identity of the Galician Diaspora on the Internet	317
	XABIER CID AND IOLANDA OGANDO	
18	Basque Diaspora Digital Nationalism: Designing "Banal" Identity	338
	PEDRO J. OIARZABAL	
	Contributors	351
	Index	357

Tables and Figures

TABLES

1.1.	Internet use by region of origin	28
1.2.	E-mail use by origin	29
1.3.	Webcam use by origin and age groups	29
1.4.	ICT penetration in countries of Spanish immigration	31
5.1.	Alexa.com traffic details for Nigeriaworld.com	96
5.2.	Alexa.com traffic details for Nairaland.com	100
5.3.	Alexa.com traffic details for Facebook.com	103
9.1.	Sample of schools by country with Caribbean student organizations	156
9.2.	Content categories, codes, and indicators	158
9.3.	Trends in Caribbean university student organization Web pages	159
9.4.	Caribbean student organization activities and images	162
9.5.	Caribbean student organization cultural mourning and transnational indicators	163
16.1.	Polls on the desirable Chinese attitude toward Xinjiang	303
16.2.	Accesses to English-Uyghur dictionaries	305
16.3.	Growth and penetration rates of Chinese Internet use	306
16.4.	Internet penetration rates, 2009: A comparative perspective	307

FIGURES

5.1.	Standardized percentages of unique names for girls born in Illinois, 1916–1989	92
14.1.	Growth of returnee-founded enterprises in Zhongguancun Science Park	249
15.1.	McArabism model	288
17.1.	Galician migrants, 1836–2010	319

Preface

The widespread use of computer-based technologies, such as the Internet and the Web, constitutes a new dimension in the study of emigrant and diasporic identities and cultures within the context of the current processes of globalization. The use of distance- and time-shrinking telecommunication technologies, such as electronic mail, cell phones, and the World Wide Web, by diaspora groups has attracted the attention of scholars from a variety of disciplines, such as anthropology, political science, philosophy, sociology, mass media, and computer-based communication.

Diasporas in the New Media Age builds on previous works while providing fresh insights into a wide range of dispersed populations and their interactions with information and communication technologies. This multifaceted book explores the richness of the intricate reality of digital diasporas as a true global phenomenon. These groups share similar concerns, anxieties, hopes, and desires, which to a certain extent go unnoticed by both their host societies and their countries of origin. The use of telecommunication technologies does more than enable diasporic communities to connect to their homelands while reinforcing their sense of collective identity, as clearly illustrated by the different studies provided here. Furthermore, *Diasporas in the New Media Age* offers theoretical discussion, implicit in each of the essays, in order to understand the diaspora phenomenon regarding the use and consumption of technology and media.

As evidenced by the findings of the contributors to this book, the impact of technology on international migration is unquestionable, as it facilitates the flow of people between regions, countries, and continents as well as the formation, growth, and maintenance of diaspora communities. In particular, the personal computer and access to the Internet have become quotidian resources among migrants who use them to develop, maintain, and re-create transnational social networks. Unquestionably, there have been major differences in the experience of migration before and since the creation of the Internet and digital communication media. As Thomas Faist argues in *The Volume and Dynamics of International Migration and Transnational Social Spaces,* "Information plays an important role for migration decision-making. It is one element that helps us to pay more attention to the bonds between movers and stayers, pioneer migrants, migration brokers, and followers.

Depending on the availability of info on transportation and opportunities for jobs and housing potential migrants can optimize their benefits. Such information may flow along various communication channels, such as mass media and friends who migrated before, but also pioneer migrants outside the inner circle of relatives and friends" (2000, 40).

This book has been conceived under the framework of the so-called Web 2.0, or social Internet network. We are entering into a new era when social networks, including diasporas, are reshaping the Internet itself. Rapid technological changes require continuous revision. In this sense, *Diasporas in the New Media Age* echoes these changes by presenting theoretical studies as well as concrete examples of the digital diaspora phenomenon.

The present volume of original essays brings together a solid and diverse selection of authors whose experience and knowledge have been key to accomplishing the goal of the book: to analyze the interrelation between diasporas and global communication media from an interdisciplinary perspective. *Diasporas in the New Media Age* aims to become a major scholarly contribution to the fields of new media and diaspora studies and, consequently, to help set the terms for future debate.

The main goals of this book are to provide a theoretical framework with which to understand the meaning of the changes introduced by technology in diasporas, as well as to understand and explain in practical terms how these changes are taking place in reality. *Diasporas in the New Media Age* presents a collection of eighteen theoretical, empirical, and rhetorical multidisciplinary essays. Twenty-one academics, writers, technologists, and cultural critics from diverse disciplines (e.g., anthropology, communications, geography, international relations, philosophy, political science, sociology, and technology) and fields (e.g., Internet, media, migration, diaspora, ethnic, cultural, and Web studies) provide fresh insights into African (Cape Verdean, Eritrean, and Nigerian), Arab and Muslim (Arab and Uyghur), Asian (Chinese and South Asian or Indian), Caribbean (Jamaican and Caribbean), European (Basque and Galician), and Latin American (Brazilian and Salvadoran) diasporas in relation to information and communication technologies.

The book is divided into two parts. Part 1, "Inside-Out the Screen: Diasporas at the Margins of Cyberspace," reflects on the Internet and other technologies within parameters of identity, politics, and culture. Four essays cover this theoretical approach. Adela Ros describes the rise of a new interconnected migration that uses the Internet, as well as other technologies such as cell phones. Connectivity is becoming a crucial issue for immigrants because it is essential not only for keeping in touch with families and friends who remained behind but also for entering the job market. Jennifer M. Brinkerhoff explores how the Internet can help create identities among diasporans and integrate them into the host society. Immigration can be a source of trouble for host lands if there is no policy to harmonize different cultures and identities. The Internet might alleviate some of those tensions, but only if there are sensitive policies that make easy and appropriate the use of telecommunications. Michel S. Laguerre clarifies to a certain extent some of the meanings of the concept "digital diaspora" by deconstructing it and relating it to other similar concepts in order to understand how diasporas interrelate with information and communication technologies. Finally, "Cybergolem" Andoni Alonso and Iñaki Arzoz explore the crucial role that digital diasporas (e.g., nations without states) can play in promoting digital activism and digital multiculturalism.

Part 2, "Dialogues Across Cyberspace," deals with the aforementioned case studies of diasporas from a theoretical or empirical perspective or both. The fourteen articles included go further than the analysis of ethnic and foreign cultures as a mere catalog of case studies. They also explore how diasporans redefine, construct, and represent notions of nation, homeland, diaspora, and identity, while considering political questions such as censorship in cyberspace and the promotion of digital transnationalism and long-distance or Internet nationalism. Fieldwork and case analysis try to show the wide scope of information and communication technologies in relation to diasporas and vice versa.

It was important in such studies to present as much diversity as possible within the limits of one book. Consequently, we have divided this second part into six generic diasporas: African (Tolu Odumosu and Ron Eglash, Gina Sánchez Gibau, and Victoria Bernal); Caribbean (Heather A. Horst and Dwaine Plaza); Latin American (Javier Bustamante and José Luis Benítez); Asian (Radhika Gajjala, Brenda Chan, and Yu Zhou); Arab and Muslim (Khalil Rinnawi and Yitzhak Shichor); and European (Xabier Cid and Iolanda Ogando as well as Pedro J. Oiarzabal). The essays combine specific issues as well as more generic approaches in order to understand the increasing interrelation between migration and information and communication technologies. Communication and new technological devices are quintessential for some of the changes in diasporic identity and its future. Technology is the key concept that relates different diasporas across the globe within a common framework.

The diverse migrants' experiences and their ways of dealing with information and communication technologies (some explored in detail) undoubtedly enrich everyone's sense of identity. Additionally, they pose new questions about ourselves; about the ways we communicate and re-create our national, ethnic, political, religious, or gender identities and our sense of territorial community and nation; and about the new manners of becoming socially, culturally, and politically motivated participants and activists in both off-line and online dimensions of reality. This book lays a common ground for future work on digital diasporas, and we hope that this first step will encourage new research and publications.

ACKNOWLEDGMENTS

Finally, *Diasporas in the New Media Age* would have not been possible without the unconditional support of those who took part in this collection of essays as well as the University of Nevada Oral History Program and the University of Extremadura, Spain. We feel especially indebted to William A. Douglass, Javier Echeverria, and Carl Mitcham, whose valued suggestions have enriched the final product, as well as the editors of the University of Nevada Press, Margaret Dalrymple, Charlotte Dihoff, and Sara Vélez Mallea.

Diasporas in the New Media Age

The Immigrant Worlds' Digital Harbors
An Introduction

ANDONI ALONSO AND PEDRO J. OIARZABAL

I couldn't take it anymore when we found ourselves alone in that small boardinghouse without love, or any friend to talk to, and release my pain.
—SANTIAGO IBARRA (1954), quoted in *Santiago Ibarra: Historia de un inmigrante vasco,* by Ángeles de Dios de Martina

Santiago Ibarra was born in Bilbao in the Basque province of Bizkaia in 1899, and at the early age of fifteen immigrated to Argentina with his seventeen-year-old brother. The chapter epigraph recounts his first day in Buenos Aires, according to a 1954 autobiography. It addresses his loneliness, nostalgia, and the overall impossibility of communicating with the loved ones who remained at home. In a sense, according to Grinberg and Grinberg, "migration requires a person to recreate the basic things he thought were already settled; he must recreate another work environment, establish affective relations with other people, reform a circle of friends, set up a new house that will not be an overnight tent but a home, and so on. These activities demand great physic effort, sacrifice, and acceptance of many changes in a short time. But to be able to carry them out gives one a sense of inner strength, an ability to dream, a capacity to build, a capacity for love" (1989, 176). One can only wonder how different it would have been for Santiago or any pre–information society immigrants, refugees or exiles, if they had had the possibility of connecting to the Internet and establishing not only instantaneous communication with parents, family members, and friends but also a digital network social world shared with others of common affinities.

At the outbreak of the Spanish Civil War in 1936, poet Antonio Machado defined Madrid as "the breakwater of all the Spains"—as the final destination of the incessant waves of refugees seeking protection as well as a solid barrier to repel attacks by Generalisimo Francisco Franco's fascist troops. In contrast to historical points of entry for immigrants, such as the emblematic Ellis Island in the United States, the Internet (along with satellite television and cellular phones and other mobile devices) is becoming the new harbor for contemporary immigrants. For many, the Internet is the first window or point of informational entry into their new destinations, prior to physical arrival, as well as a new interactive link back to their homelands. Even more, cyberspace—the communal space digitally created by the interconnection of millions of computerized machines and people—has become the virtual home for many diverse and dispersed communities across the globe. It is another space to reconnect with fellow natives around the world as well as with those remaining at home. It is a new space of hopes, desires, dreams, frustrations, and beginnings.

To appreciate the significance of diaspora creation and diaspora interaction with information and communication technologies, it is necessary to consider the spectrum of meanings of the term *diaspora*, the extent of the diaspora phenomenon, especially its political dimension, and the different ways that diasporas interact with technologies.

CONTESTED DIASPORAS

In general, the Greek term for *diaspora* (*diaspeirein*, "to sow" or "to scatter") refers to the dispersal of any population from its original land and its settlement in one or various territories. This definition originally had a positive connotation but was later redefined to include the collective expulsion of Jews from the Holy Land. The diaspora concept thus gained a negative meaning in relation to the destiny of Jewish people.

According to Tölölyan (1996), the defining elements of the Jewish diaspora conceptualization entailed the destruction of the homeland or the collective expulsion from it or both, a homeland-return movement, traumatic and coerced departure and collective trauma (victimization), a clear identity in the homeland and collective memory, and the maintenance of communications with the homeland and with coethnic members in host societies. These common elements were then applied to other realities, such as dispersed

African populations as the result of slavery and Armenians as the result of genocide in 1911, constituting along with the Greeks the so-called classical diasporas (Chaliand and Rageau 1995, 4). In other words, the "Jewish experience" became "the blue print for interpreting diaspora as a concept" (Reis 2004, 44).

Tölölyan (1996) argues that the Jewish paradigmatic definition of diaspora prevailed until the late 1960s. Since then, an emergent body of literature (see, for example, Cohen 1997a, 1997b; Laguerre 1998; Papastergiadis 1998; Braziel and Mannur 2003; Kokot, Tölölyan, and Alfonso 2004) has departed from the Jewish paradigm to explore contemporary diasporas formed after World War II. Tölölyan (1996) and Schnapper (1999) maintain that the term *diaspora* needs to retain the diverse meanings borrowed from the Jewish, Greek, Armenian, and Chinese diasporic experiences while advocating for the expansion of its classical semantic notion to enhance its effectiveness as an analytical concept and to accommodate it to new contemporary transnational realities.

On the one hand, new redefinitions of the diaspora concept have been created in order to accommodate almost all forms of dispersed minority populations scattered across the globe, including migrants, exiles, and refugees. For instance, Connor considers any "segment of people living outside the homeland" (1986) to be a diaspora, whereas Sheffer defines modern diasporas as "ethnic minority groups of migrant origins, residing and acting in host countries, but maintaining strong sentimental and material links with their countries of origin" (1986, 3). Cohen (1997b) broadens the traditional view of diasporas by introducing the following clear-cut typology of diasporas: victim (Jews, Armenians, Africans, Irish, and Palestinians), labor (Indians, Chinese, Sikhs, and Italians), trade (Venetian and Lebanese), imperial (ancient Greek, British, Spanish, and Dutch), and cultural (Caribbean).

Analysis of the diverse conceptual proposals elaborated by the aforementioned scholars provides the following basic comparative features of diasporas. There is a traumatic (forced or voluntary) dispersal to two or more locations and an active maintenance of a strong collective conscious ethnic identity, which might exist before leaving the land of origin or homeland. Tölölyan (1996, 14–15), Schnapper (1999, 249), and Butler (2001, 192) assert an extreme importance in maintaining collective transnational ties between dispersed coethnic communities, their homeland, and their host societies. These attachments and relationships are the most distinguishing aspects that differentiate diasporas from other dispersed minority ethnic groups. Vertovec refers to diaspora as a social form, as "the emphasis remains on an identity group characterized by their relationship-despite-dispersal," as well as a type of social consciousness (1997, 278). That is, there is a "particular kind of [multilocal] awareness said to be generated among contemporary transnational communities" (281). The final feature refers to a possible troubled relationship with the host society, creating dilemmas concerning dual loyalties to the host society and the homeland.

On the other hand, scholars such as Safran (1991, 1999) view this conceptualization enlargement process as a way of emptying the authentic meaning of diaspora and argue that the concept of diaspora is losing its analytical utility. Safran (1999, 278–80) states that diaspora status can be applied only to Jews and Armenians, denying Greek and Chinese dispersed community status as diasporas, whereas Sanjek (2003) does consider the African dispersed population to be a diaspora. In this regard, Schnapper raises the following concerns: "Has the almost indefinite extension of the concept emptied it of all intelligibility? . . . Concepts themselves must not be essentializing. The meaning of 'diaspora' can obviously change. The question is whether the change helps to clarify historic evolutions or whether its uncontrolled application ends up grouping together under a single term phenomena with different significance or meanings" (1999, 249).

Despite the many attempts to readjust the meaning of a concept dating back two millennia to contemporary's global realities, we do believe that the term *diaspora* still provides some useful insights into the understanding of transnational communities within a global context. Nevertheless, those attempts should go beyond determining the distinct elements that constitute a diaspora. In all cases, the term *diaspora* carries a sense of displacement. Sökefeld conclusively argues that "the multiplicity of different definitions of diaspora notwithstanding, all [are] based upon a decisive condition of space: the spatial separation of the diaspora community from 'its' homeland. Diaspora is about not being there" (2002, 111).

INTERNATIONAL MIGRATION, DIASPORA TECHNOLOGIES, AND TECHNOLOGICAL DIASPORAS

The aim of this book goes beyond the analysis of the reasons that underlie the phenomenon of contemporary international migration. In this regard, there are numerous historical, political, economical, sociological, psychological, and anthropological studies that attempt to understand the causes and effects of ongoing international flows of people (see, for example, Brettel and Hollifield 2000; Castles and Miller 2003; Faist 2000; Hirschman, Kasinitz, and DeWind 1999; Massey 1999; and Massey et al. 1993). Having said that, the great array of theories on migration focuses on three interlinked and non–mutually exclusive levels: macro, micro, and meso. The macro level deals with the structural political and economic conditions that "push" and "pull" individuals to migrate (e.g., neoclassic theories, dual labor-market theory, "new economics" of migration theory, and world system theories). The micro level deals with the decision making of individuals (e.g., rational-choice theory), whereas the meso level explores the social relations and networks that influence migrants (e.g., social network theory, institutional theory, and accumulative causation theory).

A growing body of literature uses a social network perspective for the analysis of contemporary international migration (see, for example, Boyd 1989; Brettel 2000; Kearney 1986; Portes 1995; and Vertovec and Cohen 1999). In this sense, migration is understood as a multidirectional, dynamic movement, that is, a networked building system facilitated to a great extent by information and communication technologies. Boyd states, "Studying networks, particularly those linked to family and households, permits understanding migration as a social product—not as the sole result of individual decisions made by individual actors, not as the sole result of economic or political parameters, but rather as an outcome of all these factors in interaction" (1989, 642). Tilly (1990) strongly asserts that it is not people who migrate but networks. Within the frame of this book, our particular interest lies in those theories of migration from a meso-level approach that attempt to explore interpersonal decision-making processes and migrant networks. This approach seeks the causes of such migratory movements and particularly their persistence over space and time, while highlighting the increasing role of information and communication technologies. These technologies allow migrants to create and maintain social migration networks in the context of so-called information and knowledge societies (Castells 1996; Cohen 1997a, 1997b).

A number of scholars (e.g., Adams and Ghose 2003; Anderson 1997; Dentice-Clark 2001; Hiller and Franz 2004; Ignacio 2002; Lal 1999; Mills 2002; Parham 2004; Rai 1995; Stubbs 1999; and L. Wong 2003) have focused on how emigrant communities (e.g., Arab, Chinese, Croatian, Indian, and Filipino) utilize online and mobile technologies to communicate, interact, maintain their identity, and enhance political mobilization while assessing their impact and implications on diasporic emigrants' daily lives. To a certain extent, they portray the Internet as an antidote for the assumed disjuncture or dislocation resulting from spatial and temporal distance between diasporas and their homelands. In addition, the study of virtual ethnicity and race (e.g., Diamandaki 2003; Everett 2009; Kolko, Nakamura, and Rodman 2000; and Nakamura 2002, 2007), digital diasporas, and online communities is becoming a substantial body of theoretical consideration and empirical research.

Despite the impressive number of works published on diasporas, which indicates the growing interest in this field of study, a few recent book-length works have begun to address the use and consumption of communication media (e.g., film, radio, television, video) and the Internet by diasporas. Among these collections of essays are Cunningham and Sinclair (2000), which focuses on the experiences of the Vietnamese, Fiji Indian, Thai, and Chinese communities in Australia; Allievi and Nielsen (2003), which focuses on different Muslim diasporic communities across Europe; Karim (2003), the most comprehensive collection to date of the complexities of emigrant diasporas' use of media, which highlights a few case studies (e.g., Macedonian and Rhodesian); S.-L. Wong and Lee (2003), which studies some Asian communities in the United States; and Landzelius (2006), which explores the interrelation of indigenous communities (e.g., Tongan and Assyrian) and communication technologies.

The creation and development of informal and formal transnational migrant networks among individuals, groups, and organizations from the country of origin and the country of settlement constitute webs of exchange of information and transfers of knowledge in the physical world as well as in the digital world. These networks lead to chain migration, which, in turn, helps to perpetuate migration flows between specific sending and receiving areas and among consecutive generations of immigrants (see Glick-Schiller, Basch, and Szanton 1992; and Vertovec and Cohen 1999).

The United Nations (2006) estimates that in 2005 worldwide international migration involved approximately 191 million people. Contrasted with a world population of more than 6 billion, international migration is not a large proportion of the total and in fact could be considered marginal. Moreover, only a few countries, such as the United States, the United Kingdom, France, and Spain, among others, receive any significant share of international immigrants.

However, over the past few decades immigration has doubled, and socioeconomic tensions and political conflict have grown at local, national, and international levels, forcing many to leave their home countries in search of new opportunities. Immigration is a phenomenon that spans time, generations, and geographies; it has a history. According to Appadurai (1996), historical and political changes have reshaped our notion of immigration. As seen, the term *diaspora* conveys different meanings and includes historical phenomena such as globalization, translocalities, and the crisis of the traditional state. Thus, in many circumstances, immigration becomes a question of identity, a diasporic process. A diaspora transcends, though is distinct from, immigration and has a clear political connotation that is reshaped by economy, politics, and technology in the era of globalization.

Statistics show that only 20 percent of the world population uses the Internet, which leaves a large number of people "off-line" (Internet World Stats 2007). Moreover, countries with less Internet usage have larger emigration rates than wealthy countries, which, in turn, become the main destinations of the majority of immigrants. Not all immigrants have equal access to information and communication technologies, even in their new host countries, and consequently there is a potential danger for many to remain behind or become increasingly excluded from their host societies while also becoming detached from the ongoing changes taking place in their homeland.

Historically, there has been a close correlation between technology and migration. Technological advancement of communication and transportation systems and infrastructures has facilitated both population movements and the formation of diasporas. For instance, the use and knowledge of technologies are major forces in motivating scientists and skilled workers to leave their homelands, as can be seen in India, South Korea, China, and Russia. Additionally, the diffusion of awareness of better lifestyles and wealth, as spread by global media and the Internet and by immigrants already settled in "first world countries," is increasingly becoming a stimulator to migration.

Certain aspects of contemporary globalization, such as neoliberal capitalism or the development of so-called global cities and technological advances in information systems, telecommunications, and transportation, are also, according to all the evidence, accelerating diaspora formation, growth, and maintenance. This was also true in past eras. For example, the nineteenth and twentieth centuries witnessed the articulation of national media and systems of communication such as newspaper, radio, and television. For the past two centuries an info-sphere (information environment) has profoundly changed the way national identities are created and reproduced. The speed and the outreach of that info-sphere have exponentially expanded. National media contribute to the imagining of a nation as a shared territorial community of nationals, through the production of homogeneous discourses of identity and culture (Anderson 1991). In this sense, the existence of that info-sphere—first newspapers, later radio and cinema, and, finally, television—helped to articulate and reinforce much of the romantic nationalist discourses. Although there was bitter criticism of such influence, the info-sphere quickly became a very significant and necessary part of everyone's daily life, with both negative and positive effects.

The appearance of new devices such as satellite communications, the Internet, and cell phones has introduced substantial changes within this info-sphere. Now, not only does communication obey a vertical axis like television, radio, and the newspaper, but we are also witnessing an increasing implementation of horizontal communications, which include more active roles by recipients, transforming identity processes into something more complex and diverse. Small communities, isolated individuals, and marginalized groups can use a platform such as the Internet to easily raise their voices and increase their possibility of being heard. For example, cell phones are now used in the effort to change regimes, as happened in the Philippines, where President Joseph Estrada was peacefully overthrown in January 2001, or to alter national opinion, as in Spain after the March 11, 2004, al-Qaeda bombings, or to bypass government control over media and censorship, as happened during the so-called Green Revolution in Iran in June 2009. This mobile technology has become somehow a more affordable and easy-to-use commodity whose ramifications are yet to be fully appreciated.

In addition, the emergence of the so-called Web 2.0 points out that something profound is changing the way that we relate to the Web. New proposals like "cloud computing" open the possibility of a network where users and their knowledge reshape not only content but also economic forces, allowing for the possibility of new business. It is said that 80 percent of the content on the Internet will come from users, from social networks. If this is true, diasporas will play a central role in that process because more and more new generations of immigrants are already embedded in the telecommunication system.

This info-sphere constitutes a postnational or global media that transcends national boundaries, creating a deterritorialized space or cyberspace. The old idea of deterritorialized communities, bounded by common interests and not by space or time, is now real (Licklider and Taylor 1968). This idea echoes the "republic of letters" during the Enlightenment that described the exchange of private correspondence between Western philosophers and other influential intellectuals. Transnational media reach a borderless audience of nationals and nonnationals and disrupt that romantic notion of a single territory for each "race" and one national media for each national culture. Now different nationalisms are confronted with a globalized landscape. Thus, information and communication technologies that were once confined to producing national cultures no longer conform to these fixed territorial boundaries.

At the same time, there is a real political challenge in coping with the new media while making sense of their power to convince, persuade, and re-create the vision of countries and nationalities that might differ from what is posited by existing governments. Part of the effort to develop an information society addresses directly how identity and nation are expressed. For example, in the case of the Basque Country the Basque Autonomous Government in Spain pursues the articulation of a digitally networked nationalist ideology within the Basque diaspora (Oiarzabal 2006). The digital nation could adopt a different, or even contradictory, image to that of the one in the "real" world. That possibility deserves careful attention.

Diasporans map an atlas of identity that occupies multiple geographical locations, construct different ideological discourses, speak different languages and dialects, represent various degrees of assimilation into their countries of residence, and maintain various degrees of transnational connections among themselves and with the homeland. Nationalism becomes a multilayered or multifaceted set of discourses allowing for great diversity. Concepts such as nation, identity, and belonging take on new meanings. Diasporans re-create psychological or emotional communities that "inhibit" an interstitial space between the land of origin and the land of settlement.

In this regard, diasporans have historically utilized a variety of means of communication—from newspapers, newsletters, and radio and television programs to the Internet—as ways to overcome barriers of temporal, spatial, and psychological distance, which exist among diverse codiasporic nodes and their countries of origin. For example, the Internet as a post–geographically bounded global communication system has significantly provided the ability for dispersed groups such as diasporas to connect, maintain, create, and re-create social ties and networks with both their homeland and their codispersed communities. The Internet offers the ability for diasporas to exchange instant factual information regardless of geographical distance and time zones. Again time and space shift meanings; there are no constraints on synchronicity or locality. That is, the Internet offers the possibility to sustain and re-create diasporas as globally imagined communities.

Technology affects human movements in decisive ways as time and space "shrink." Thus, technological devices are continuously reshaping these concepts. In a sense, our own identity is also redefined by information and communication technologies as it is embedded and contextualized in our perception of time and space. Technologies allow us to re-create our own reality that even employs time that is long gone or space that is far distant, transforming both within an imaginary landscape. *Translocality* is a term closely related to information and communication technologies. Already in the mid-1960s, Joseph C. R. Licklider and Robert Taylor, both of whom worked in the Advanced Research Projects Agency of the Pentagon (headquarters of the U.S. Department of Defense in Virginia) and were pioneers in promoting the development of the Internet, had begun to conceive of the computer as a communication device more than a calculating machine. That is, Licklider and Taylor (1968) forecast computers as machines able to create communities beyond time and space. The ties among users would be a community of interests and affinities. In a similar vein, these authors referred to identity in an indirect way.

Expansion of new technologies has deepened the question of identity even further. From Haraway's *Cyborg Manifesto* (1985) to Hayles's *How We Became Posthuman* (1999), the question of human identity has produced thousands of scholarly works based on a different array of practical experiences, manifestos, testimonies, and reports floating around the Internet. In addition, Internet phenomena such as Second Life (a virtual-world game created in 2003) evidence how online identity (diasporic or otherwise) is a hot topic for some discourses in our present technological society.

"Virtual life," "virtual community," and "digital diasporas" are concepts that need to be handled with care, as they display an immense range of connotations. We briefly focus on two of those meanings, which we refer to as "weak" or "soft" and "strong" or "hard," depending on the involvement with the real world. Within the first meaning, immigration has been used as a metaphor to explain the passage of an individual from real life into a digital world—that is, new users who acquaint themselves with new technologies. Related terms are *digital nomadism, cyborg identity,* and *virtual community.* At times, questions such as involvement, responsibility, and identity are difficult to resolve in such online communities. Second Life is a good example of this weak, or soft, meaning.

The idea of real people using virtual technologies to interact with the real world relates to the strong, or hard, connotation of the aforementioned concepts. Online activists—people defending the Internet as a way to achieve fair globalization, nongovernmental organizations, independent journalists, and alternative agencies—are an example of such use of the Internet as a way to improve the *real* world. In this category digital diasporans bring to the Internet a sense of identity and community prior to modern technology. So technology either reinforces or transforms their previous meanings and attitudes.

This book attempts to explore "digital diasporas" and their interaction with technology as part of our common daily politics, not as an aloof element.

Thus, we define digital diasporas as the distinct online networks that diasporic people use to re-create identities, share opportunities, spread their culture, influence homeland and host-land policy, or create debate about common-interest issues by means of electronic devices. Digital diasporas differ from virtual communities and nations because in digital diasporas there are strong ties with real nations before creating or re-creating the digital community, thus differing in some ways from Licklider and Taylor's idea of a virtual community. On the Internet, all of us are "immigrants" who simultaneously share a common space called cyberspace. That is to say, cyberspace does not belong to any particular nation, state, or diasporic group. This is essential to the understanding of what we mean by digital diaspora.

There is very little doubt that there exists an increasing interest in the fields of diasporas and information and communication technologies, not only in the academic world but also in society, as reflected by the mass media. It is not a coincidence that *Time* magazine chose the "Internet User" as the "Person of 2006." It is widely common to say that the Internet has completely changed our lives (although the details of those changes are not so well known). If it true that the Internet has transformed our lives, then the Internet must influence how immigrants (or all people, for that matter) use the new media in relation to different social aspects, including the interrelation with their homelands, identity processes, and roots. Thus, there is a need to understand diaspora communities and their dynamic process of political, cultural, and financial online networking on national, transnational, and global scales. In the near future, the Internet could be a good reflection of what happens with diasporas, their aspirations, rights, and responsibilities.

REFERENCES

The reader should notice that ephemerality is an intrinsic nature of the World Wide Web, and despite the effort made to update the URLs of many of the Web sites mentioned throughout the book, some might have disappeared by the time of its publication.

Adams, Paul C., and Rina Ghose. 2003. India.com: The construction of a space between. *Progress in Human Geography* 27, no. 4: 414–37.
Allievi, Stefano, and Jørgen Nielsen, eds. 2003. *Muslim networks and transnational communities in and across Europe.* Leiden and Boston: Brill.

Anderson, Benedict. 1991. *Imagined communities: Reflections on the origin and spread of nationalism.* London: Verso.

Anderson, Jon W. 1997. Cybernauts of the Arab diaspora: Electronic mediation in transnational cultural identities. Couch-Stone symposium "Postmodern Culture, Global Capitalism, and Democratic Action." University of Maryland, April 10–12. Available at http://www.bsos.umd.edu/css97/papers/anderson.html.

Appadurai, Arjun. 1996. *Modernity at large: Cultural dimensions of globalization.* Minneapolis: University of Minnesota Press.

Boyd, Monica. 1989. Family and personal networks in international migration: Recent developments and new agendas. *International Migration Review* 23, no. 3: 638–70.

Braziel, Jana Evans, and Anita Mannur. 2003. *Theorizing diaspora: A reader.* Malden, Mass.: Blackwell.

Brettel, Caroline B. 2000. Theorizing migration in anthropology: The social construction of networks, identities, communities, and globalscapes. In *Migration theory: Talking across disciplines,* ed. Caroline B. Brettel and James F. Hollifield. London: Routledge.

Brettel, Caroline B., and James F. Hollifield, eds. 2000. *Migration theory: Talking across disciplines.* London: Routledge.

Butler, Kim D. 2001. Defining diaspora, refining discourse. *Diaspora* 10, no. 2 (Fall): 189–219.

Castells, Manuel. 1996. *The rise of the network society.* Malden, Mass.: Blackwell.

Castles, Stephen, and Mark J. Miller. 2003. *The age of migration: Int ernational population movements in the modern world.* 3d ed. New York: Guilford Press.

Chaliand, Gérard, and Jean-Pierre Rageau. 1995. *The Penguin atlas of diasporas.* New York: Penguin.

Cohen, Robin. 1997a. Diaspora, the nation-state, and globalization. In *Global history and migrations,* ed. Wang Gungwu. Boulder: Westview Press.

———. 1997b. *Global diasporas: An introduction.* Seattle: University of Washington Press.

Connor, Walker. 1986. The impact of homelands upon diasporas. In *Modern diasporas in international politics,* ed. Gabriel Sheffer. New York: St. Martin's Press.

Cunningham, Stuart, and John Sinclair, eds. 2000. *Floating lives: The media and Asian diasporas; Negotiating cultural identity through media.* Brisbane: University of Queensland.

de Martina, Ángeles de Dios. 2004. *Santiago Ibarra: Historia de un inmigrante vasco.* Vitoria-Gasteiz: Servicio Central de Publicaciones del Gobierno Vasco.

Dentice-Clark, Lucia. 2001. My home town is a URL in cyberspace: The Internet, Italian ethnic identities, and the European Union. *Cultural Survival Quarterly* 24 (January).

Diamandaki, Katerina. 2003. Virtual ethnicity and digital diasporas: Identity construction in cyberspace. *Global Media Journal* 1, no. 2 (Spring). Available at http://lass.calumet.purdue.edu/cca/gmj/SubmittedDocuments/archivedpapers/Spring 2003/diamondaki.htm.

Everett, Anna. 2009. *Digital diaspora: A race for cyberspace.* Albany: State University of New York Press.
Faist, Thomas. 2000. *The volume and dynamics of international migration and transnational social spaces.* New York: Oxford University Press.
Glick-Schiller, Nina, L. Basch, and C. Blanc Szanton. 1992. Transnationalism: A new analysis framework for understanding migration. *Annals of the New York Academy of Sciences* 645: 1–24.
Grinberg, Leon, and Rebecca Grinberg. 1989. *Psychoanalytic perspectives on migration and exile.* New Haven: Yale University Press.
Haraway, Donna. 1985. A cyborg manifesto: Science, technology, and socialist-feminism in the late twentieth century. In *Simians, cyborgs, and women: The reinvention of nature.* New York: Routledge.
Hayles, Kathelyn. 1999. *How we became posthuman: Virtual bodies in cybernetics, literature, and informatics.* Chicago: University of Chicago Press.
Hiller, Harry H., and Tara M. Franz. 2004. New ties, old ties, and lost ties: The use of the Internet in diaspora. *New Media & Society* 6, no. 6: 731–52.
Hirschman, Charles, Philip Kasinitz, and Josh DeWind, eds. 1999. *The handbook of international migration: The American experience.* New York: Russell Sage.
Ignacio, Emily Noelle. 2002. Filipino ka ba? Internet discussions in the Filipino community. In *Contemporary Asian American communities,* ed. Linda Trinh Võ and Rick Bonus. Philadelphia: Temple University Press.
Internet World Stats. 2007. World Internet users, December. Available at http://www.internetworldstats.com/stats.htm.
Karim, H. Karim, ed. 2003. *The Media of diaspora: Mapping the globe.* London: Routledge.
Kearney, Michael. 1986. From the invisible hand to visible feet: Anthropological studies of migration and development. *Annual Review of Anthropology* 15: 331–61.
Kokot, Waltraud, Khachig Tölölyan, and Carolin Alfonso, eds. 2004. *Diaspora, identity, and religion: New directions in theory and research.* New York: Routledge.
Kolko, Beth, Lisa Nakamura, and Gilbert B. Rodman, eds. 2000. *Race in cyberspace.* New York: Routledge.
Laguerre, Michel S. 1998. *Diasporic citizenship.* London: Macmillan.
Lal, Vinay. 1999. The politics of history on the Internet: Cyber-diasporic Hinduism and the North America Hindu diaspora. *Diaspora* 8, no. 2 (Fall): 137–72.
Landzelius, Kyra, ed. 2006. *Native on the Net: Indigenous and diasporic peoples in the virtual age.* London: Routledge.
Licklider, Joseph C. R., and Robert Taylor. 1968. The computer as a communication device. *Science and Technology: For the Men in Management,* no. 76 (April): 21–31.
Massey, Douglas S. 1999. Why does immigration occur? A theoretical synthesis. In *The handbook of international migration: The American experience,* ed. Charles Hirschman, Philip Kasinitz, and Josh DeWind. New York: Russell Sage.
Massey, Douglas S., Joaquin Arango, Graeme Hugo, Ali Kouaouci, Adela Pellegrino, and J. Edward Taylor. 1993. Theories of international migration: A

review and appraisal. *Population and Development Review* 19, no. 3 (September): 431–66.

Mills, Kurt. 2002. Cybernations: Identity, self-determination, democracy, and the "Internet effect" in the emerging information order. *Global Society* 16, no. 1: 69–87.

Nakamura, Lisa. 2002. *Cybertypes: Race, ethnicity, and identity on the Internet.* New York: Routledge.

———. 2007. *Digitizing race: Visual cultures on the Internet.* Minneapolis: University of Minnesota Press.

Oiarzabal, Pedro J. 2006. The Basque diaspora Webscape: Online discourses of Basque diaspora identity, nationhood, and homeland. Ph.D. diss., University of Nevada, Reno.

Papastergiadis, Nikos. 1998. *Dialogues in the diasporas: Essays and conversations on cultural identity.* London: Rivers Oram Press.

Parham, Angel A. 2004. Diaspora, community, and communication: Internet use in transnational Haiti. *Global Networks* 4, no. 2: 199–217.

Portes, Alejandro. 1995. Economic sociology and the sociology of immigration: A conceptual overview. In *The economic sociology of immigration,* ed. Alejandro Portes. New York: Russell Sage.

Rai, Amit S. 1995. India on-line: Electronic bulletin boards and the construction of a diasporic Hindu identity. *Diaspora* 4, no. 1 (Spring): 31–58.

Reis, Michele. 2004. Theorizing diaspora: Perspectives on "classical" and "contemporary" diaspora. *International Migration* 42, no. 2: 41–60.

Safran, William. 1991. Diaspora in modern societies: Myths of homeland and return. *Diaspora* 1, no. 1 (Spring): 83–99.

———. 1999. Comparing diasporas: A review essay. *Diaspora* 8, no. 3 (Winter): 255–89.

Sanjek, Roger. 2003. Rethinking migration, ancient to future. *Global Networks* 3, no. 3: 315–36.

Schnapper, Dominique. 1999. From the nation-state to the transnational world: On the meaning and usefulness of diaspora as a concept. *Diaspora* 8, no. 3 (Winter): 225–55.

Sheffer, Gabriel. 1986. A new field of study: Modern diasporas in international politics. In *Modern diasporas in international politics,* ed. Gabriel Sheffer. London: Croom Helm.

Sökefeld, Martin. 2002. Alevism online: Re-imagining a community in virtual space. *Diaspora* 11, no. 1 (Spring): 85–123.

Stubbs, P. 1999. Virtual diaspora? Imaging Croatia on-line. *Sociological Research Online* 4, no. 2. Available at http://www.socresonline.org.uk/socresonline/4/2/stubbs.html.

Tilly, Charles. 1990. Transplanted networks. In *Immigration reconsidered: History, sociology, and politics,* ed. Virginia Yans-McLaughlin. New York: Oxford University Press.

Tölölyan, Khachig. 1996. Rethinking diaspora(s): Stateless power in the transnational moment. *Diaspora* 5, no. 1 (Spring): 3–36.

United Nations, Population Division of the Department of Economic and Social Affairs of the United Nations Secretariat. 2006. *Trends in total migrant stock: The 2005 revision.* Available at http://esa.un.org/migration/.

Vertovec, Steven. 1997. Three meanings of "diaspora," exemplified among South Asian religions. *Diaspora* 6, no. 3 (Winter): 277–300.

Vertovec, Steven, and Robin Cohen. 1999. Introduction to *Migration and transnationalism,* ed. Steven Vertovec and Robin Cohen. Aldershot: Edward Elgar.

Wong, Loong. 2003. Belonging and diaspora: The Chinese and the Internet. *First Monday* 8, no. 4 (April). Available at http://firstmonday.org/issues/issue8_4/wong/index.html.

Wong, Sau-Ling Cynthia, and Rachel C. Lee, eds. 2003. *Asian America.Net: Ethnicity, nationalism, and cyberspace.* London: Routledge.

PART I

INSIDE-OUT THE SCREEN

DIASPORAS AT THE MARGINS OF CYBERSPACE

1 Interconnected Immigrants in the Information Society

ADELA ROS

> *Owing to the communications and transportation revolution, today's international migrants are, more than ever before, a dynamic human link between cultures, economies and societies. Penny-a-minute phone cards keep migrants in close touch with family and friends at home, and just a few seconds are needed for the global financial system to transmit their earnings to remote corners of the developing world, where they buy food, clothing, shelter, pay for education or healthcare, and can relieve debt. The Internet and satellite technology allow a constant exchange of news and information between migrants and their home countries. Affordable airfares permit more frequent trips home, easing the way for a more fluid, back-and-forth pattern of mobility.*
>
> —KOFI ANNAN, report of the secretary-general of the United Nations, 2006

Fatima is a twenty-five-year-old woman who works as a journalist at a local television station in Barcelona. The first thing she does every day when she arrives home at eight o'clock is go to the computer and connect her Webcam and Skype. A few minutes later, her parents and grandmother get connected from Casablanca. Then, her sister appears on the screen from Romania. A few minutes later, her cousin in Belgium shows up. "Not only do we speak every day, we stay together," she says. Fatima is one of the almost one million immigrants who have arrived in the past decade in the autonomous region of Catalonia, in northeastern Spain. Many of them are using the world of information and communication technologies to meet their communication and information needs. The effects and implications of such technological adoption still remain unknown.

The example of Fatima may illustrate a new situation. Castles and Miller (2003) have shown that one of the outcomes of global interdependence and interconnection is the expansion of the effect of international migration. Now immigrants may be taking advantage of that same interconnection that generated their move and use it to overcome the main difficulties that they face.

One of the new elements differentiating the present migration domain from that of other periods in history comes from the possibilities that communication and information technologies offer everybody, including immigrants and their families, to keep in constant contact through communication and information. Migrating in the age of mobile technologies—multimodal communication from anywhere to anywhere, SMS interchange, fast e-mail, virtual communities, chats and forums, video conferences by telephone and Internet—introduces new dimensions. Today, immigrants have many more opportunities to live interconnected with situations and people in their country of origin. According to the consultancy firm Telegeography, "There is a high increase in the growth of telephone traffic between countries with strong migration connections; looking at the number of minutes of teletraffic between specific countries with strong migration connections in the years 1995 and 2001 data, they suggest a remarkable growth in traffic" (Vertovec 2004b, 10).

Migration is probably one of the oldest ways of interconnection between different, distant parts of the world (Held et al. 1999, 283). Migration has always constituted a "natural" flow of products, ideas, cultures, and languages. Millions of social, economic, and political interconnections have been produced after migration has linked two or more regions in the world. Once established in host societies, migrants' links with families and friends in the country of origin have always remained. In Held's terminology, the pattern of global interconnection in the domain of migration has always been high.

But the recognition of a long history of interconnections, even the affirmation that connection constitutes an intrinsic element of migration, does not alter the fact that the increase and changes in interconnection in the Information Society may be transforming the nature, meanings, and logic of immigration. Immigrant interconnection has now taken on a new dimension, with new physical spaces reserved for it and, what is even more decisive for the future, associated economic benefits. In the context of Catalonia, the presence of Internet cafés and telephone centers in immigrant neighborhoods in major cities and in medium towns and the hundreds of advertisements everywhere from private companies trying to make immigrants call and send money home are constant reminders that something new is happening. Now it is time to pass from an intuitive observation and use all these pieces of reality to start making something of them more important: interconnection can be used as the basis for a new look at immigration.

This chapter has two main purposes. First, the chapter attempts to go a step further in the analysis of interconnections in immigration. Although immigrants follow a general trend of increased interconnection within globalization, there is not much data on how patterns of communication and information develop different levels and forms of interconnection and the identification of the main factors that intervene. As patterns of communication have always existed in migrant contexts, the question is whether there is something new in the nature of interconnection in the Information Society and what the main consequences are. Second, this work examines a closer interrelationship between immigration and a whole set of elements that are transforming our way of life that have been characterized as the Information Society. High interconnection is only one of the traits of the Information Society paradigm. In the Information Society, immigrants are experiencing historical transformations, too.

The organization of this chapter reflects these goals. Each section corresponds to a different part. First, I will frame immigration within the Information Society context. Second, I will take a look at the interaction between immigrants and technologies of information and communication. Third, there is an analytical proposal to analyze interconnection patterns of communication and information. This work poses more questions rather than giving quick responses. Future answers will be the result of a long-term initiative taken by the Immigration and Information Society Research Programme.[1] Data shown in these pages are the result of ongoing research that employs various techniques such as in-depth interviews, statistical analysis, participant observation, and other more specific, innovative techniques to register the use of technological tools.[2]

The study of interconnection is nothing new in the immigration literature. But today it is becoming even more necessary than ever to track all that transnational studies have helped to recognize. That is, immigration implies a new relationship with both the host societies and those of origin simultaneously. But immigration also produces continuous contact with the local reality and needs. A deep interest in the relationship of the local to the global, or in the continuous processes of compression and decompression in which immigrants are immersed, lies behind the following pages. We need to ask how this constant movement is produced in day-to-day practice and how it is changing the social organization of immigrants. More particularly, nowadays, could people who are separated due to migratory movement be organizing their social lives and filling them with meaning, overcoming the factor of distance? And are we moving from a pattern of immigration to another of mobility as a new model that implies a lesser will to transform one's life since life could easily be developed within one's own newly built, technologically maintained networks? The consequences of a real recognition of interrelation from a local perspective are still unknown.

Nothing of what has been said up to this point seems to have been understood by policy makers yet. In Europe, for instance, the dominant model these days reminds us more of a scheme of integration as disconnection than of any possible advantage of having interconnected immigrants. The influence of international geopolitics does not help at all.

THE INFORMATION SOCIETY IN CONTEMPORARY IMMIGRATION

The emergence of information and communication technologies in immigrant communities in Western societies represents a new element in migration contexts that may be transforming different elements of the nature of migration. The introduction of information and communication technologies (ICTs) in the world of immigration could be incorporating new elements into the reality of immigration that, far from being complementary, could imply a transcendental move toward the emergence of new issues in the social organization of immigration.

Contemporary migration occurs in the Information Society, and it is intimately interwoven with it. However, Information Society and migration are two realities that have not yet been combined. Although the presence of technologies of communication in migration settings has already been pointed out as a new trend, a full integration of the Information Society paradigm in migration remains to be studied. First, the way in which politics and business have reacted responds quite well to the scheme of "projecting dreams and fears of the kind of society that will result" (Castells et al. 2007, 2). Mobile telephone companies shape a new ideal world for immigrants, where communication becomes "the closest thing to being together" and helps one avoid homesickness.[3] From politics, the incorporation of ICT into the immigrant's world is interpreted more as a difficulty than as an opportunity.

What does this look consist of? This approach involves considering contemporary migration as part of the reality of the Information Society, which implies a new model of social organization of migration made possible by the revolution of technologies. An approximation of immigration to the Information Society would allow new elements and new questions to arise. The Information Society is thus the necessary framework within which we will be able to find emerging patterns of immigrant incorporation into host societies.

The Information Society is the result of a technological paradigm that engenders an augmentation of human capacity of information processing and communication made possible by the revolutions in microelectronic-based information and communication technology such as computers and digital communications. Such a paradigm, which was shaped in the United States in the 1970s in interaction with the global economy and with world geopolitics, was organized around information technology; globalization—as a process of increasing interpenetration and interdependence of activities—has accelerated systems of interaction.

Technological tools play a central role in the organization of society and in the shaping of the opportunities and constraints, meanings and ways, of life. People and groups adapt technology to their needs and interests, producing transformations in the organization of social life and profoundly changing the structures of current society in such a way that "the rhythms of everyday life are being transformed in ways which, by any historical comparison, are remarkable" (Webster 2001, 1). Thus, information and communication technologies imply three new features: more capacity for processing information, more capacity for interaction of people, and more capacity for flexibility in continuous fields of presence. In recent years, analytical work has been carried out on the meaning that these three features have in different spheres of social organization. We propose immigration as a very appropriate site to empirically analyze how a technologically based paradigm is expressed in a specific reality and group. To look into migration using an informational perspective, that is, using the main traits, transformations, and challenges identified for the Information Society, makes for an interesting analytical exercise.

Information Processing

That migration networks distribute information among potential migrants and offer assistance to recently arrived immigrants is in itself nothing new. Networks (i.e., interconnected nodes) are the pattern of social life and of dominant functions and processes in the rise of the network society (Castells 1996). Although the relationship between migration and information is closely tied, the conditions under which networks of information and communication function in migrant contexts are new. Contemporary migration expresses a new model of mobility with high levels of information and communication networks, powered by informationalism as a new capacity for processing information. This implies a closer look at new informants and at new sources of information, multiple nodes and new actors, and the possible capacity that migrants have to make more informed decisions. The central actors are the creators, designers, and disseminators of information flows. These new actors result in more autonomy from power centers, since this autonomy requires multidirectionality and continuous flows of interactive information processing.

Revolutions in transportation and new technologies of information and communication are the main factors explaining current patterns of interconnectedness and the possibility to have wider access to information anywhere in the world. The impact of high interconnectedness on contemporary international migration could be offering new information and communication potentialities to migrants. Migrant networks can distribute information and coordinate efforts in a better way than ever before. They can connect with different nodes around the world and process information with an impact and velocity never seen before (Massey 1998; Fawcett 1989; Tilly 1990). Horst observes that increased communication in Jamaica "enabled through the presence of house phones and especially the ownership of mobile phones has led Jamaicans to more realistic expectations of the migration experience and opportunities associated with living abroad" (2006, 155).

The penetration of ICTs in migration contexts, in both the country of origin and the country of destination, directly benefits migrants, as they have wider access to information. As a Senegalese immigrant reported, "Information, we have it all." However, these new potentialities are associated with a lack of structured information, opening the door to asymmetric information, unreliable sources, false expectations, and more. Although there are new potentials for information transmission in migration contexts, there is still a lack of information among most migrants. The information they have is often insufficient and leads migrants into situations of misinformation or confusion, opening the way for frauds or intermediaries to intervene. A resident in Catalonia from Benin was very explicit in reference to this lack of useful information: "People see images, but they do not know what lies behind them." A woman from Algeria added that "there are many things in communication, telephone, the Internet. But this information never arrives; it doesn't cross the Mediterranean Sea."

Finally, at the level of migration policies, the new tools of the Information Society have not yet been used to solve the basic problem of a lack of information, both before and after the arrival into a new country, which is still prevalent in many migration situations. It does not seem, therefore, that contemporary migration in the information age has contributed to overcoming a situation of unequal distribution of knowledge.

More Interaction

High connectivity in migrant contexts shapes a new space for interaction and simultaneity, both at a distance and in the local context. First, new patterns of distant interaction and sociability based on new communication tools would be emerging in immigrant contexts and allow them to maintain ties to a distant community while supporting face-to-face ties closer to home. As new technologies give more opportunities of communication, existing family relationships may get stronger and much more coordinated (Castells et al. 2007). Second, more interaction may also imply more and new contacts in the host society.

Immigrants are using the possibilities created by mobile technology, working in diffused networks, in order to coordinate work across long or short distances. Immigrant workers are also mobile workers who keep in constant contact with their managers to get last-minute information while working anywhere or even when they are not working (Castells et al. 2007, 78), thus blurring the boundary between work and the private sphere. In addition, personal communication from thousands of miles away penetrates the formal boundaries of work. In Catalonia a train is an excellent place to observe how immigrants—who are regular users of public transportation—use mobile phones to apply for jobs, receive work offers, or simply let somebody know of a delay.

In addition, immigrants living with different levels of uncertainty keep their mobile phone as the only certain way to be contacted. Having a mobile phone is quickly perceived as a labor need and requirement. I have had the opportunity to meet many illegal immigrants. Even though they may live in poor conditions, they still own a mobile phone. They all understand perfectly that in the Information Society, the capacity of interacting means an open door for better opportunities.

Immigrants have realized that a new sociability means permanent, ubiquitous forms of connectivity—staying in touch anytime, from anywhere, and keeping multiple channels of communication open. And they are practicing it with their new friends and family members. The first results of experimental "diaries of technology" show that the mobile phone is basically used for linking up communication (Horst 2005, 2006) with family and peers in the proximity. From the same diaries, we will try to analyze any possible effects that the increase in interaction may have on shaping new community networks and what kind of new interactions with the host society—if any—are produced.

The dynamics of increased interaction mean immigrants also suffer from the same common problems of contemporary communicative practices: constant microadjustments, excess of control, and invasions of privacy. Fatima, the young women who uses Skype every day to stay in touch with her family abroad, complains about her father exercising a traditional role of a controlling father from Casablanca—he gets upset when his daughter arrives home late for their daily "e-meeting." She also recalls when the manager of the telephone center that she used to use one day dialed her phone number from memory. Many immigrants complain about the noise and lack of privacy conditions in the phone centers. These negative effects need to be further analyzed in the future.

Space of Flows

Although migrants respond to traditional—physical—patterns of mobility, contiguity may be defined in new terms. In the Information Society, the time-space context is transcended by information and communication technologies and blurred in individual practice in a new space of flows (Castells 1996). Such time-space compression, as a process of redefinition of time and space as constraints on the organization of human activity (Inda and Rosaldo 2002), needs to be applied to the reality of immigration. As the meaning of place changes, the meaning of living in a place, separated from the original one, is changing for immigrants as well. Now, more than ever before, it is possible to overcome the restrictions of space.

One of the most important consequences of these changes in the meaning of time and space in the domain of migration is the reconceptualization of distance. New, different approaches to the distance factor—a key element for migrant communities—are required. Indeed, as Diminescu (2007) has suggested, the whole category of "space," which is deeply embedded in the logic of migration analysis, requires redefinition. Migrating in a "space of flows" supposes a much more continuous reality where the meanings of "origin" and "destination" are blurred. Diminescu has pointed out that the definition of a migrant based on physical criteria and in varying degrees of difficulty is being challenged. Thus, she proposes to use the concept of "the connected migrant."

How can distance be understood vis-à-vis migrants' continuous presence through phone calls, SMS, and videoconferences? Can people and communities living apart now build an elastic, flexible terrain of time and place? Are time and space being transcended in social practice through the capacity of constant "from anywhere to anywhere" contact?

In interviews with immigrants, most of them agree that "distance has been shortened" compared to the past and that technologies of communication produce a positive effect of closeness. The feeling of proximity is even greater when image is incorporated into communication—through VOID technologies and mobile teleconferencing. Many immigrants describe an unbelievable feeling of "being there" or "being with them" when talking with their families in the countries of origin.

The capacity to overcome time, space, and even sociocultural constraints provided to immigrants by new ways of interaction could be expanding individualism and increasing self-consciousness among migrants. Now that they are free to relate to the world, it would be a way of enhancing their autonomy. In the case of immigrants, they may be able to set up their own connections, thus bypassing traditional channels of communication of states and other institutions and organizations (Castells et al. 2007), constructing personal communities that provide a sense of belonging with people who may not live nearby. Following Wellman's idea of community as networks of communication (1979), the sense of belonging to a community in immigrant contexts needs to incorporate a more communicative perspective. One question remains open: how do these flows interact with local references and needs?

Power Distribution

The possibility that migrants, and the sociogeographic contexts where they live in the countries of origin and destination, have taken technology and used it in ways that developers never expected raises the question of the relationship between power and technology for powerless social groups and regions (Wellman 2004). Presence and absence in the network are critical sources of domination and change in any society (Castells 1996). But are recent migrants really taking advantage of being part of a network to gain empowerment? Or, on the contrary, is the process of disintegration for powerless groups, such as migrants, ongoing, even if they are accessing ICTs? Who does amass the resources for migrating and successfully entering and integrating into other societies (Held et al. 1999)?

IMMIGRANTS AS ICT USERS

Because one of the necessary conditions to become a digital immigrant is the use of information and communication technologies, it is important to identify how the penetration of ICTs occurs in the immigrant community. As a follow-up to the usage and interaction of immigrants with ICT, it becomes necessary to analyze whether a specific way of using ICT by immigrants exists and determine what its main features are. That will provide an excellent picture of the intersection of immigration and the Information Society.

However, neither official statistics nor academic research has been much help in producing data on ICT use by the immigrant population. Telecom companies do not facilitate data on "their customers." In the case of Catalonia, where 17 percent of the population are immigrants, none of the most recent data on technologies uses and home ICT equipment have considered origin as a main sociodemographic variable. A combination of pragmatism—that is, immigration adds complexity to surveys—and a lack of social sensibility is probably the reason for such "disregard."

Although the existing data on ICT use by immigrants in Catalonia until now are not very representative, they do allow us to carry out an initial evaluation comparing the immigrant and native populations. Taking into account its limitations,[4] the Survey on New Technology Use (Direcció General de Atenció Ciutadana de La Generalitat de Catalunya 2006) reveals some interesting data on the use of information and communication technologies in Catalonia.[5] The results show that, of the different ethnic groups, the "EU and rest of Europe" collective and the Latin American collective use the Internet (see table 1.1) and e-mail (see table 1.2) more than the native population. The "rest of the world" collective, which includes immigrants from Asia and Africa, uses the technology less. However, the use of communication technology is higher in all three immigrant communities in comparison to the native population.

One could argue that age could be affecting these results, because the immigrant population is younger than the native population. However, an initial comparative analysis of each age group concludes that immigrants—especially those from Latin America—have usage patterns more similar to young natives (between sixteen and twenty-nine) than to older ones. This is very clear in the case of Webcam use (see table 1.3).

ICTs are used in all the different phases of the migration process. The immigrant population makes the most of information and communication systems

TABLE 1.1 | INTERNET USE BY REGION OF ORIGIN

REGION	TOTAL NUMBER OF ABSOLUTE CASES	YES (%)	NO (%)
Native population	1,329	56.6	43.4
EU and the rest of Europe	36	78	22
Latin America	113	77	23
Rest of the world	40	45	55

SOURCE: Our elaboration. Data from the Survey on New Technology Use (2006) (Direcció General 2006).

TABLE 1.2 | E-MAIL USE BY ORIGIN

REGION	TOTAL NUMBER OF ABSOLUTE CASES	YES (%)	NO (%)
Native population	1,329	43.4	43.4
EU and the rest of Europe	36	72	28
Latin America	113	68	32
Rest of the world	40	30	70

SOURCE: Our elaboration. Data from the Survey on New Technology Use (2006) (Direcció General 2006).

TABLE 1.3 | WEBCAM USE BY ORIGIN AND AGE GROUPS

AGE	REGION (%)	DAILY (%)	A FEW TIMES A WEEK (%)	A FEW TIMES A MONTH (%)	LESS FREQUENTLY (%)	NEVER (%)
16–29	EU and the rest of Europe	13	0	13	6	69
	Latin America	15	25	10	8	43
	Rest of the world	0	6	6	6	81
	Native population	4	11	5	9	70
	Total	6	12	6	8	68
30–44	EU and the rest of Europe	27	9	0	9	55
	Latin America	11	11	8	8	62
	Rest of the world	5	0	5	5	84
	Native population	2	5	3	3	86
	Total	4	6	4	4	83

SOURCE: Our elaboration. Data from Survey on New Technology Use (Direcció General 2006).

during the different phases and situations in the migration process: in making the decision to leave home, in preparation for the journey, on the journey, upon arrival, during postarrival, and also in their relationships with people in their countries of origin as well as in other parts of the world. At each moment, according to different needs, technology plays a different role. The telecom-usage pattern in immigrant contexts includes chatting, contacting friends, entertainment, work-related activities, obtaining news, and the like.

Thus, for example, in the process of the decision to migrate, technology plays an important role in generating information and also in creating images of the destination. Horst (2006) points out that, thanks to ICT, more realistic expectations are generated. However, the issue of a substantial difference between the image seen on international television and reality emerged in the interviews.

In preparation for the journey, information is a key factor. We can state that here ICT plays a fundamental role in obtaining information. From our interviews, it does not seem that those who have arrived in Catalonia in recent years did so with very much formal information about what to expect. However, we need to know more about what the current strategic information sources are and what role informal as opposed to formal (or institutional) agents play. Further investigation is needed into how the information process occurs and if such information has improved.

For a journey full of uncertainties—and human drama—access to technologies is a fundamental way to inform family abroad of the arrival. There are some reports in the interviews of getting a SIM (subscriber identity module) card at the same moment of arrival. One interviewee even refers to *cayuqueros* (those in charge of the boats that transport illegal immigrants from the coasts of Senegal and Mauritania to Spain) using mobile phones.

If we move on to another key phase in the migration process, the moment of arrival, ICT becomes a basic resource for social and employment integration. The mobile phone, in this sense, is a basic instrument in the social relationships of immigrants when they arrive. Through the mobile phone they contact other people for support and company. As previously stated, the mobile phone is also the main form of contact for employment purposes.

When immigrants arrive in societies with high levels of ICT penetration, they respond to general patterns and adapt and use them according to their own needs. Since levels of ICT penetration differ greatly between countries of origin and destination, many immigrants have their first contact with the Information Society upon initial arrival. Some immigrants recall that they were not acquainted with the communication facilities until they arrived. Others brought a mobile phone with them and then bought a new SIM card once they arrived.

Differential rates of ICT adaptation are a basic factor in this terrain. Of course, interconnection is not possible without a counterpart. For the Spanish case, even if Spain maintains one of the lower rates of ICT penetration of Europe, the differential rates in relation to the principal countries of origin are still enormous (see table 1.4).

However, as technology diffusion penetrates the regions of origin, more immigrants incorporate some technological tools, such as mobile phone and e-mail, before leaving. More and more, immigrants' regions of origin are filling with technological devices (mobile phones) and Internet access (in cybercafes). A Senegalese woman explained that, after visiting her country last year, she realized that "there is a cyber[cafe] on every corner." A Moroccan woman observed that cybercafes in Morocco have become gathering places where people go to receive news from those abroad.

The analysis of ICT use by immigrants in Catalonia shows that ICT is part of what it means to be a migrant today. From the interviews it became

TABLE 1.4 | ICT PENETRATION IN COUNTRIES OF SPANISH IMMIGRATION

COUNTRY	OPERATIONAL TELEPHONE LINES	DWELLING WITH INTERNAT PER 100 INHABITANTS	BROADBAND INTERNET PER 100 INHABITANTS	PERSONAL COMPUTERS PER 100 INHABITANTS	SUBSCRIBERS TO MOBILE TELEPHONE SERVICES PER 100 INHABITANTS	TOTAL TELEPHONE SUBSCRIBERS PER 100 INHABITANTS	INTERNET USERS PER 100 INHABITANTS
Spain	43.16	2.28	0.29	26.64	93.91	137.07	34.85
Argentina	22.38	2.38	0.03	7.72	34.76	57.14	13.17
Bolivia	6.97	0.09	0.00	2.12	20.07	27.04	3.90
Colombia	19.52	0.43	0.01	6.67	23.16	42.68	7.98
Ecuador	12.22	0.07	0.00	5.49	34.44	46.66	4.73
Morocco	4.38	0.01	0.00	2.07	31.23	35.60	11.71
Pakistan	2.96	0.02	0.00	—	3.30	6.27	1.32
Peru	7.44	0.40	0.02	9.75	14.85	22.28	11.68
Romania	19.70	0.22	0.02	11.00	45.85	65.55	20.20
Senegal	2.21	0.01	0.00	2.34	9.94	7.77	4.66
China	23.79	0.01	0.01	4.03	25.49	49.29	7.16

SOURCE: Our own data based on data from the International Telecommunication Union (2006).

apparent that the interviewees were very familiar with information and communication technologies and felt at ease and comfortable talking about them. In many cases, a change of "tone" was noticeable when more difficult subjects were raised.

An important characteristic is that immigrants are knowledgeable about and familiar with ICT. The immigrant population's digital culture seems to be heavily consolidated around communication systems. We have verified this familiar, daily contact with ICT in various ways. One way was through the volume of information they manage concerning prices, offers, products, and the combination of strategies they use. For example, in the "mobile culture" among immigrants, a young Senegalese woman said that she, like many other young women from Senegal, had two mobiles as a strategy for controlling spending. Immigrants spend time and money on mobile phones and other communication devices. They often spend a larger amount of their income and time compared with other groups (Qiu 2004). Immigrants develop strategies to reduce communication costs to a minimum with a combination that allows them to create their own ad hoc supply. In the case of phone calls, they use combinations of landlines and mobile and Internet calls.

In addition to knowing about prices and supply, the immigrants interviewed knew all about state-of-the-art technology, the latest in mobiles, in Web 2.0, or in programs like Skype. Limitations on ICT use appear to be associated with a lack of technical knowledge (Internet), a lack of time (Internet), and the cost (mobile phone).

The increasing importance of ICT, as a general feature in the Information Society, is also found among immigrants, particularly in relation to mobile phones. A young African put it this way: "If I lose my phone, I lose everything." Immigrants are attached to their cell phones and cell phone numbers. In many cases, the address book is stored only in the mobile, as the mobile phone develops as a status symbol, and as part of the urban dream.

Technologies appear to be associated with making life easier—for example, staying in touch with one's roots, making distance more bearable, facilitating contact for job offers. But on the other hand, more negative values emerge linked to the use of technologies. In this sense, technology emerges as a new instrument of power that cannot always be completely trusted. In particular, there are fear and mistrust of the power that can now be exercised by those controlling the technology, whether they are large companies or small telephone call shops.

IMMIGRANT INTERCONNECTION PATTERNS

To state, as has been done previously, that there is a greater flow of information and communication is not particularly informative. It is necessary to look into the new patterns of interconnection among immigrants and their families and friends abroad to discover if there is anything new—and what that is—in the way in which information and communication flows are being organized. With the purpose of understanding how this interconnection is being produced, there are some analytical categories that may prove helpful. To build a model of migration interconnection, the categories of flows in globalization established by Held et al. (1999)—extensity (that is, range), intensity, velocity, and impact—are a convenient starting point. As Vertovec (2004b) has pointed out, these categories are highly applicable to migration environments. With the help of interviews, we have operationalized these dimensions as follows.

Extensity: Interconnection extensity means the current capacity of new technologies to reach more places than ever and involve many actors, thus creating high capacity for action between spaces and people who remain far from each other. Interconnection network extensity, as Castells (1996) points out, is strengthened considerably in a context of informationalism that increases the capacity to process information. The strategies for expanding access to technologies are mainly in the hands of regulators, namely, governmental and telecommunication interests.

When we apply the category of extensity to interconnection in migration, we find different elements that need to be further analyzed. First, who are the actors communicating with each other in the current migration context, and how do variables such as age, sex, habitat, economic level, and education affect the connection space? Second, how does mobile communication technology, which breaks the variable space, as opposed to fixed communications affect the connection, and where are the agents who enter into communication?

Intensity: One of the basic questions in connection and communication flows is the regularity or rate at which they occur. It would seem logical to think that the changes that have taken place in communications, especially ease of access, have increased the patterns of frequency and rate.

If we apply all these categories to immigration contexts, a more frequent, continuous, and prolonged pattern of interaction could be very significant and could represent one of the most important new aspects in relation to previous migratory movements. In short, we could be looking at a change in the pattern of interactions that might be moving from occasional to regular. Here, we need to respond to different sets of questions:

1. In relation to communicative frequency:
   - How often does communication occur?
   - How is it affected by cost? Do those with greater economic capacity communicate more? Or are there other variables at stake?
   - Is there a communicative strategy, or is it improvised?
   - How do temporal circumstances affect communication frequency?
   - How and who sets the frequency patterns?
2. In relation to communication duration:
   - How long does contact last?
   - What factors affect duration?
3. In relation to budget:
   - What kind of budget is set aside for communication?

According to recent research, migrants use information and communication technologies on a frequent basis, thus allowing them to maintain a high level of interconnection within networks and keep connected to their families and other loved ones thousands of miles away (Qiu 2004; Barendregt 2005; Cartier, Castells, and Qiu 2005; Thompson 2005, all of them in Castells et al. 2007). In the interviews, different references to different levels of intensity were found (from more than three times a week to once every two weeks). It is particularly interesting to note that some immigrants explain that they changed their frequency in interconnection as they became acquainted with the increasing facilities and different options available, each one with a different cost. Now, as an Algerian boy explained, instead of calling once a week, for the same price, he can make three calls a week. Immigrants seem to prefer more frequency in communication. Their strategies seem to be shorter but frequent rather than longer but infrequent contact.

Continuity in interconnection is also present in specific moments of time. Some people refer to periods of problems (health problems, for instance) or of religious celebration. This is the case for Fatima, who, when Ramadan arrives, is in constant contact with her family and friends in Morocco through telephone, Internet, and text messages (of religious texts) that help her feel "as if I were there, with them, although it is not real."

Velocity: ICT introduces and modifies the time factor in communication in the form of high-speed communicative interactions. ICT allows for the instant transmission of information and instant communication. This simultaneity introduces a very significant change in comparison to earlier stages of communication in migration contexts and raises key questions, such as: What is known of the realities, problems, and mutual concerns in origin and destination communities? Are decisions made from a distance? At what times do interconnections take place? Simultaneity in communication is now possible thanks to telephone and VOID communication. A woman from Senegal explained her family's strategy: First, she sends them a text message or an e-mail notifying them that in an hour's time, she will be calling them. Then her family goes to a cybercafe, and they chat via the computer.

Flow impact: As an effect of the other categories in the immigration experience—extensity, intensity, and velocity—regarding key issues such as a new distance or long-distance interrelation, is a sense of proximity created? What are the negative effects of the combination of distance and frequent contact? Is there more mutual control? Does the increased communication flow lead to conflictive situations or to new duties for immigrant communities?

One of the main impacts that we should attempt to analyze are the new possibilities of connecting lives over here and over there. As an Ecuadoran immigrant woman explained to me, there are some elements such as remittances, house tenancy, knowledge of politics, and knowledge of personal matters that link her "two lives." To what extent are these "two lives" experienced as a whole reality in continuous interaction? How do factors of distance influence immigrants' feelings, projects, and the general vision of their lives and future? And how do they manage to resolve tensions and conflicts that the volume of so much interconnection may produce? The words of the Ecuadoran woman are very clear in that respect: "I have another life there, and I have also duties and responsibilities there. I have a mother, nephews, a house, and a dog there. Sometimes, they think we have a lot of money here, and they continuously ask for money and help. This summer, I had to go [over] there to organize things, and I had to explain to them how much we earn, and [about] all the expenses we have here, and how little is left."

Interviews with Senegalese immigrants showed an existing tension, as they experience a high level of constant demands from the country of origin. Family relations are strengthened, and transnational families become more possible. For many immigrants who have left part of their families in the country of origin, to be connected with them is an advantage that produces satisfaction and a feeling of proximity and helps them avoid depression. Some mothers report that, by calling every other day, they are able to follow their children's homework. They greatly appreciate spending time talking about everyday things and not having to be too concerned about time or costs.

LOOKING AHEAD

Recognizing interconnection as a new feature of immigration could be a positive step toward a much better capacity for shaping integration processes in the Information Society in which they already play an active role. Immigrants have entered the world of technology through a mechanism that seems to be working well: the need for communication and information. Because they have incorporated new rules of interaction within a very short period of time, now is the time to take advantage of the new possibilities to accelerate their incorporation and stop wasting opportunities. A new context of facilities to interconnection permits immigrants to be embedded in their home communities while building lives in their new environments.

However, there are some aspects in the relationship between immigration and the Information Society that need to be addressed if we want to prepare a better terrain for incorporation. First, the use of communication and information technologies and the assimilation of new interaction patterns by immigrants could be the preconditions for a second and necessary stage in which immigrants develop the capacity to apply their knowledge of technology to the world of work, adding human capital value. Otherwise, we may encounter a new imbalance in the "digital divide" process.

Second, immigrant interconnection as a terrain of opportunities needs better social infrastructures. Technical infrastructures that immigrants have appropriated present many problems of access, cost, and intimacy. That makes cybercafes and mobile phones the only solutions, while service providers and local governments have not activated any strategy for supporting mobile phone usage among migrants. Immigrants usually report receiving unsatisfactory services and being badly served, including access problems to fixed-line telephones (Qiu 2004). Part of the reason for the high demand for mobile phones among immigrants has to do with these constraints. What this usually means is that they cannot follow a rational system of paying and often find themselves paying more, getting less, and depending on corporate strategies while no public policy exists (Castells et al. 2007).

Third, one of the new challenges that the immigrant population is facing in the Information Society is how to manage the infinite possibilities of interconnection. Recent immigrants are presented with all kinds of tools and opportunities to be able to transport themselves, within a few seconds, into a different reality of which they don't have complete control. In addition, demands and questions coming from countries of origin could be so continuous that they would need to learn how to respond and manage such inputs. We need to be very aware that more interconnection does not always equate to an expression of domination but, on the contrary, may mean a new expression of unequal distribution of resources and, therefore, of power.

NOTES

1. The Immigration and Information Society Research Programme was created in September 2006 at the Internet Interdisciplinary Institute (http://in3.uoc.edu/), thanks to the financial help of the Generalitat de Catalunya, the Autonomous Community Government of Catalonia.

2. I want to thank Elisabet González, Papa Sow, and Graciela de la Fuente, the research team of the Immigration and Information Society Research Programme, for all their contributions to this work.

3. In the last months of 2007, one of the Spanish mobile phone companies introduced a new television publicity campaign targeting the immigrant sector. The lyrics to the campaign's signature song, which accompanied images of immigrant people working, were: "Nostalgia cero, o sea nada" (Zero nostalgia, that is nothing). The advertisement announced a special price called "Tarifa juntos" (the Together Rate) and finished by saying, "Lo más parecido a estar juntos" (The closest thing to being together).

4. The methodological limitations of this survey must be taken into account in the interpretation of the results. First, case selection was carried out by telephone. The sample was limited to the foreign population with access to a telephone at home, that is, individuals with a certain level of stability, which is not representative of the immigrant population as a whole. Second, the sample is not random. Third, the sample of immigrant population was insufficient—189 individuals from a sample of 1,500. Fourth, the native population sample represents a wider cross section

than the immigrant population sample. This makes comparison between the native and immigrant populations more difficult.

5. I would also like to thank Elsa Urrutia at the Direcció d'Atenció Ciutadana of the Generalitat of Catalonia for her kind help and the Centre d'Estudis d'Opinió for making the data available to us for this study. Thanks also to Antoni Marín Saldo for his advice on statistical work.

REFERENCES

Barendregt, B. 2005. *The ghost in the phone and other tales of Indonesian modernity.* Proceedings of the International Conference on Mobile Communication and Asian Modernities. Hong Kong, June 7–8.

Cartier, C., Manuel Castells, and J. L. Qiu. 2005. The information have-less: Inequality, mobility, and translocal networks in Chinese cities. *Studies in Comparative International Development* 40, no. 2: 9–34.

Castells, Manuel. 1996. *The information age: Economy, society, and culture.* Oxford and Malden, Mass.: Blackwell.

Castells, Manuel, M. Fernandez-Ardevol, J. L. Qiu, and A. Sey. 2007. *Mobile communication and society: A global perspective.* Cambridge: MIT Press.

Castles, S., and M. Miller. 2003. *The age of migration: International population movements in the modern world.* 3d ed. New York and London: Guilford Press.

Diminescu, D. 2007. Le migrant connecté. *Migrations/Société* 17, no. 102: 275–92. Available at http://www.ticm.msh-paris.fr/spip.php?article32.

Direcció General de Atenció Ciutadana de La Generalitat de Catalunya. 2006. *Enquesta sobre l'ús de les noves tecnologies.* Catalunya: Idescat. Available at http://www.idescat.net/cat/idescat/estudisopinio/rpeo/R-364.pdf.

Fawcett, J. 1989. Networks, linkages, and migration systems. *International Migration Review* 23, no. 3.

Held, D., A. McGrew, D. Goldblatt, and J. Perraton. 1999. *Global transformations.* Cambridge, Mass.: Polity Press.

Horst, H. 2005. From kinship to link-up. *Current Anthropology* 46, no. 5 (December).

———. 2006. The blessings and burdens of communication: Cell phones in Jamaican transnational social fields. *Global Networks* 6, no. 2: 143–59.

Horst, H., and D. Miller. 2006. *The cell phone: An anthropology of communication.* Oxford: Berg.

Inda, J. X., and R. Rosaldo. 2002. *The anthropology of globalization: A reader.* Oxford: Blackwell.

International Telecommunication Union. 2006. *World telecommunication ICT: Development report.* Canada: International Telecommunication Union.

Massey, D. 1998. *Worlds in motion: Understanding international migration at the end of the millennium.* Oxford: Oxford University Press.

Project Internet Catalonia. n.d. Universitat Oberta de Catalunya (IN3: Internet Interdisciplinary Institute). Available at http://www.manuelcastells.info/en/invesdoc_index.htm.

Qiu, J. L. 2004. (Dis)connecting the Pearl River Delta: Transformation of a regional telecommunication infrastructure, 1978–2003. Ph.D. diss., University of Southern California.

Ros, A., E. González, A. Marín, and P. Sow. 2007. Migration and information flows: A new lens for the study of contemporary international migration. Working paper, Internet Interdisciplinary Institute, UOC. Available at http://www.uoc.edu/in3/dt/eng/ros_gonzalez_marin_sow.html.

Telegeography. 2006. *Telegeography executive summary, 2007.* N.p.

Thompson, J. 2005. Mobile phones and ration cards: Refugee resettlement and the Sudanese experience. Paper presented at the Third Annual Forced Migration Student Conference, Oxford Brookes University, Oxford, May 13–14.

Tilly, C. 1990. Transplanted networks. In *Immigration reconsidered: History, sociology, and politics,* ed. V. Yans-Mclanghin. Oxford: Oxford University Press.

United Nations. 2006. *International migration and development.* Report of the UN secretary-general. New York: United Nations.

Vertovec, S. 2004a. Cheap calls: The social glue of migrant transnationalism. *Global Networks* 4, no. 2: 219–24.

———. 2004b. Trends and impacts of migrant transnationalism. Working Paper no. 3. Oxford: Oxford University, Centre on Migration, Policy, and Society. Available at http://www.compas.ox.ac.uk/publications/papers/wp0403.pdf.

Webster, F. 2001. *Culture and politics in the information age.* London and New York: Routledge.

Wellman, B. 1979. The community question: The intimate networks of East Yorkers. *American Journal of Sociology* 84, no. 5: 1201–31.

Wellman, B., et al. 2005. Connected lives: The project. In *Networked Neighbourhoods,* ed. P. Purcell. Berlin: Springer.

2 Migration, Information Technology, and International Policy

JENNIFER M. BRINKERHOFF

As other chapters in this volume illustrate, diasporas' use of the Internet yields important consequences beyond cyberspace. Diasporas are increasingly recognized for their potential contributions to homeland socioeconomic development. These contributions include economic remittances, diaspora philanthropy, knowledge transfer, diaspora investment and business development, and policy influence (see Brinkerhoff 2008b). Diasporas may also threaten global security, as they promote and may participate in regime change and contribute to and sustain conflict in their home countries, increasingly with spillover effects into neighboring countries and the global arena. In their host societies, diasporans integrate to varying degrees of success. Those who fail to integrate into the host culture and become socially, economically, and politically marginalized may be vulnerable to recruitment into violent activities (Hernandez, Montgomery, and Kurtines 2006; Galtung 1996). Those who lack a collective identity have been described as "psychologically desperate" and easy prey for terrorist organizations who seek to fill this psychological void (Taylor and Lewis 2004, 184).

The Internet plays a role in each of these potential activities. It can provide an important foundation for the collective identity necessary for purposive action and the prevention of recruitment into violent activities (Brinkerhoff

2009). The interactive components of the Internet enable the creation of cyber communities that connect dispersed populations and provide solidarity among members. Members use discussion forums to disseminate information about the homeland faith or culture or both, to reinforce or re-create identity to make it more relevant and sustainable across generations in diaspora, and to connect to and participate in homeland relationships, festivals, and socioeconomic development. Members' discussions may reflect diasporas' embrace and experimentation with liberal values, which inform conflict mitigation, political agendas, and homeland socioeconomic development contributions (ibid.). Furthermore, the Internet is a tool for mobilizing this collective identity into action. It is an organizational and networking resource for assembling and communicating among individuals and groups, for providing information and referrals to other actors; furthermore, it facilitates issue framing and confidence building. Results of mobilization agendas can be posted and disseminated to inspire continued commitment and subsequent mobilization.

Diaspora identity expression and transnational engagement occur with or without policy interventions, and in those areas where diasporas' economic and political contributions are less noticeable, we are likely to see fewer, if any, government policies and programs. One such area is in the digital arena. The purpose of this chapter is to build upon our cumulative understanding of digital diasporas to generate a series of policy recommendations for homeland governments, adopted country governments, and international development actors. These recommendations build largely from my own comparative analysis of nine digital diaspora organizations (ibid.).

ADOPTED COUNTRY GOVERNMENTS

Migrant integration can be eased when diasporans have opportunities to express their hybrid identities collectively. The Internet provides important opportunities for creating a sense of identity and solidarity around a shared cultural heritage and diaspora experience. The communities created serve to combat feelings of marginalization among diasporans, providing them identity and other forms of support as they cope with the diaspora experience. This identity support, in turn, enables diasporans to integrate new ideas, values, and experience into their identity frame of reference, testing the boundaries for what it means to their homeland identity as well as a potentially more modern and individualistic adopted country identity. At the same time, digital diasporas are important vehicles for disseminating information and advice for accessing public goods and services and easing the transition from newly arrived migrant to productive member of the host society.

The Internet's conducive features for identity negotiation and representation of liberal values (in both form and function) are important contributors that should be sustained and protected. The motivation to express identity is natural and common to all human beings. Whether addressed in cyberspace or in the physical world, diasporans face a psychosocial need to develop and express their hybrid identities and to experiment with the integration of alternative and additional values and conceptions, including those representing liberal values. Given hybridity, affirming the homeland identity and even talk of return do not mean that diasporans do not value or embrace their adopted country and society. An improved public awareness of this complexity could encourage tolerance and, ideally, incentives to seek improved understanding of particular individuals and communities.

Understanding hybridity is crucial in the consideration of social policies. Portes (2006) examines three practical theoretical stances or policy orientation options vis-à-vis migrants in receiving societies. Hard assimilation seeks to completely assimilate migrants to the exclusion of a homeland identity by the second generation. Permanent culturalism emphasizes the homeland identity, sometimes to the exclusion of an adopted country identity. It can lead to ethnic enclaving, which, in turn, can isolate immigrant communities, potentially sending them down a path of downward assimilation, where the second generation and beyond experience a downward spiral of economic and educational opportunities (see Rumbaut and Portes 2001; and Waldinger and Feliciano 2004). Soft assimilation tolerates ethnicity but promotes the ultimate goal of integration. It encourages knowledge and respect of mainstream culture, laws, education, and language. It typically leads to gradual and voluntary integration. Minimal requirements for its success include knowledge of the adoptive country language and observance of its rule of law. It is this model, Portes argues, that leads to the American tolerance and even encouragement of hyphenation. Soft assimilation celebrates hybridity, and digital diasporas are an important vehicle for expressing hybridity. Digital diasporas afford migrants opportunities to embrace their American selves while still maintaining a homeland identity; in other words, digital diasporas contribute to soft assimilation and should thus be allowed to function unencumbered.

Diasporans make implicit cost-benefit analyses with respect to both identity orientation (see, for example, Portes and Zhou 1993; Waters 1999; and Sánchez Gibau 2005) and identity mobilization and its direction (Brinkerhoff 2008a). Policy makers can influence the terms of this analysis by attending to the "context of reception." This is typically defined according to three factors: "the state policies directed at a specific migrant group; the reactions to and perceptions of the immigrants by public opinion; and the presence or absence of an established ethnic community to receive the immigrants" (Grosfoguel and Cordero-Guzmán 1998, 357). The importance of a receiving ethnic community confirms the need for solidarity benefits stemming from a shared, collective identity; it further highlights the salience of digital diasporas' contributions.

Policy makers can focus on changing the stakes for identity mobilization, enhancing the possibility of constructive directions by creating an enabling environment for a high quality of life and community for diasporas. Such an enabling environment might include, for example, access to small business and home ownership loans and support for access to vocational training and higher education. For example, Grosfoguel and Cordero-Guzmán's review of the Cuban diaspora experience confirms the importance of both social networks—internal community solidarity—and "broader social structures that constrain or enable access to capital, information, and resources by members of a specific community's micro-networks" for integration and success in the receiving country (1998, 355).

In short, digital diasporas support integration by providing solidarity, combating marginalization, facilitating access to public goods and services, and embracing hybridity and soft assimilation. They are particularly salient to the context of reception, which further supports migrant integration. Two policy recommendations emerge from this discussion. First, information technology (IT) regulation should maintain privacy and access in order not to interfere with opportunities for exploring identity and representing liberal values. Second, and relevant to diasporas beyond cyberspace, policy makers should focus on changing the stakes for identity mobilization by creating an enabling environment for a high quality of life and community for diasporas, thus discouraging ethnic isolation and mobilization for destructive aims.

Host governments are also concerned with the security implications of diasporas. With the recent war on terrorism, migration to industrialized countries has sparked new concerns ranging from the migration of terrorists or the harboring of terrorists by immigrant communities to the financial support of terrorist activities through charity foundations and the recruitment of terrorists from within seemingly assimilated diaspora communities. These concerns can be partially addressed through a cooperative strategy with diaspora organizations. Loose partnerships or cooperative arrangements can create structures for anonymous reporting by concerned diasporans and associated response procedures by the receiving governing bodies. Not all communities will equally embrace this idea, and the parameters of what is acceptable will need to be negotiated on a case-by-case basis. The Internet can be a useful tool in such cooperative reporting.

HOMELAND GOVERNMENTS

This discussion of homeland government policy options focuses on how to interface with established diasporas rather than prevent them in the first place (i.e., through migration management policies). There is a broad scope for governments to engage with their diasporas to facilitate their constructive political, social, and economic contributions to the homeland. Proactive efforts and explicit policies have been slow in coming. Portes, Escobar, and Radford reviewed the experience and research to date and summarized it as follows: "All empirical evidence indicates that economic, political, and socio-cultural activities linking expatriate communities with their countries of origin emerged by initiative of the immigrants themselves, with governments jumping onto the bandwagon only when their importance and economic potential became evident" (2007, 258; see also Portes 2003).

Proactive diaspora engagement has been labeled the "diaspora option" and encompasses several potential approaches. This policy framework conceives the skilled diaspora as an asset to be captured (Meyer et al. 1997), and encompasses three types of related diaspora engagement strategies: remittance capture, diaspora networking, and diaspora integration (Gamlen 2005). Each of these arenas holds policy implications for digital diasporas.

Regarding remittance capture, home governments are increasingly soliciting remittances and offering policy incentives (e.g., tax-free investment opportunities and matching) and investment options (e.g., remittance-backed bonds and foreign currency accounts) to encourage diaspora contributions (see, for example, Orozco with Lapointe 2004; Lowell and De la Garza 2000; and Pires-Hester 1999). Less attention is given to the importance of an enabling IT infrastructure for remittance transfer, banking, and other transactions. For example, Thamel.com has evolved a sophisticated set of remittance services that enable diasporans to make monthly payments directly to bank accounts with specified purposes, such as making monthly payments to support private education or home equity loans. Reliable infrastructure and a supportive IT regulatory environment are crucial to the effectiveness of such services.

Diaspora networking refers to networking that links the homeland to the diaspora. This includes, for example, fulfilling intermediary functions, such as acting as a coordinating body between the supply and demand of potential contributions (see Meyer and Brown 1999), facilitating the migration process, and ensuring transportability of qualifications (Vertovec 2002). Diaspora networking can support the capture of diaspora socioeconomic contributions to the homeland. Outreach and communications, including visiting delegations between home- and host land, contribute both to networking and to integration strategies. IT is often used to support these efforts; for example, intermediary organizations (including government agencies) may set up online databases to fulfill matching functions, governments may set up Web portals specifically to interface with their diasporas, and e-newsletters may be distributed to diaspora organizations and individuals abroad.

The diaspora integration strategy recognizes the diaspora as a constituency that is marginalized from the homeland. Thus, related policies include the extension of citizen rights such as voting and the organization of diaspora summits and diplomatic visits to diaspora organizations in their host countries. Diaspora integration policies confer social status, political influence, and legitimacy to the diaspora and its potential efforts to contribute to the homeland. Combined with diaspora networking, this strategy may be used to seek endorsement and legitimacy for homeland political agendas.

Digital diasporas are a relatively untapped resource for supporting each of these policy strategies. Beyond databases and their own e-mail Listservs, homeland governments can more systematically use digital diasporas to solicit and possibly influence diaspora contributions, disseminate information about homeland developments, perhaps seeking endorsement (e.g., for postconflict draft constitutions, as in the case of Afghanistan), and stay connected to their diasporas, acknowledging them in ways that may encourage diasporas' further engagement in support of the homeland.

THE INTERNATIONAL DEVELOPMENT INDUSTRY

As a practice, international development can be viewed as an industry, driven by international donors, multilateral and bilateral, whose resources are represented by official development assistance (ODA). Universities, private consulting firms, and nongovernmental organizations (NGOs) are the major producers of international goods and services, purchased through ODA. Where do diaspora contributions fit in the formal policies and programs of official development assistance?

Globalization has exacerbated many of the challenges to effective development assistance (see Florini 2000), including a declining capacity of national governments, changing private sector and NGO roles, and weak and outmoded global institutions (Brinkerhoff 2004). In response, Lindenberg and Bryant call for a new global NGO architecture, including networks and virtual organizations. However, they conclude, somewhat pessimistically, that "none of the globalizing organizations from any part of the world have a fresh vision of how more genuine, multi-directional global networks for poverty alleviation and conflict reduction might be most realistically developed" (2001, 245). On his part, Dichter (2003) notes some positive advancement outside of the development industry, notably the growing promise of telecommunications and economic remittances from diaspora communities. Digital diasporas hold great potential to meaningfully contribute to the socioeconomic development of their homelands in ways that supplement and may even enhance current industry efforts.

To date, most diaspora-related initiatives originating from the donor community concentrate on research and enhancing the productive benefits of remittances. Isolated experimentation and experiences are accumulating, wherein diaspora organizations are more engaged in the interface with formal development institutions, including providing technical assistance and intermediary services and even as implementing partners (see Brinkerhoff 2004). Diaspora members have individually been engaged in formal development assistance for some time. A more systematic review of options and experience is needed, particularly with the increased role of official development assistance in postconflict reconstruction, where societies have largely lost their human capital to war and forced and elective migration.

Whether through formal partnerships or informally, international development practitioners can tap digital diasporas to solicit information and cultural and technical expertise, disseminate information about their programming for the purpose of constituency building and coordination, and seek intermediary support in order better to reach the diaspora and target communities and those with requisite skills and experience. Independently, diasporas are pioneering their own development models, often coordinated through online communities. The development industry could enhance the effectiveness and impact of these efforts, helping these organizations to scale up these activities, building the organizational capacity of these diaspora organizations, and providing forums for these groups to share experiences and learn from each other, as well as industry development efforts.

SUMMARY OF POLICY IMPLICATIONS

The previous discussion generates six policy recommendations for host- and homeland governments, as well as international development policy makers and analysts. First, IT regulation should maintain privacy and access in order not to interfere with opportunities for exploring identity and representing liberal values. Second, policy makers should focus on changing the stakes for identity mobilization by creating an enabling environment for a high quality of life and community for diasporas, thus discouraging ethnic isolation and mobilization for destructive aims. Third, host country governments can use the Internet as a tool for fostering cooperative strategies with diaspora communities to ensure security. Fourth, homeland governments should use digital diasporas to solicit and possibly influence diaspora contributions, disseminate information about homeland developments, perhaps seeking endorsement, and stay connected to their diasporas, acknowledging them in ways that may encourage diasporas' further engagement in support of the homeland. Fifth, international development practitioners should tap digital diasporas to exchange information and solicit support and expertise for their development programming. Sixth, international development actors should consider assisting digital diasporas to improve the effectiveness and impact of their efforts and support their learning from each other as well as from the development industry.

These recommendations focus primarily on enabling or at least not inhibiting the potential constructive contributions of digital diasporas. The phenomenon of digital diasporas is broad and diverse, necessitating flexible responses that address a much wider range of implications than those noted here. This chapter has sought to provide the initial groundwork for further investigation and experimentation. As diasporas' significance in global affairs is increasingly recognized, and as their use of the Internet becomes justifiably more noticeable, we are likely to witness an expansion of real-world issue identification and policy experimentation, far beyond what has been conjectured here. The fact remains that few policies targeted specifically to digital diasporas exist, and those that do have yet to be systematically inventoried and studied. In the future, it is likely that chapters such as this will necessarily focus on a more limited scope in terms of policy targets and with a much higher volume of examples from experience.

REFERENCES

Brinkerhoff, Jennifer M. 2004. Digital diasporas and international development: Afghan-Americans and the reconstruction of Afghanistan. *Public Administration and Development* 24: 397–413.

———. 2008a. Diaspora identity and the potential for violence: Toward an identity-mobilization framework. *Identity: An International Journal of Theory and Research* 8.

———, ed. 2008b. *Diasporas and international development: Exploring the potential.* Boulder: Lynne Rienner.

———. 2009. *Digital diasporas: Identity and transnational engagement.* Cambridge: Cambridge University Press.

Dichter, Thomas W. 2003. *Despite good intentions: Why development assistance to the third world has failed.* Amherst: University of Massachusetts Press.

Florini, Ann, ed. 2000. *The third force: The rise of transnational civil society.* Washington, D.C.: Carnegie Endowment for International Peace.

Galtung, Johan. 1996. *Peace by peaceful means: Peace and conflict, development and civilization.* Thousand Oaks, Calif.: Sage.

Gamlen, Alan. 2005. The brain drain is dead, long live the New Zealand diaspora. Working Paper no. 10. Oxford: Oxford University, Centre on Migration, Policy, and Society.

Grosfoguel, Ramón, and Héctor Cordero-Guzmán. 1998. International migration in a global context: Recent approaches to migration theory. *Diaspora* 7: 351–68.

Hernandez, Lynn, Marilyn J. Montgomery, and William M. Kurtines. 2006. Identity distress and adjustment problems in at-risk adolescents. *Identity: An International Journal of Theory and Research* 6: 27–33.

Lindenberg, Marc, and Coralie Bryant. 2001. *Going global: Transforming relief and development NGOs.* Bloomfield, Conn.: Kumarian Press.

Lowell, B. Lindsay, and Rodolofo O. De la Garza. 2000. *The developmental role of remittances in U.S. Latino communities and in Latin American countries: A final project report.* Washington, D.C.: Inter-American Dialogue and the Tomas Rivera Policy Institute.

Meyer, John-Baptiste, and Mercy Brown. 1999. Scientific diasporas: A new approach to the brain drain. Prepared for the World Conference on Science, UNESCO-ICSU. Budapest, Hungary, June 26–July 1.

Meyer, John-Baptiste, Jorge Charum, Dora Bernal, Jacques Gaillard, José Granés, John Leon, Alvaro Montenegro, Alvaro Morales, Carlos Murcia, Nora Narvaez-Berthelemot, Luz Stella Parrado, and Bernard Schlemmer. 1997. Turning brain drain into brain gain: The Colombian experience of the diaspora option. *Science-Technology and Society* 2. Available at http://sansa.nrf.ac.za/documents/stsjbm.pdf.

Orozco, Manuel, with Michelle Lapointe. 2004. Mexican hometown associations and development opportunities. *Journal of International Affairs* 57: 31–49.

Pires-Hester, Laura. 1999. The emergence of bilateral diaspora ethnicity among Cape Verdean–Americans. In *The African diaspora: African origins and New World identities,* ed. Isidore Okpewho, Carole Boyce Davies, and Ali A. Mazrui. Bloomington: Indiana University Press.

Portes, Alejandro. 2003. Conclusion: Theoretical convergencies and empirical evidence in the study of immigrant transnationalism. *International Migration Review* 37: 874–92.

———. 2006. Remarks at the Bellagio dialogue on migration, closing conference. Bellagio, Italy: German Marshall Fund and the Rockefeller Foundation, July 14–15.

Portes, Alejandro, Cristina Escobar, and Alexandria Walton Radford. 2007. Immigrant transnational organizations and development: A comparative study. *Immigration Migration Review* 41: 242–81.

Portes, Alejandro, and Mi Zhou. 1993. The new second generation: Segmented assimilation and its variants (interminority affairs in the U.S.: Pluralism at the crossroads). *Annals of the American Academy of Political and Social Science* 30: 74–97.

Rumbaut, Rubén, and Alejandro Portes, eds. 2001. *Ethnicities: Children of immigrants in America.* Berkeley and Los Angeles: University of California Press.

Sánchez Gibau, Gina. 2005. Contested identities: Narratives of race and ethnicity in the Cape Verdean diaspora. *Identities: Global Studies in Culture and Power* 12: 405–38.

Taylor, Donald M., and Winnifred Louis. 2004. Terrorism and the quest for identity. In *Understanding terrorism: Psychosocial roots, consequences, and interventions,* ed. Fathali M. Moghaddam and Anthony J. Marsella. Washington, D.C.: American Psychological Association.

Vertovec, Steven. 2002. Transnation and skilled labour migration. Presentation at the conference "Ladenburger Diskurs 'Migration' Gottlieb Daimier-und Karl Benz-Stiftung." Ladenburg, February 14–15.

Waldinger, Roger, and Cynthia Feliciano. 2004. Will the new second generation experience "downward assimilation"? Segmented assimilation re-assessed. *Ethnic and Racial Studies* 27: 376–402.

Waters, Mary C. 1999. *Black identities: West Indian immigrant dreams and American realities.* Cambridge: Harvard University Press.

3 Digital Diaspora
Definition and Models

MICHEL S. LAGUERRE

> *By virtual diaspora, I mean the use of cyberspace by immigrants or descendants of an immigrant group for the purpose of participating or engaging in online interactional transactions. Such virtual interaction can be with members of the diasporic group living in the same foreign country or in other countries, with individuals or entities in the homeland, or with nonmembers of the group in the hostland and elsewhere. By extension, virtual diaspora is the cyberexpansion of real diaspora. No virtual diaspora can be sustained without real-life diasporas, and in this sense it is not a separate entity, but rather a pole of continuum.*
>
> —MICHEL S. LAGUERRE, "Virtual Diasporas: A New Frontier of National Security," cited in "Incipient Soviet Diaspora: Encounters in Cyberspace," by Larisa Fialkova and Maria N. Yelenevskaya

The goal of this essay is to assess how information technology has affected subaltern diasporic communities in the Silicon Valley and San Francisco metropolitan area and to develop a theoretical understanding of the diverse manifestations of the problem so that genuine public policy can be engineered.[1] The emphasis of the debate on the digital divide has overshadowed the other problems that the interface between information technology and diasporic communities has generated. This essay intends to provide a more adequate conceptualization of the issue that both reflects and sheds light on the conditions of these digital diasporic communities.

DEFINING THE DIGITAL DIASPORA CONCEPT

The notion of diaspora, which is used to refer to an immigrant group outside its homeland, has evolved to that of "digital diaspora," which reflects the engagement of its members in activities related to information technology (IT) (Laguerre 2005). Since their virtual performances may involve more than just the homeland and any specific enclave, it is important to sort out what is included under digital diaspora and to provide an operational definition so that its identitary characteristics can be spelled out.

Three building blocks are required as social infrastructure in order for a digital diaspora to emerge as part of the fabric of society. They are "immigration," "information technology connectivity," and "networking." A digital diaspora cannot exist without immigration. Therefore, immigration is necessary to make the individual or group *diasporic*. The only exception to this general rule is when a diaspora comes about as a result of the redesigning of the national borders of a country. Some residents of the traditional homeland may then become citizens of a new national territory.

IT connectivity is another requirement because it is the backbone of digitization. A diaspora becomes digital to the extent members of the group can access and use telecommunication instruments as a mode of information and communication to reach local and distant contacts. In other words, IT connectivity makes it possible for the diaspora to express and perform its *digital* identity.

Networking is also a condition to attain because it requires one to make or entertain contacts with others in order for the "netizens" to function as a virtual community. Online community status is realized through networking, which makes the entire process *operational*. A digital diaspora then becomes operative when members of the immigrant group engage in cyber communications with other participants in their networks of contacts.

The criteria discussed above provide a frame of reference for the elaboration of a definition of digital diaspora. In this light, I argue that *a digital diaspora is an immigrant group or descendant of an immigrant population that uses IT connectivity to participate in virtual networks of contacts for a variety of political, economic, social, religious, and communicational purposes that, for the most part, may concern either the homeland, the host land, or both, including its own trajectory abroad.*

The interface of a diaspora with IT that turns it into a digital diaspora has different incarnations. For example, a digital diaspora may be identified as a *virtual community* because multiple individuals with the same ethnic background or national origin belonging to a multitude of host-land sites converge their online activities in the creation of one or more interactive virtual sites. However, to the extent that these individuals participate in the everyday life of a diasporic enclave, their online activities necessarily contribute and influence the shape of their community of residence. In this sense, the virtual diasporic community is a cyber expansion and the other façade of the community of residence. For this reason, I conceive of the digital diaspora as the interweaving of the virtual and the real in the hybrid production of everyday life in an immigrant enclave.

In a second incarnation, a diaspora becomes digital because of both the positive and the negative impacts of the IT revolution on the immigrant enclave. While in the first instance, the community is in control of the way it uses this new technology, in the second example, the IT revolution is seen as responsible for the gentrification of the place of residence and the displacement of longtime dwellers on fixed incomes. The lower use of IT among this group qualifies them for being unconnected, but not nondigital, since they are immersed in a digital environment on which they depend to meet some of their needs. The IT revolution swept them willingly or unwillingly into the orbit of the digital environment.

In a third incarnation, a diaspora becomes digital because of the global penetration of the IT revolution into the community and the global outreach of residents to meet their local needs and improve their lifestyles. One can think of telemarketers based in India, for example, that reach these local communities for the purpose of selling manufactured products or providing services on behalf of American firms. Here IT is used to globalize markets and to deghettoize diasporic communities. In this case, external agents play an important role in the way in which community residents are brought to participate in specific sectors of the global digital economy.

DECONSTRUCTING THE DIGITAL DIASPORA QUESTION

The IT literature on diasporic communities has focused on a number of themes including the *digital divide* to explain questions of access between the haves and the have-nots in order to show the increasing or decreasing size of the gap (Compaine 2001; Wresch 1996); *race* to explain its invisibility online or its manifestations in different shapes in cyberspace (Nakamura 2002; Kolko, Nakamura, and Rodman 2000); *gender* to discuss how it has impacted the division of labor among minorities in terms of job displacement, telecommuting, and promotion at the workplace (Matthews 2002); *community informatics* to explain how projects undertaken by the government, nongovernmental organizations, and established philanthropic organizations have contributed to upgrading the level of Internet participation and empowerment of these communities (Servon 2002; Keeble and Loader 2001); *virtual communities* to explain how diasporans use the Net to communicate with the homeland, to maintain relationships with coethnics, and to access information and services in the private and public sectors (Matei and Ball-Rokeach 2002); *public policy* to discuss ways in which the local or federal government can intervene to alleviate the burden of these communities (Pellow and Park 2002); *online religion* to discuss how they acquire information about their religion via the Internet, how the religious netizens sustain each other in the practice of their faith, and how some churches provide services in virtuality to their members (Dawson and Cowan 2004); *new media* to explain how diasporans are able to access and produce online media for the consumption of members of their ethnic group and the homeland (Tynes 2007); *development* to examine how they can contribute to the modernization and democratization of their country of origin (Brinkerhoff 2004); and *national security* to determine how they may constitute through their online interaction a challenge to the preservation of peace and order in their country of residence (Latham and Sassen 2005).

To explain how IT has imploded inside subaltern diasporic communities in the Bay Area, this essay focuses on five topics: *digital marginality* in reference to African Americans who have suffered from a double process of exclusion online and off-line, showing how one type contributes to reinforcing the other and vice versa; *empowerment* as individuals access the Net or are introduced to it by government or grassroots organizations to ameliorate their economic conditions and social circumstances; *displacement* in reference to the flight of San Francisco Mission District Latinos to other quarters as a consequence of the invasion of the dot-commers that raised the cost of renting and contributed to the tightening of the housing market; *technopolization* in reference to Indian high-tech workers' transformation of their neighborhood in Fremont into a technopolis, a success that other immigrant communities have been unable to achieve; and *globalization* in reference to both the penetration of ethnic enclaves by online marketers and the use of the Net by individuals to deghettoize their local conditions.

The interplay between IT and subaltern diasporas has not produced homogeneous results, and in this essay I explain the heterogeneity of these outcomes. The digital modulation of these communities has different contours because of their different IT practices and experiences. It is shaped by a lack of access to digital devices for some (digital marginality), by gentrification for others (displacement), by a concentration of technical expertise in a specific enclave (technopolis), by individual upward mobility (empowerment), and by transnational practices (globalization) for still others. The bulk of the analysis operationalizes and elaborates these five models in order to assess the modus operandi of the recent transformations of these communities.

THE DIGITAL DIASPORIC MARGINALITY MODEL

If the digital divide is the answer, what is the question? Social exclusion is a system of meanings and practices that operates at various levels and in various domains and tends to reproduce itself over time. It is important therefore to identify and analyze the sites where digitization is the mechanism that separates one side (those who have access to the Internet) from the other (the unconnected). I think that there are many digital divides in society, and poverty is only one aspect of this multifaceted problem (Wresch 1996; Loader 1998; Zlolniski 2006). In regard to digitization, social exclusion is the result of a number of processes, among them *exclusion-embedded design, appropriation, access, usage, policy,* and *reproduction.*

Exclusion-embedded design: The design of the software and hardware is made for commercial profit and not for social inclusiveness. So it is a profit-driven design. Profit-driven design is made for the common user without reference to race, ethnicity, national origin, or gender that would make it more specific and perhaps more complicated to produce. In any case, it would require the hiring of those with this kind of expertise who can identify what the cultural assumptions and expectations are. But a specialist on race or gender may not have the knowledge of assumptions for different cultural groups. That's why sensitivity to these issues is likely to lead to a far more complicated production process.

Thus, exclusion works here at the conceptual level during the designing process itself and before the design is implemented. Design presupposes a master idea since the product is the embodiment of that idea or view. What is it good for, who is going to use it, and for what purposes? The making of a tool requires and presupposes a number of alternative choices as design is a choice of one shape against other possibilities: a design that will suit some people (educated elite) more than others (illiterates), those who can speak some languages (standard European languages) more than others (patois speakers), and those who have a need for the machine (the need to communicate with friends who have a computer or the need to access information from elsewhere) more than others (the poor and homeless who are more involved in securing their daily food and shelter).

Exclusion-embedded appropriation: Since the tool is made for specific use, the publicity for its sale will target specific groups more than others. Exclusion is inscribed in the process at the publicity level. Exclusion at the publicity level privileges one group at the expense of the others. The tool is made and sold to meet the taste and tasks of specific users. One may speak of exclusion in taste making and task making. Appropriation presupposes the marketing (availability of tools) and purchasing (availability of money). Exclusion is embedded in both processes: the tool is made available in some sites and markets and is less expensive there, while it is not available or more expensive in others. Income disparity further explains the divide: those who can and those who cannot acquire the item because of poverty.

Exclusion-embedded access: There are various routes to access a computer. Those who can afford one, buy one; those who work for a company may get one from their employer; or one may access a computer in public places that make such tools available to the public for free or for a fee such as the municipal library, the neighborhood kiosk, or the Internet café. In addition, the teaching of certain skills is necessary to allow people to use it. The state's or city's goal in teaching these skills is to allow individuals to be good citizens, to participate in public debate and civic life, and to be employable and capable of paying taxes. Access means also access to the language of the machine. Some groups and neighborhoods have faster and better access than others. For example, universities make available to students, staff, and faculty a much larger menu of databases than does the local neighborhood branch of the municipal library. In contrast, the ordinary individual who uses his neighborhood library must pay for the use of these same for-profit databases if they are available at all.

Exclusion-embedded usage: Lack of availability may restrict or limit usage. The person who must pay to use a database may decide not to use it. Usage for learning can be done at the neighborhood library, but this is not the place to do economic transactions, not only because the line may not be secure but also because a public place simply does not facilitate these kinds of activities. Short of owning a home computer, one may be restricted from engaging in such economic transactional activities.

Exclusion-embedded policy: Social exclusion can be generated by the *public design* of public policies. City governments design IT policies that discriminate against certain groups and favor others, such as the business community. The business sector is seen as generating income, while the poor neighborhoods are viewed as asking for entitlements. These two agendas of the city must be adjusted in relation to each other. Policies facilitate companies to place underground cables in one district (business area) but not in others (poor residential area) or more cable capacities for broadband in one area than in the others. This *infrastructural discrimination* provides the context for structural discrimination that embeds cultural discrimination and the reproduction of one aspect of the divide.

Exclusion-embedded reproduction: The digital divide can be seen as producing inequality because of lack of access, but it can also be seen as reinforcing inequality in the case of those who have already been marginalized. IT therefore creates the divide where it did not exist before, as its introduction marginalizes those who cannot access it. One witnesses a double process of marginality whereby social exclusion reinforces digital marginality and digital marginality in turn reinforces social exclusion. One strengthens and thereby reproduces the other.

The architecture of digital exclusion is made possible by a double process of vertical and horizontal marginality. One must then distinguish the process by which economic poverty engenders digital exclusion because of a lack of access (horizontal marginality) from the process by which digital marginality is self-imposed (lack of motivation) or because of government policies (unavailability of digital infrastructure or vertical marginality).

THE DIGITAL DIASPORIC EMPOWERMENT MODEL

Digitalization has caused, on the one hand, the marginalization of some sectors of the population, and, on the other hand, it has been beneficial to others (Jordan 1999; Mehra, Merkel, and Bishop 2004). So the *empowerment model* is the other side of the *digital divide model.* In other words, there are positive aspects to digitalization that benefit diasporic communities. It must be said at the outset that the same community may experience both aspects at the same time, or may experience one aspect at one time and the other at another time.

Digitization empowers marginal peoples in many different ways, as the initiative may come either from within or from without. Those who must stay at home because of family obligations (taking care of children, elderly parents, or handicapped persons) have been isolated from the labor force, but with computer access they are able to look for jobs that do not require physical presence in the workplace but simply a telepresence (telecommuting jobs); they can compete with others (nonhomebound individuals), and they can even be members of a labor union (Huws 1984, 6). What the Internet does here is to help someone who is homebound to return to the labor force via online mechanisms.

IT demarginalizes home workers in their own location, turns a disadvantageous position into a competitive one, and turns isolation into participation. Empowerment is effected not only in the realm of employment but also in other aspects of social life. Through computer access and use, one may participate in the political process, the Internet economy, and civic activities. So empowerment is social, communicational, religious, economic, and political.

IT undivides what was previously divided. The divide and undivide are part of the transformational process. The computer produces both ends. It makes it possible for marginalized people to penetrate and access middle- and upper-class circuits. For example, as members of virtual communities their true lower-class identities are not revealed, something they could not hide in physical-interface interaction. Social mobility online can lead to social mobility off-line the same way that marginality online can lead to marginality off-line. If knowledge is power, the disenfranchised have access to the same information that is available to everyone online. Part of their poverty was related to a lack of access to information. They could not access this information because of lack of free time to go to the library or take classes, inability to afford the cost involved, transportation issues, and so on. The computer eliminates some of these problems that play to the disadvantage of the poor.

This empowerment can also be engineered by outside institutions and may take various forms depending on what is targeted and the outcome that is projected. The creation of IT neighborhood centers to train the unemployed is a specific practice whose goal is to return some individuals back to being productive members of society. Placing computers in public places, such as branch libraries, allows individuals access for whatever purpose they wish to use these digital devices. Sometimes demarginalization works better at the social (use of e-mail or the Internet for personal growth) than at the economic (use of the Internet to buy things online) or political level (to contact city hall agencies, file tax returns online, or read political platforms of candidates for public offices); sometimes it works with equal effectiveness in all of these domains.

Empowerment also occurs as the use of the computer allows a global space for social interaction. Here again this access to global institutions and practices can be beneficial in some domains like telecommuting for an overseas firm more so than in the domain of social interaction. In any case, this form of practice deisolates the individual and relocates him or her in a more global social universe of interaction.

This empowerment may also come about as a reinforcement of local practices. The computer is used not only to contact people in faraway places but also to interact with neighbors. Sometimes these are individuals, who already know each other, who use the computer to maintain contacts or advance a community or church cause. In other cases, these are people who meet online. Neighborhoods that have online newsletters provide a forum where community members can discuss matters of common interest and in the process can become friends. *Community empowerment* may enhance individual empowerment. One may have *individual empowerment* without community empowerment as individuals may use their access to the computer for social mobility, as a means that allows one's flight from the neighborhood. Sometimes this empowerment is *generational,* as youngsters become the beneficiaries of the process and not the parents who may not be able to take advantage of the Internet. In some cases, community empowerment is highlighted when the Internet is used as an organizing tool to recruit and assemble people so as to stage a protest and protect one's rights (Mele 1999).

THE DIASPORIC DISPLACEMENT AND GENTRIFICATION MODEL

This model explains that the rise of the high-tech industry in Silicon Valley has affected some lower-income communities more than others, which has resulted in the flight of community members to other city quarters. This displacement is seen in the area of housing caused by the rise in the cost of purchasing homes and of renting in general because middle-class newcomers were ready to pay more for these living facilities.

Sometimes displacement is caused by the loss of jobs, and therefore residents become unable to keep the property since they have no means to pay for it. More often than not, this form of displacement affects some households but not others and is not likely to leave a visible mark on the community because some individuals are likely to rely on their savings until they land a job. Its characteristic is that it tends to occur at a downturn of the economy.

The displacement model developed here is linked to three associated phenomena: gentrification caused by the incoming dot-commers, the rise of the cost of housing, and the flight of residents to a place whose monthly cost they could afford. The peculiarity of this displacement is that it did not happen at a downturn of the economy but rather when the high-tech economy was at its peak. These individuals were recruited to work in the high-tech industry of Silicon Valley or the services facilities it generated. This displacement takes the following forms: replacement of apartment users by dot-commers and the flight of previous residents to other quarters. This occurs either without the alteration of the structure of physical environment or with the bulldozing of structures by developers in order to erect new office spaces for the incoming dot-commers.

Displacement obviously affects those who must go, but also the old residents, as they must interact with the newcomers. So there is an *upturn* impact (flight of some old-timers) and also a *downturn* impact (adaptation of the remaining population to dot-commers). Those who must go experience the discomfort that accompanies displacement and relocation. Those who stay are impacted by the loss of affordable apartments, new regulations for street parking, and the inability to access places that were previously accessible.

New jobs were made available for the upkeep of buildings (janitors, maids, security personnel), for providing services (secretaries, cooks, and so on), and to staff new facilities financed by philanthropic organizations (computer training for youth, for example). Nearby restaurants and other facilities (shops, dry cleaners, and banks) derived some profit from the patronage of newcomers.

Gentrification developed and evolved as newcomers, mostly middle-class Anglo individuals and families, settled in the Mission District neighborhood. These individuals were often of different social and ethnic backgrounds but took advantage of newer, perhaps cheaper, apartments than what they could find in other quarters, made available to them because they could afford the increasing cost of rent. It is disciplined gentrification in that those who come do so exclusively for the purpose of working in computer-related industries or ventures. This homogeneity of the enclave IT workforce contrasted with the diversity of the surrounding panethnopolis. Since some of the newcomers work but do not live there or live but do not work there, this entails that a group of commuters will appear every day in the neighborhood. Their presence is felt on a part-time basis in the neighborhood. The eventual mushrooming of *upscale* bars, restaurants, and clubs develops to serve this exclusive group. As a consequence of this silent invasion, more police are hired to patrol this area and protect the dot-commers and their property.

The early cultural, aesthetic, and spatial separation of old-timers from newcomers can be seen in the way in which gentrification took place in the Mission District of San Francisco. The old residential community was Spanish speaking, while the newcomers were mostly monolingual English speakers; new building structures that welcomed the newcomers contrasted with the old building structures; and the newcomers lived in a specific ghettoized portion of the neighborhood vacated by the flight of the former residents, while the older residents were more dispersed and covered a more extensive geographical area.

THE DIASPORIC TECHNOPOLIS MODEL

The diasporic technopolis is conceived as a high-tech enclave that is produced by elite labor migration, which displays a concentration of expertise in a given locale and reflects the technological work in which a substantial segment of the group is engaged. Migration to the technopolis is wholly influenced by the high-tech activities that attract them, including their family members. What this definition stresses is the importance of high tech in the construction and daily life of the community, how it shapes migration patterns, how it influences transnational relations with homeland experts and technological centers, and how and why such an enclave is different from other diasporic enclaves. A diasporic technopolis is fundamentally an urban enclave that develops its identity because of the concentration of expertise it houses, its visible contribution to the regional industrial system, and its vibrant transnational connections to the homeland.

The majority of the people are either directly involved in high tech or are related to individuals who are or who have influenced the direction of their migration. The enclave is a magnet for the secondary migration of members of the group either for family reunification, for schooling and training, or for labor purposes. The enclave tends to hold a status higher than that of other ethnic enclaves because of the educated-elite status of its residents, its superior economic position, and its political influence because of its contribution to the regional and national economy. The enclave is a hotbed of entrepreneurial activities that further reinforce the visibility and superior economic status of the residents.

However, a concentration of expertise is not enough to make it a viable technopolis. It is important to have a sustainable community made possible by the social institutions it develops (church, school, newspapers) and the social organizations it forms (professional, recreational, religious, political, economic). High tech tends to influence the shape of these institutions, from membership to function significance. Its enclave status means it is related to a larger technopolis that it feeds by its labor and that also influences its shape. It is so because diasporic technopolises do not emerge by themselves but rather in relation to a larger urban system. These experts do not necessarily work at first for themselves in firms they create, but often they begin by working for others. But over the years, some may end up establishing their own companies either by themselves or in association with noncompatriots.

The technopolis comes about as a result of a number of firms and state policies and practices. First, a shortage of computer specialists and the growth of the information technology industry led to the recruitment of experts who migrated to the area (industry practices). Second, once established, these experts recruited their friends and classmates, and otherwise made the area known to prospective international migrants in their homeland. And third, the U.S. government—through the H-1B visa program—facilitated the migration of these experts (public policy). These combined factors have contributed to the development of the Indian technopolis in Fremont, Silicon Valley. The diasporic technopolis is one specific form of the technopolis and oftentimes emerges as a singular feature of the multicultural metropolis.

A technopolis is then a city that has as its primary focus the production, application, and commercialization of technical knowledge and fundamentally reproduces itself through these activities, intranational and international migration. Failing to reproduce itself either because of generational change of orientation or the transformation of the territorial space, the technopolis is likely to be transformed into an ethnopolis or panethnopolis.

THE DIASPORIC GLOBALIZATION MODEL

This model presents the local and the global as two interacting aspects of the same continuum. It argues that IT allows residents to deghettoize themselves in using the Internet as a way of interacting with people outside their neighborhood. This expansion of interaction has economic, political, and social dimensions that can be transcribed in terms of profits accruing to these individuals. The economic aspect of neighborhood globalization can be seen either in the neighborhood serving as the initiating process of the phenomenon or in its point of destination. By that I mean that individuals do take advantage of that connection to global networks in order to search for employment, to relay output, or to telecommute as it happens in some forms of outsourcing where it is used to gain employment or to sustain it. As a site of destination, external entities (telemarketers, overseas firms) penetrate the neighborhood to use its talent or to sell things as a way of expanding the global market environment.

Digitalization divides or fragments the neighborhood into the local and the global residents. This segmentation is influenced by Internet access, but also by time (immigrant generation), social positions, and one's incorporation into new social networks. Through the Internet, one may participate in various forms of homeland politics or lobby on behalf of the homeland.

The use of the Internet to communicate with the homeland fragments the neighborhood by ethnicity. Online practices following ethnic lines undo the panethnopolis character of the neighborhood at the virtual level and turn it into a mix of juxtaposed ethnopoles. So the online practices, far from reinforcing the character of the off-line spatially contiguous community, project in virtuality its ethnic diversity. While the community may be united over local goals, it may not be so over global goals because they derive from contingencies of different homeland needs. The Internet also reinforces the ethnopole features of each cultural segment because of the online connections with compatriots who live in the city or nation, but not in the panethnopolis.

How does the community manage the online versus off-line fracture? Could this be a source of tension? Can the community live with different rhythms caused by different ethnic holidays, for example, and different political crises in the homeland that resonate with different rhythms in the panethnopolis? Do the subalterns become concerted to the rhythms of the hegemonic ethnic group in the panethnopolis to the extent that Cinco de Mayo is celebrated by all? Or do ethnic holidays (a Puerto Rican parade or Guatemala Independence Day) become celebrated by all? What kind of ethnic marginalization (different from marginalization caused by the mainstream) is at play? Is Cinco de Mayo hegemonic and other Hispanic festivals (Salvadoran, Guatemalan, Chilean) subalternized in the Mission District by the Mexican American majority? Does the community represent itself, or is it represented by others as hegemonic Hispanic or rather as Mexican American, thereby giving more weight to the Mexican connection than other South American links?

Asynchronization of ethnic festivities is reinforced by the Internet in the neighborhood because of denser interactions among coethnics and the homeland. Different holidays are peak periods for some ethnic groups while not celebrated by other ethnics. So peak periods of interaction on the Net may be influenced by peak periods off the Net and vice versa.

A major consequence of the revolution in information technologies is the transnationalization of diasporic practices (Van den Bos 2006). This transformation of diasporic communities occurs because of the transnational and global practices it generates and facilitates. These practices are used as a factor in the translocation of aspects of diasporic life.

In this sense, the use of the computer and the Internet has contributed to the delocation and respatialization of practices. It gives access to jobs located elsewhere, bypassing physical presence in the search, interview, presentation of the self, contractual engagement, assignments, and fulfillment of tasks. It allows individuals who cannot find jobs in their locality to seek employment beyond their place of residence. It allows the cyber expansion of social interaction beyond one's neighborhood through participation in virtual communities and online discussions. It allows the request for services beyond one's locality, as in the case of online sales and transactions. It allows diasporans to maintain transnational relations with the homeland, thereby keeping alive the host-land and homeland circuit and the transnationalization of the diasporic community. This globalization of practices seems to be limited to individuals who have computer skills and have access to computers. In this sense, IT redivides the population into global and local people.

This global interaction facilitates migration to the diasporic site. It facilitates the ongoing connections between and among various diasporic sites in more than one country. It makes it possible for diasporans to follow the vagaries of everyday politics in the homeland. For example, homeland newspapers are available online, letters to the editor can be sent via e-mail, and requests for information are done by e-mail as well. It contributes to the vibrancy of the online public sphere. It allows the invasion of the enclave by outsiders, like telemarketers and others in terms of opening the enclave to market penetration (extension of markets), but it also allows diasporans to look for less expensive things to buy online. It makes it possible for residents to maintain a physical presence in the neighborhood while keeping a virtual presence globally. It leads to a disjunctive physical ghettoization at the local level and social deghettoization at the global level.

It facilitates the global networking of neighborhood activists as they link up with activists worldwide. This globalizes neighborhood problems, thereby inducing local actors or officials to pay more attention to these demands.

It also creates new divisions in the neighborhood. In a panethnopolis, various ethnic groups link up differently with their groups and the homeland. Ideological divisions become transnationalized as well, as each group expands its sphere of interaction in support of a homeland group with similar orientation.

CONCLUSION

This analysis of the interface of diasporic communities with IT allows us an opportunity to explicate the modalities of the production and engendering of the digital diaspora phenomenon. It is not enough to speak of the digital diaspora as a given, yet it is important to delineate a definition that both derives from and reflects everyday social and transnational practices. Digitization has enhanced the sustenance of global interactions in immigrant enclaves, provided tools for the creation or participation in virtual public spheres to discuss matters of common concerns to the group, and generated new perceptions of social reality and contributed to new forms of organizational practices.

From a practice standpoint, the digital diaspora is characterized by the *mobility* that molds the flows of its internal, external, and cross-border interactions; by the multimedia environment that feeds its communicational and informational modes of interventions and performances; and by its *global engagement,* which reflects its ongoing relations with multiple diasporic sites and the overseas homeland.

Mobility is a constitutive feature of digital diasporic practices not simply because of the immigrant characteristic of the community, which entails movement from one place to another, but also because of the cross-border interactions that contribute to the constant social construction and deconstruction of virtual nomadic structures. One refers here to not only the mobility of virtual institutions but also that of agents, tools, and interactions.

The digital diaspora evolves within the context of a multimedia environment. This leads to a flexible regime of practice, which makes it possible for the activities diasporans engage in to take different virtual forms. Individuals may use different media at different times to accomplish different things, or at the same time for different reasons, such as personal preferences or as a task requires. For example, they may use a multipurpose tool, a landline or a mobile device, or an interoperable gadget. The virtual identity of the digital diaspora is revealed through the multiform use of these multiple types of digital media.

Finally, globality, in reference to the diverse locations of sites in different host lands, projects the geographical parameters within which virtual interventions of diasporans take place. While transnationalization tends to highlight diaspora-homeland relations, globalization enlarges the parameters by including diaspora-diaspora relations as well. The digital diaspora navigates its international arena of practice in a universe dotted with homeland and diverse host-land sites as the primary foci of its digital interventions.

NOTES

1. Revised version of a paper prepared for delivery at "The Networked World: Information Technology and Globalization: An International Conference," sponsored by the Center for Science, Technology, and Society, Santa Clara University, April 23, 2003. An abridged French version of an earlier draft of this paper was presented at the Centre de Recherche sur l'Immigration, l'Ethnicité et la Citoyenneté at the University of Québec–Montréal, June 10, 2003.

REFERENCES

Brinkerhoff, Jennifer. 2004. Digital diasporas and international development: Afghan-Americans and the reconstruction of Afghanistan. *Public Administration and Development* 24: 397–413.
Compaine, Benjamin. 2001. *The digital divide: Facing a crisis or creating a myth?* Cambridge: MIT Press.

Dawson, Lorne L., and Douglas E. Cowan. 2004. *Religion online: Finding faith on the Internet.* New York: Routledge.

Fialkova, Larisa, and Maria N. Yelenevskaya. 2005. Incipient Soviet diaspora: Encounters in cyberspace. *Croatian Journal of Ethnology and Folklore Research* 42, no. 1.

Huws, Ursula. 1984. *The new homeworkers: New technology and the changing location of the white collar workers.* London: Low Pay Unit.

Jordan, Tim. 1999. *Cyberpower: The culture and politics of cyberspace and the Internet.* New York: Routledge.

Keeble, Leigh, and Brian D. Loader, eds. 2001. *Community informatics: Shaping computer-mediated social relations.* New York: Routledge.

Kolko, Beth E., Lisa Nakamura, and Gilbert B. Rodman, eds. 2000. *Race in cyberspace.* New York: Routledge.

Laguerre, Michel S. 2005. *The digital city: The American metropolis and information technology.* New York: Palgrave Macmillan.

Latham, Robert, and Saskia Sassen, eds. 2005. *Digital formations: IT and new architecture in the global realm.* Princeton: Princeton University Press.

Loader, Brian D. 1998. *Cyberspace divide: Equality, agency, and policy in the information society.* New York: Routledge.

Matei, Sorin, and Sandra J. Ball-Rokeach. 2002. Belonging in geographic, ethnic, and Internet spaces. In *The Internet in everyday life,* ed. Barry Wellman and Caroline Haythornthwaite. Malden, Mass.: Blackwell.

Matthews, Glenna. 2002. *Silicon Valley, women, and the California dream: Gender, class, and opportunity in the twentieth century.* Stanford: Stanford University Press.

Mehra, B., C. Merkel, and A. P. Bishop. 2004. Internet for empowerment of minority and marginalized communities. *New Media & Society* 6, no. 5: 781–802.

Mele, Christopher. 1999. Cyberspace and disadvantaged communities. In *Communities in cyberspace,* ed. Marc A. Smith and Peter Kollock. New York: Routledge.

Nakamura, Lisa. 2002. *Cybertypes: Race, ethnicity, and identity on the Internet.* New York: Routledge.

Pellow, David Naguib, and Lisa Sun-Hee Park. 2002. *Silicon Valley of dreams: Environmental injustice, immigrant workers, and the high-tech global economy.* New York: New York University Press.

Servon, Lisa J. 2002. *Bridging the digital divide: Technology, community, and public policy.* Malden, Mass.: Blackwell.

Tynes, Robert. 2007. Nation-building and the diaspora on Leonenet: A case of Sierra Leone in cyberspace. *New Media & Society* 9, no. 3: 497–518.

Van den Bos, Matthijs. 2006. Hyperlinked Dutch-Iranian cyberspace. *International Sociology* 21, no. 1: 83–99.

Wresch, William. 1996. *Disconnected: Haves and have-nots in the information age.* New Brunswick: Rutgers University Press.

Zlolniski, Christian. 2006. *Janitors, street vendors, and activists: The lives of Mexican immigrants in Silicon Valley.* Berkeley and Los Angeles: University of California Press.

4   An Activist Commons for People Without States by Cybergolem

ANDONI ALONSO AND IÑAKI ARZOZ

CYBERGOLEM AS A DIASPORIC AND ACTIVIST ENTITY

We must begin this article writing about the deep connection between the purpose of this paper and its dual authorship.[1] The "we" refers to an activist and intellectual node composed by artists and scholars who have joined together to generate theoretical work under the label of *Cybergolem*. Thus, the "we" used in this paper is something more than a rhetorical strategy. Computer networks have produced a new understanding of authorship such as the one shared by two or more authors. Wu Ming,[2] also known as Luther Blisset, is a good example of the recent changes concerning the very idea of authorship. This new understanding of authorship is not arbitrary.

Hypertext and hypertextualism are new writing technological devices that require fresh strategies. For instance, communication technologies erase the spatial distance among the different authors who belong to Cybergolem. In this particular case, the authors who write this paper have been working for more than fifteen years under different pseudonyms. Cybergolem has become something more than a nickname for collective authorship. It is an entity that changes, acquires different forms, and tries to reelaborate the idea of authorship in each work. Sometimes the authors involved are two, and sometimes there are more than two.

Cybergolem began as a digital hetero-identity composed of a Basque philosopher and artist. Later on, other authors and readers from the Basque Country and elsewhere joined Cybergolem. What we are trying to underline is that Cybergolem is also diasporic, in the sense that it is a digital entity. Therefore, we need to refer to its diasporic "life history" in order to be able to explain our peculiar focus on digital diasporas. One of the authors works in his Basque homeland; the other has worked as a nomad in different places as a Basque diasporan: Madrid, Cáceres, Pennsylvania, and Reno. According to Vance Stevens (2007), computer manager for the Petroleum Institute, those types of authors would be *digital immigrants,* because they were born in the *analogical* or *pre-Internet* era. Paradoxically, the Cybergolem entity would be considered a *digital native,* because it was born much later and inside cyberspace. Cybergolem as a Basque diasporic entity as well as a digital native entity would determine our experiences and our focus of research on the Basque digital diaspora. That is to say, it is not about using new technologies to close the existing geographical gaps among the different Cybergolem authors. Our focus can be defined as the effort to build a complex identity that assumes multiple perspectives according to different works and activities.

The beginning of Cybergolem was more conventional and focused primordially on Basque culture and STS (science, technology, and society) approaches. Later on, Cybergolem went through a period of criticism about cyberculture that took the form of books and Web projects such as *La nueva ciudad de dios* (*The New City of God*) (Alonso and Arzoz 2003).[3] That period underlined the birth of a new pseudoreligion composed by new technologies as the ultimate promise (Bloom 1996). A posterior work, *La quinta columna digital* (*The Fifth Digital Column*; Alonso and Arzoz 2005), addressed activist actions inside cyberculture itself. The Basque issue influenced that book very much, as it dealt with questions such as Basque identity and the place for peoples without political states. It is clear that the long, violent conflict experienced in the Basque homeland was an important question that has not yet been fully explored. This activist approach has also marked other Cybergolem interests. Contributions to other collective works such as Tester[4] (2006) or the Free Knowledge Foundation proved the need for a more involved and activist positioning. Some works were published in Basque, English, and Spanish: *Euskal Herria digital 1.0* (The Basque Country Digital 1.0) (Alonso and Arzoz 1996), *The Electronic Forest* (Alonso and

Arzoz 1999), *Destino cyborg* (Destiny Cyborg) (Alonso and Arzoz 2007), and above all *Basque Cyberculture* (Alonso and Arzoz 2002).

Cybergolem is a collective venture that understands research as a diasporic and activist Basque-entity enterprise. Digital diasporas and sociocultural activism become a substantial part of the contemporary Basque diaspora.[5] Basque identity in the era of cyberculture and globalization is a central issue that requires a significant number of careful studies. Although our philosophic and activist approach tries to provide a more global insight—something that could be useful for other diasporas around the world—our experiences come from diverse diasporic and activist Basque positions. Such experiences shape our writing.

DIGITAL DIASPORAS AS (TRANS)VERNACULAR COMMONS

As stated in previous works (e.g., Alonso and Arzoz 2003), we have suggested that cyberculture did not fully emerge at the end of the twentieth century or at the beginning of the twentieth-first century. Latest technologies often hide the origins of an earlier digital milieu that we have placed in a previous technological culture. Somehow culture produced in past eras has been determinant for the rise of the digital culture as we know it today. One of the roots of cyberculture is a Western vernacular culture that preceded industrialization.

Past vernacular cultures dealt with many types of techniques and technologies, but their survival was difficult. The vernacular culture shows different degrees of disappearance around the world. Vernacular milieu, as Illich (1991) described it, consists of a distinct way to relate to objects, customs, ways to produce, or articulation of gender. Illich argued that *homestead, homemade, and homespun* are the main features of that vernacular culture. Somehow new technologies show some analogies with that culture.

For instance, the history of the Internet teaches us how scholars and universities took over a military technology by adding new values and uses and transforming *Arpanet* into something different. The first virtual communities and communication technologies in history accepted egalitarian and communal knowledge (Rheingold 1993). It seems that technologies existed in order to be appropriated by different people. Vernacular culture was transformed by the new digital technological culture, but at the same time the vernacular culture took over the recently born digital culture and added some of its features. The subtitle of Rheingold's seminal work, *Homesteading on the Electronic Frontier,* is not casual.

Activists and libertarians began to consider the Internet as a possible vernacular domain. Somehow the first declarations and manifests about the independence of cyberspace went along this line, because they thought that cyberspace should be considered a commons (Barlow, Toffler, and Dyson 1994). That is, they considered cyberculture an open and free network for communicating, getting and exchanging information, or creating identities. Of course, we argue that cyberspace is not the same as a commons. The wide scope and variety of cyberspace do not allow for such a thing. So very soon such a paradise for a free and an open culture was confronted by many challenges in the same way that traditional commons were confronted in the past.

We have identified various opinions regarding such identification between cyberspace and commons. For example, new Luddites such as Zerzan (1994) seem to describe vernacular culture as belonging to a primitive and wild society. So the new utopia is to regain that primitiveness. However, technologies are the main obstacle to return to that romantic past. Zerzan's position oversimplifies the content of commons. Indeed, there are more ways to rethink the vernacular. Studies on ethnography and architecture propose more detailed and complex interpretations (e.g., see Alonso, Arzoz, and Ursua 1996).

Vernacular culture is also an alternative economic model based on shared property, subsistence, and solidarity (Wright 2001). Some of the European vernacular cultural characteristics have survived in certain historical institutions and customs, as the commons applied to the property of certain lands. In fact, the commons was the kernel for a social, cultural, and economic system in many places. For example, it explains communitarian lands and community work (*auzolan* in Basque) in today's Basque culture.

In other words, one of the best-known features of a commons is the model of property shared by a community. Commoners share benefits while engaging the shareholders in caring and sustaining that common property. Private property excludes communities either from benefit or from responsibility. Many things like air, water, and language still belong to a commons. Recent debates about public goods bring back the model proposed by traditional commons.[6] It could be useful to understand commons as a complete political and economic philosophy as well as an alternative model to our current society, which is based on services and private property. That philosophy could be akin to some libertarian and communitarian models, because they ask for a participative democracy instead of a model based on representatives (Bookchin 2002).

It is easy to find hidden historical ties between the vernacular commons and cyberculture. The worldwide alternative cyberculture movement has relied on a model based on the commons. Free software and free knowledge movements[7] claim such a relation because it underlines their usefulness in order to understand software and knowledge as shared property or public good. The commons reflects different ways of organizing economy, as Wright (1991) evidences. The social, economic, and cultural fabric of diasporas can be included under the commons. Now technology transforms vernacular into trans- or cyber-vernacular. Digital diasporas are a commons for the exchange of information, dissemination of general or personal news, re-creation of cultural memories, and new cultural activities. Therefore, diasporas have acquired the same network structure as the Internet thanks to e-mail, Listservs, chats, blogs, virtual communities, and cell phones. Cyberspaces have become a digital and communal space.

Even more, the pre-Internet-era diasporas that used the weak threads of letters and traveling were not communal until their transformation into digital. Social and communicative commons were renewed due to the digital commons of cyberculture. Cyberculture creates a commons that had almost disappeared and builds ethnic national commons where they never existed. Diasporas can be supported, recovered, reformulated, and even developed as commons due to the existence of a technology with vernacular roots. The interesting issue, therefore, is not to favor the development of a particular digital diaspora but to digitalize diasporas themselves. This could be one of the most powerful ways to connect, unite, and re-create diasporas of people without states, such as the Basques.

The goal is not to revitalize a conventional nationalism using the Internet but to promote a new social and cultural identity in the context of globalization. There was a particular practice among Basques who immigrated into the American West of the United States that exemplifies our argument. Throughout the territory of the American West, Basque shepherds (their main occupation for nearly a century) used to carve with their knives messages, slogans, and rough drawings onto the white bark of poplars (Mallea-Olaetxe 2001). Those carvings showed their condition as Basques, their dreams, and their nostalgia for their homeland. Trees were a primitive way of communication. Understood as a commons, the trees were a means used to address people from the Basque diaspora. The Internet can be seen as an evolved *electronic forest* (Alonso and Arzoz 1998) and an extremely powerful tool for the communicative commons.

THE RADICAL (HETERO-)IDENTITY OF DIGITAL DIASPORAS

Activist researchers on cyberculture defend an alternative globalization able to respect vernacular cultures. Instead of globalization, there is a possibility for a *glocalization* where the global gets its meaning from the local, and from the local we reach the global (Castells 2001). In fact, the global is understood as something that assimilates vernacular values and mixes elements of a global culture (sharing with other cultures) within the local. We argue that glocalization is the only way of survival for vernacular cultures in this century. Cyberculture and globalization have established a new framework for an elective and complex identity (Zurawski 1998). This identity allows for an ethnic self-representation and self-determination, because the user can recreate his or her identity.

But beyond adopting or simulating identity as a fixed picture, cyberculture allows us to interact thoughtfully with our own cultural context and also with other cultures. Virtual life allows access to richer, more flexible, and more diverse identities than a conventional native identity (no matter how complex or intercultural that identity might be). Users' identities experience transformations by living the otherness of other users who might belong to different cultures. Limits and rules for the online world on ethnic origin, religion, gender, even (vernacular) languages are not as determinant as they are in the off-line world. Very often, the off-line world creates ghettoes for nonstandard identities. Although cybercultural utopias on identity have limits, this process becomes a hetero-identity with specific features such as radical, multiform, changing, open, progressive, or elective.

In a sense, an online identity alleviates traditional ethnic requirements and transforms it into something different more suitable for cyberspace. At the same time, this online identity allows for the recovering of vernacular values and ties. That plastic quality of online identity is extremely important for diaspora users of a particular community. It allows coming back to their culture without being considered *second-class* citizens (first class supposedly would require living in the homeland). Being immigrants or natives no longer matters, because living in the digital era transforms everybody into immigrants. The access to cyberspace extends the limits of immigration. Paradoxically, the diasporan is basically a digital native. Therefore, to the same extent, he or she is not a real diasporan because his or her culture is digital. Everybody adopts a diasporan identity in what Bey (2001) qualifies as a *temporally autonomous zone*. That zone offers a freer, more creative cultural identity, a virtual *no-man's-land* that has been populated by different avatars, virtual communities, and digital nations.

Internet and cyberculture have allowed the birth of that radical (hetero-) identity leading to the cyborg identity. As Haraway (1991) claims, the cyborg identity allows for the reinvention of an ethnic identity as a symbiosis of vernacular culture and technological devices, which, by the way, are both of vernacular origin. In other words, this means the return to the vernacular and to an ethnic local universe. At the same time, it creates a universal space that every single country can share.

Understanding digital identity in this way reflects a pragmatic character to explore technical possibilities and not some neo-Luddite attitude of the vernacular. Cyborg identity of ethnic users (those who live their ethnicity on the Net) allows us to talk about a cyborg diaspora as a collective entity (Gajjala 2005). In this context, "collective" means an interconnected entity and not an insectlike hive (Kerckhove 1997), while the term *cyborg* is not used as vague science fiction rhetoric. A cyborg is not a virtual angel, as extropians propose (Hayles 1999). A cyborg is a real identity in Haraway's terminology; a flesh-and-bones individual who uses or interacts with machines in a very close way. Speaking today about cyborgs implies the symbiotic quality of flesh and machines, or, rather, the interaction between the off-line world and the online cyberspatial planet. The cyborg diaspora of diasporic users with their own codes must be the main actor of a possible activist and communalist virus.

DIASPORA AS A COMMUNAL ACTIVISM IN THE NET SOCIETY

Digital diaspora or cyborg diaspora is something more. It cannot only be a (trans)vernacular commons. If it were, it would be transformed into folkloric virtual tourism. For nationalities without states such as the Basques, political institutions are not fully able to develop their own identities, so they must adopt the position of the activist commons. Possibilities opened up by an alternative cyberculture and a cyborg diaspora go much further than simple communications among natives and immigrants in search of genealogical trees inside Mormon databases (e.g., the ScanStone software in Utah) or in building a melancholic diasporic culture. A digital diaspora can transform into a political community and incite the uprising of more engaged political communities by means of activism. Direct contribution to newsgroups and blogs on politics could be one of the ways to reach such a community.

However, for this to happen, it would be required to generate an activist awareness inside the virtual communities of the majority of diasporas without states (such as the Basques). Information nodes and digital services would make people more aware of their political possibilities. Prior to a direct political discourse, there is a realm for activist work consisting of social and cultural realities. That realm has already been used as a first step toward politics in the off-line world. For instance, debates, political campaigns, and polls constitute the fabric for the political off-line world. This could also be the space to generate political awareness in the online world.

Different homeland parties are trying to incorporate diasporans to homeland polls. The first step for peoples without a state could be to establish contact and exchanges among natives and diasporans. That would help to preserve cultural identity features such as folklore and vernacular languages. That effort could allow diasporas to be part of that homeland political life.

Additionally, homeland politicians are beginning to understand that a cyborg community is a vital part of today's world. Little by little many diasporas witness the support and implication of their homeland online. Recognition of the significance of digital diasporas opens up two possibilities for an activist commons. The first possibility could be the transformation of that commons into a loudspeaker for nationalism in order to achieve a state for diasporas. The second is the generation of a virtual nation as the sum of the different activist nodes.

The first option is too conventional and relies on an *analogical logic.* Digital diasporas transform themselves into a companion for the set of nationalistic strategies proposed from the homeland. That goal undermines the real importance of digital diasporas and the possibilities of the digital commons. The second option goes further, because it relies on the potentialities given by digital activism and virtual diasporas. It presents an innovation on the traditional political discourses by reconsidering old nationalistic goals and discourses. Nevertheless, this does not imply the idea of generating a virtual nation as a way to fulfill the need of a state for those nations without one. Ethnic users would create a cybernation or free e-nation based on hetero-identities and on the cyborg identity of digital diasporas. That e-nation does not exclude the off-line world, but it is reinforced and enriched by a cyberspatial commons.

As John Perry Barlow declared the independence of cyberspace with particular features, digital diasporas can declare the independence of digital ethnic communities. They can rely on that discourse but not as a neocolonial struggle for independence. That struggle took place during the twentieth century in the off-line world and ended there. It is akin to the opening of a new space with a libertarian flavor open to different political sensitivities. It should be a nation able to be interpreted in different ways and limited only by due respect according to netiquette and the human rights discourse. Sometimes the off-line world defines the boundaries of states with armies, but a progressive Net society is the best option, the most realistic, and the most fruitful for peoples without states, and not the option of military or terrorist actions.

The question does not imply the denial of a traditional nationalistic point of view. Independence and state are probably overvalued goals in today's politics. Conversely, the idea is to open the ethnic identity to the richness of globalization and cyberculture in an activist and transvernacular commons. A virtual nation, inhabited by cyborg peoples, should remain independent from problems in the off-line state. A national community could survive whether it achieves statehood or not. Maybe, paraphrasing Wittgenstein (2007), questions about states should be dissolved. A cybernation as an activist commons must enrich nations and keep them active and creative whether or not they exist as states. The e-nation is the absolute best diasporic option. The digital diaspora could be an active part of the nation and contribute to its development in the blogosphere or in a democracy, in which diasporans progressively become citizens with all the rights that entails. That is to say, the identity for peoples without a state acquires its full sense due to the diaspora. Only peoples without a state would prevail if they transform themselves into an e-nation or cyborg nation. In this sense, digital diasporas are the off-line nations' secret resource.

DIGITAL DIASPORA IN THE CONTEXT OF A HISTORY OF THE FUTURE

Up to now we have described the current situation of the digital diasporas. However, our rapidly changing times make any statement merely temporary. There are an increasing number of technological studies, under the label of "The History of the Future," devoted to forecasting the future of digital diasporas (Wagar 2001; Kaku 1998). Thinking of the near future has become a crucial necessity for technology as well as for planning the economy. That kind of study bears little resemblance to science fiction fantasies. Maybe the so-called near future or the current cyberpunk may have some common ground. Therefore, it is necessary to speak about the digital diaspora from the point of view of a near future defined by technological and digital communication development.

In this sense, there are new devices (e.g., cell phones or personal digital assistants) not as widely used as the Internet that can become customary. On the one hand, diasporans connected by telephone or e-mail would transform their communications into a sensorial, interactive, and visual connection. The simple existence of the telephone and e-mail changed profoundly the daily life of different diasporas. Before those devices, diasporans had to get used to long delays for letters sent by boat or plane that, many times, were outdated when they arrived. Chats and phone calls over IPs (Internet protocols) have transformed instant familiar communications, as evidenced by many different studies (some of them present in this book). Telepresence devices such as Webcams are becoming extremely affordable for everyone.

Internet systems including videoconferencing are already notably changing interpersonal communications among relatives, friends, groups, and associations within the diverse diaspora nodes and with the homeland.

On the other hand, interactive worlds like Second Life could also create a new kind of diaspora in the long term. It is true that Second Life has not been able to generate a parallel universe, a *metaverse,* as Stephenson (1992) or Egan (1995) have described it. Even the eight million active users on that site are still little involved in Second Life. *Wired* magazine ironically qualifies Second Life as an Empty Planet (Anderson 2007). The forty thousand users who connect habitually are part of a strange and motley community mostly in search of virtual sex or risky business. The rest are virtual tourists who soon become bored of grotesque avatars and primitive interaction.

In fact, Second Life and other similar sites such as Habbo Hotel[8] are the first outcomes of a virtual tendency inside cyberculture. The most important fact is not aesthetics or economy; rather, void advertising is the main goal. In spite of that, those sites show important tendencies, such as online video games. A better technological development is required to become a good opportunity for the diasporas. Global communications among diasporans are based on the simplicity and efficacy of e-mail, chats, and digital phone services such as Skype. Those technologies do not require much money or technical expertise, so they can fit the requirements of the poorest diasporas. Second Life hardly represents a little part of present digital diasporas.

Phenomena like the blogosphere, MySpace, and YouTube are the future for communication in cyberspace, and, of course, those new tools are influencing present diasporas. In the near future convergence of different technologies—cell phones, Wifi and Wimax connections, and laptop computers—will offer an interactive world to be used by diasporans in a utilitarian way. Nevertheless, in another sense there is an activist commons where diasporas can take part as well. Peer-to-peer exchange, "copyleft licensing," free and libre software, communal Wifi sharing, communal production of content like the Wiki phenomenon, and more offer a new and richer scenario for a cultural and politically aware diaspora.

As a result, this scenario for digital diasporas should change in different ways. First, diasporas would be progressively alphabetized in these technologies like cyberspace, common systems, telepresence, and interactivity. Consequently, the current existing digital divide would be eradicated in the long term. Then diasporans belonging to underdeveloped countries would become more aware of their situation and strengthen the content of their diasporas. Connection among different digital diasporas would be a tendency, and the connection with the homeland would be more tight due to these technologies. Connection to homeland and production of particular contents mean a deep awareness of belonging to a diaspora. But in a world where natives and diasporans are equally digital diasporans, there would be an opposite effect. Communications have two ends. Therefore, connection with the homeland implies a progressive homogenization among diasporans and natives. The increase of contact would produce a common ground.

The use of new technologies and the contact with different cultures could have many other effects for recent diasporas. Those new political or economic diasporas would be in touch with strange technologies. That fact could be either positive, as in a mestizo culture, or negative, producing cultural and social shock. For old diasporas, like the European, new technologies can be a blessing because the diasporans are more used to technology. However, all diasporas would be able to enrich and contribute to activist commons in politics and culture. They would innovate in cyberactivism by cultural production, Net art, solidarity projects, political debate in the blogosphere, or collaborative projects on science and culture.

From a realistic point of view there are different prognoses on the History of the Future, but not all of them are equally positive. Climate change and its consequences, the energy crisis, and war could deeply and negatively influence the world economy or the political hegemony of the United States. Therefore, technological developments could be at stake. Migration movements could become massive, which would make difficult the formation of new diasporas in a strict sense. "Invasions" of large numbers of desperate people could appear in Rio Grande or in the Mediterranean area, for instance. The possibilities to create a real diaspora could collapse. But in any case, in any possible scenario, the empowerment of diasporas as activist commons is crucial. An undesired outcome from massive immigration would be a scenario of nationalistic cyberguerrillas. On the contrary, a digital commons could be the result of shared efforts toward social, economic, and political development for those peoples without a state in our likely diasporic planet.

THE BASQUE DIGITAL DIASPORA

Digital diasporas can be one of the best available tools to re-create a social, cultural, and political community for Basques in the short term. Transformation of a Basque identity into a digital identity with a new sense of "Basqueness" would allow re-creating a Cyber–Euskal Herria, a Cyber–Basque Country.[9] That recreation would empower and unite all Basque people even without a state. That would mean respect for singularity, but at the same time it would favor an evolution to accommodate such a culture in times of globalization.

The Basques are an old European diaspora that resulted from a weak economy, which forced many Basques to immigrate to countries such as the United States, Argentina, Mexico, Uruguay, and Australia. In addition, other Basques, exiles such as ETA (Basque Country and Freedom in its Basque acronym) members, joined already established diaspora communities as a result of the Spanish civil war and political persecution during Franco's regime. More recently, there have been other Basques who have been persecuted by ETA and have left the homeland and moved to other parts of Spain or abroad. In general, the Basque diaspora is stable and structured, with present low growth but able to consolidate as an activist commons, especially due to the European, South American, and U.S. nodes. The Fourth World Congress on Basque Collectivities held in Euskal Herria and organized by the Basque government had the following significant motto in Basque: "Zubigintzan" (Building Bridges). That congress has made clear the crucial need to count on the digitalization of the Basque diaspora.

In other words, on the one hand, the Basque government would support and help maintain and innovate informative and participative services for the Basque diaspora. On the other hand, the Basque diaspora itself would generate its own autonomous digital life. Transformation of the Basque digital diaspora into a true activist commons in favor of Cyber–Euskal Herria depends more on users and communities than on the Basque local government, because diasporan users confront the evident fact that the Basque homeland is fragmented into two countries (Spain and France) and into three different administrative territories—the Basque Autonomous Community, Navarre, and Iparralde.

The growing diasporic blogosphere could be the best place to develop a more communitarian and participative e-democracy, particularly within a homeland context where antiterrorist policies undercut civic and political rights. The democratic experience gained by the Basque digital diaspora in their host countries could be useful for the resolution of the homeland violent conflict that spans more than fifty years. We believe that this is the main challenge that activist commons have to undertake. One hopes the digital commons is not as divided as the off-line political sectarianism seems to be. In this sense, the digital diaspora can offer a more conciliatory and embracing attitude toward the conflict. Peace could become the main communal purpose for the establishment of a digital *auzolan*[10] of the Basque diaspora and the Basque digital community as a whole. The Basque (digital) diaspora's active involvement in participative initiatives (e.g., supporting nonviolent and human rights movements)[11] to provide a resolution to the Basque conflict is crucial. There should be no difference among initiatives from "above," such as the Basque government–sponsored network, "Konpondu," or from bottom-up activist nodes such as the "Batzart! Initiative."[12] If the conflict is not resolved, the Basque diaspora and its homeland are at risk of being in a permanent ideological and social fracture, which would most likely block capacities in the short run.

Consolidating Cyber–Euskal Herria beyond the circumstances of the homeland responds to the logic of survival of an old nation without a state. In the end, it has to survive with or without its own state. Former *lehendakari* (president) of the Basque Autonomous Community government, Juan José Ibarretxe, said, "The great linguist Koldo Mitxelena used to say that 2,000 years ago there were no state or nation but the Basque people existed. And we do not know if in 2,000 years there will be states of nations but we know the Basque people will exist" (2007). In this sense, new laws reinforcing ties with different communities in the Basque Country should embrace the fate of the diaspora as a digital entity with its own features and developments, including those that are political, cultural, and social.[13]

The main task is, therefore, to develop a social, cultural, and activist Basque commons. There have been some proposals, such as "KZ Gunea," on tourism in the Basque Country.[14] Also, there is a Basque community in Second Life known as the "Euskarians." The meeting point for this virtual community is the Euskarian Etxea (Home of the Euskarians).[15] Nevertheless, among those proposals, as important as they are, the most urgent one is to develop a free cyberculture that can combine homeland and diaspora Basques alike. That implies the existence of a good connection with networks of people without states and their respective diasporas. This is the real ground where Basque culture can play an important role. If the Basque diaspora, even lacking institutional support and having deep ideological internal divisions, could achieve this goal, it could become a reference for other digital diasporas of peoples without states.

As part of the European Union homeland, Basques are familiar with new technologies, and historically this has not harmed their ethnic idiosyncrasy. On the contrary, Basque culture has survived and evolved in a quite prosperous socioeconomic environment. The preservation of idiosyncratic features of diaspora Basques, well integrated into their host lands, may become an interesting contribution to multicultural societies in cyberspace. If it was possible for one of the most ancient European ethnic groups to survive in the middle of the European Union due to an activist diaspora, and if it was possible to solve or at least put up with its own political struggles, could not this example be used as a model for other diasporas without states?

ACTIVATING DIGITAL DIASPORAS

From our limited experience on the Basque digital diaspora and on the free cyberculture, we do not dare to suggest a set of proposals on how to constitute a digital diaspora. Surely, our recommendations would not be shared for all existing diasporas. Instead, we are proposing a set of basic insights that can summarize and guide our vision on diasporas as an activist commons. For instance, it is clear that there is a need to create and support digital information services and mass media (radio, TV, newspapers) in both directions—from homelands and from diasporic nodes around the world.

Of course, homeland institutions must support the basic digitalization of diasporas as well as the different social and cultural projects that diasporas might create in cyberspace. Digital diasporans should establish some resident status in the newly established e-nations. There must be an effort to create an e-democracy with the participation of diasporans. Also, digital diasporans should create an activist commons while attempting to confront the existing digital divide among their members and communities, since not everybody has access to new technologies. The articulation of different activist nodes would be the first path to take. Then a central node, or main agency, could be set up in order to study the reality of peoples without states and help their digital diasporas to coordinate social, intercultural, economic, and pacifist actions.

In *First Diasporist Manifesto,* Kitaj (2007) proposes that diasporas should widen their meaning and embrace exiled or marginalized people, Africans, homosexuals, Gypsies, or political refugees. That widening indicates the possibility of a large interwoven network with different intersections. Those crossing threads could become a resistant fabric for solidarity and creativity. Consequently, diasporas must explore that common or communal space for resistance and activist creativity. As Leung points out, "The net [should be understood] as means for minorities . . . tactics and technologies of resistance" (2006, 60). Digital diasporas for peoples without states must join the diasporic and global movement, because that is one of the best ways to support a strategy of solidarity. Whether we like it or not, our world is a diasporic place both in reality and on the Net.

NOTES

1. Text revised by Bernardo Santano at the University of Extremadura, Spain.
2. The official site, http://www.wumingfoundation.com/, offers an explanation and history of this group. They used a previous name, Luther Blisset, producing the novel *Q,* which is available at http://www.wumingfoundation.com/italiano/downloads.shtml.
3. See http://www.siruela.com/ncd and http://www.quintacolumna.org.
4. See http://www.e-tester.net/eng/index.html.
5. We owe a deep debt to Pedro J. Oiarzabal's works on diaspora and homeland identity (Oiarzabal and Oiarzabal 2005) and the Basque diaspora Webscape (forthcoming). His pioneering work has started the kind of debate and research that any scholar interested in diasporas should follow. His analysis and applied methodology on the Basque Webscape are crucial not only for the Basque diaspora but for any other diaspora in the intersection with the digital world.
6. UNESCO has considered of great importance the preservation of "intangible heritage" as a public good. In this sense, there are more public goods than water and air. Other groups like Platonic (http://www.Platonic.org) have organized different projects to preserve knowledge, such as the World Bank of Knowledge. See http://www.unesco.org/culture/heritage/intangible/html_sp/index_sp.sh.
7. See http://www.gnu.org/philosophy/free-sw.html and http://www.libre.org.
8. Second Life can be reached at http://SecondLife.com/ and Habbo Hotel at http://www.habbo.com/. See also Ana Malagón's blog, "Bigarren Bizitza" (Second Life) at SecondLife.com.
9. Gurutz Jáuregui (2000) coined the term *Basqueness.*
10. This is the word for "communal work" in the Basque language.
11. There are active sites like http://www.bideahelburu.org and http://www.artamugarriak.org.
12. See http://www.konpondu.net/ for the blog Batzart!
13. The Law on Basque Communities, of August 1994, was formulated by the Basque Autonomous Community government.
14. See http://www.kzgunea.org.
15. See http://9cdr.blogia.com/temas/revistas-libros.php.

REFERENCES

Alonso, Andoni, and Iñaki Arzoz. 1996. *Euskal Herria digital.* San Sebastián: Gaiak.
———. 1998. *Baserri eraitzia.* San Sebastián: Gaiak.
———. 2002. *La nueva ciudad de dios.* Madrid: Siruela.
———. 2003. *Basque cyberculture.* Reno: Basque Studies Program, University of Nevada, Reno.
———. 2005. *La quinta columna digital.* Barcelona: Gedisa.
———. 2007. "Destino cyborg." In *Estudios sobre cuerpo, tecnología y cultura,* ed. Jesús Arpal Poblador and Ignacio Mendiola. Leioa, Bilbao: Universidad del País Vasco, Servicio Editorial.

Anderson, Chris. 2007. *Wired,* July.

Andoni Alonso, Iñaki Arzoz, and Nicanor Ursua. 1996. *Cemento vernáculo: El impacto del cemento en la arquitectura vernácula de Navarra.* San Sebastián: Real Sociedad de Estudios Vascos.

Barlow, John Perry, Alvin Toffler, and Esther Dyson. 1994. Declaration of independence of the cyberspace. Available at http://www.eff.org/~barlow/Declaration-Final.html.

Bey, Hakim. 2001. Temporary autonomous zone. Available at http://evolutionzone.com/kulturezone/bey/taz/taz.html.

Bloom, Harold. 1996. *Omens of millennium: The gnosis of angels, dreams, and resurrection.* New York: Riverhead Books.

Bookchin, Murray. 2002. *The communalist project.* Harbinger: Communalism.

Castells, Manuel. 2001. *The Internet galaxy.* New York: Oxford University Press.

Egan, Greg. 1995. *Permutation city.* New York: Eos.

Gajjala, Rhadika. 2005. Cyborg-diaspora: Observations from the cyber-field. In *On the move: Mobility and identity,* ed. Kris Knauer and T. Rachwal. Bielsko-Biala: Akademii Techniczno-Humanistycznej.

Haraway, Donna. 1991. Cyborg manifesto. In *Simians, cyborgs, and women: The reinvention of nature.* New York: Routledge. Available at http://www.stanford.edu/dept/HPS/Haraway/CyborgManifesto.html.

Hayles, Katherine. 1999. *How we became post-humans.* Chicago: University of Chicago Press.

Ibarretxe, Juan José. 2007. Preguntas difíciles para una reflexión autocrítica sobre las Euskal Etxea. *Gara,* July 11.

Illich, Ivan. 1991. *Shadow work.* London: Marion Boyards. Available at http://www.preservenet.com/theory/Illich/Vernacular.html.

Jáuregui, Gurutz. 2000. *La democracia planetaria.* Oviedo: Nobel.

Kaku, Michio. 1998. *Visions: How science will revolutionize the 21st century and beyond.* New York: Oxford University Press.

Kerckhove, Derrick. 1997. *Connected intelligences.* Toronto: Somerville Books.

Kitaj, R. B. 2007. *Second diasporist manifesto.* Cambridge: Yale University Press.

Leung, Linda. 2006. *Virtual ethnicity: Race, resistance, and the World Wide Web.* London: Ashgate Publishing.

Mallea-Olaetxe, Joxe. 2001. *Speaking through the aspens: Basque tree carvings in California and Nevada.* Reno: University of Nevada Press.

Oiarzabal, Agustin, and Pedro J. Oiarzabal. 2005. *La identidad vasca en el mundo: Narrativas sobre la identidad más allá de las fronteras.* Bilbao: Erroteta.

Oiarzabal, Pedro J. Forthcoming. The Basque diaspora Webscape. Reno: University of Nevada Press.

Rheingold, Howard. 1993. *The virtual community: Homesteading on the electronic frontier.* New York: Harper Perennial. Also available at http://www.rheingold.com/vc/book/.

Stephenson, Neal. 1992. *Snow crash.* New York: Bantam Books.

Stevens, Vance. 2007. Second Life and online collaboration through peer to peer distributed learning networks. Proceedings of the METSMaC Conference, Abu Dhabi. Available at http://prosites-vstevens.homestead.com/files/efi/papers/metsmac/metsmac_secondlife.htm.

Tester. 2006. *Trabajos de nodos.* San Sebastián: Arteleku. Available at http://www.e-tester.net/cast/proyectos/book.html.

Wagar, W. Warren. 2001. *Memoirs of the future.* New York: SUNY–Binghamton.

Wittgenstein, Lugwig. 2007. *Tractatus logico philosophicus.* Available at http://www.kfs.org/~jonathan/witt/tlph.html.

Wright, Philip. 2001. *Nonzero: The logic of human destiny.* New York: Vintage.

Zerzan, John. 1994. *Future primitive.* New York: Autonomedia.

Zurawski, Nils. 1998. Culture, identity, and the Internet. Available at http://www.uni-muenster.de/PeaCon/zurawski/Identity.htm.

PART II

DIALOGUES ACROSS CYBERSPACE

5 Oprah, 419, and DNA
Warning! Identity Under Construction

TOLU ODUMOSU AND RON EGLASH

The term *African diaspora* is relatively new, having become popular after World War II and first defined formally in a 1965 essay by George Shepperson. As noted by Edwards (2001) the specific phrase *African diaspora* contrasts with prior terms such as *Pan-Africanism* in ways that convey its orientation toward a more decentralized, heterogeneous, and antiessentialist meaning, an orientation that is made even more explicit in Gilroy's framework of "the Black Atlantic." Here we investigate the formation of diasporic identity through digital media among two different groups of African Americans: those with a heritage in the United States and those who are recent immigrants (i.e., first generation) from Africa. Strange though it may sound at first, we found recent controversies involving television show host Oprah Winfrey to be a common intersection by which diasporic identity in both groups could be elucidated.

OPRAH WINFREY'S DNA ANCESTRY TRACING

On the Public Broadcasting System program *African American Lives,* which first aired in February 2006, host Henry Louis Gates Jr. traced the ancestry of eight prominent African Americans, including Oprah Winfrey. Winfrey had previously announced her ancestry to be Zulu, but that was—according to the mitochondrial DNA test offered by Gates—incorrect; her closest match was the Kpelle ethnic group in Liberia.[1] Winfrey was clearly taken aback by this news; she "had to take a breather." Her personal, emotional, and financial investment (forty million dollars for a girl's school in South Africa) in Zulu identity had been intense, despite prior warnings from historians that there was no record of the slave trade delivering people from the Zulu ethnic group.

While Winfrey's conflicting ancestral identity was the subject of a great deal of popular press, it was by no means unique. All migrants create stories, composed of facts, guesswork, and outright imagination, about their relation to heritage—indeed, even people still occupying their own ancestral lands do so (Anderson 1983). But this act of identity self-construction is particularly intense in the case of involuntary migrants such as the descendants of enslaved Africans in America. Ogbu (1978), investigating the rejection of academic success by some African American students, argues that African Americans (as well as those other involuntary inhabitants of the United States, Latinos and Native Americans) created an "oppositional identity" to the mainstream culture, making the rejection of what Jim Crow laws and more informal racist systems had held them from part of an active form of self-creation. Ogbu (1991) later clarified this concept using the term *cultural inversion* to refer to the rejection of symbols (dress, language, behaviors, and so forth) associated with a dominant culture. Fisher (2005) amends this model to stress oppositional identity as more about an attraction to alternatives than a rejection of the mainstream: in her view the students were not so much rejecting academics as they were placing a higher priority on becoming a rapper or an athlete.

Of course, the African Americans of Ogbu's and Fisher's studies—high school students in the age of hip-hop, where "keepin' it real" often references a glorified criminality—have a different construction of identity than the generation represented by Winfrey, who came of age during the 1960s, when dashiki shirts, Afros, and other symbols of a displaced African homeland often constituted black authenticity. As Dent (1992), Gilroy (1993), George (2001), and others have stressed, these strategic modes of individual and collective identity—Garveyism, Negritude, Black Power, Buppies, B-Boys, Bohos, Rastafarians, 5 percenters, Gangstas, AfroFuturists, and so on—embody a wide variety of intersections among political strategy, heritage, and social position. From assimilation to separatism, from the promotion of tolerance to resistance and revolution, from civil rights to repatriation in the motherland, these various ideologies (and in some cases accompanied theologies) align themselves with specific cultural expressions. The use of particular elements of African continental culture is thus at least as much a strategic decision as it is a recognition of historical realities. In Molefi Asante's Afrocentric framework, for example (cf. Asante and Asante 1985), ancient Egypt is reimagined as the original black homeland, with sub-Saharan African cultures as secondary derivatives. Visualizations of pyramids and Isis, not kente cloth and Swahili, became signifiers of black ethnic origins in this Afrocentric movement.

That's not to say one is authentic and the other false. Swahili itself was not a language indigenous to any inhabitants of the "slave coast," and kente cloth was influenced by cloths from India brought to Ghana through trade (Perani and Wolff 1999), nor is this peculiar to the African diaspora. There is, for example, no mention of the Jewish Exodus in the ancient Egyptian records, a fact that has led many historians to doubt its authenticity. But the decision to fashion a diasporic identity from various cultural elements often finds the historical debates less important than their political and social implications. For Asanti's followers, ancient Egypt held disproof of black intellectual inferiority (both genetically and culturally) and offered an alternative "classical civilization"—parallel to but independent from ancient Greece—as a wellspring of black cultural and spiritual origins. The tension between the desire to sustain this particular construction of the African diaspora and the need for mainstream academic support is an extremely important conflict, to be sure, but the point here is that it has not been a strong obstacle for the Afrocentrists themselves.

Thus, what is striking about Winfrey's construction of diasporic identity was not so much, as many critics leaped on, the lack of authenticity in citing a South African heritage rather than a West African one but rather how readily she relinquished it when confronted with the results of her genetic testing. One never hears of an Afrocentrist who, when confronted with evidence[2] contrary to the thesis of a black ancient Egypt, simply gives up. This may in part be due to the particular situation: Winfrey's genetic test results were delivered to her by Harvard professor Henry Louis Gates Jr. But equal authority has been leveled against Afrocentrists, and none of them had invested forty million dollars in a school for ancient Egyptians, as Winfrey had for South Africans.

PUBLIC EPISTEMOLOGICAL AND ETHICAL ASSOCIATIONS OF DNA

A better explanation might be that Winfrey felt she had to submit to the authority that has been invested in DNA. This authority has been strongly enshrined thanks to the multibillion-dollar biotechnology industry and its interpenetration with science education (a network of connectivity that runs from the funding of university laboratories and in some cases whole departments to the scientists who sit on both boards of educational institutions and boards of genetics industry corporations and educators who wish to prepare students for the reality of careers in an academe or industry centered on genetics) as well as popular media (ranging from press coverage of cloning and claims for biological determinism—that is, the "discovery" of genes for sexual orientation, criminality, IQ, and the like—to popular fiction such as the X-men trilogy, the Spiderman trilogy, and Harry Potter ["purebloods" versus "mudbloods" and so on]).[3] However, it is here that we encounter an illuminating contradiction to the concept of "digital diaspora."

Much of the literature on digital communities and identities stresses their flexibility and tenuous ties to physical realities: the famous *New Yorker* cartoon of the beagle in front of a keyboard—"In cyberspace no one knows you're a dog"—is merely the extreme for a wide variety of racial and gender "tourism"; whether one wishes to represent black identity with references to ancient Egypt, urban ghettos, Caribbean beaches, or Senegambian rice paddies, there is no easier place than virtual landscapes and communications. Digital communities based on the aforementioned cultural groupings, from Garveyism to AfroFuturists, flourish in virtual diasporic spaces where cultural elements such as Egyptian pyramid screen savers, kente-cloth Web-page borders, rap music audio backgrounds, and simple textual elements ("Ho-tep" at the end of an e-mail) can be recombined and reinvented at will. If that flexibility is the epitome of digital representation—if the digital diaspora is offering a blank canvas for ethnic self-construction—then how is it that the digital technology of DNA can be so rigid that it can force a powerful individual like Winfrey to give up the very diasporic identity that she had worked so hard to construct?

There are two possible answers here. A biological realist would say that, unlike art and narrative, DNA is an objective physical reality, one that has a powerful influence on our bodies and their lineage. A social constructivist would say that DNA exists as just another discourse, and the fact that its narratives, visualizations, and other representations are ensconced in more elite institutions is no reason to grant it a greater ontological status. Our stance on that controversy is somewhat of a synthesis (see Eglash 2007). We need not feel forced to decide between the two; it is enough to know that DNA is more than just another digital media representation. Its powerful claim on our bodies, genders, family histories, ethnic identities, and even the very concept of race is mediated not by the aesthetics of style and interpretation of historical record but by its location in a high-tech science that garners some of the highest epistemological status in our world.

Moreover, it is not merely a matter of having the authority of science behind it; DNA has also dramatically altered its moral dimension in recent decades. Although DNA still carries negative associations from racism (cf. Gould 1981), it has recently taken on a more complex series of ethical associations, ranging from its role in freeing black prisoners on death row to its glorification in the popular *CSI* television series. The confirmation of the "out of Africa" thesis (that all humanity emerged from Africa a short fifty thousand years ago) and its implications for genetic diversity (that all humans are 99.9 percent genetically identical, that there is not enough diversity for humans' racial differences to be considered biologically significant, that there is more genetic difference between individuals than between races) have only been increasingly confirmed by DNA evidence from the Human Genome Diversity Project and other sources.

The extent to which those results have been taken up by popular understanding in the African American community is uneven. Belief in evolution among the lay public is much lower in the United States than in many other industrialized nations due to its mobilization of religious fundamentalism and accompanied politicization (Miller, Scott, and Okamoto 2006); Gallup polls in 1997 showed that only about 49 percent of the lay public (in comparison to 95 percent of scientists) believe that "man has developed over millions of years from less advanced forms of life" (this includes both those who believe that God intervenes in evolution [39 percent] and those who do not [19 percent]). Factors that positively correlate with belief in evolution include atheism, education, and income.

One would think that given the strong role of religion in the African American community, and the community's lower average education and income level, surely far fewer African Americans would believe in evolution. But the drop is minuscule: the Gallup poll reported 45 percent of African Americans believe in evolution in comparison to 49 percent of Caucasians. Clearly, there is a counterbalance to these negative factors; a reasonable hypothesis is that this counterbalance has to do with the positive associations as outlined above. Moreover, the "out of Africa" thesis has an enormous following in the black community. Black cultural movements have appropriated this concept for a wide variety of purposes, some allied with theological concepts (black woman as the "the mother of all races" in relation to mythology), some with political concepts (as seen in the Afrocentric critique, or in a less racially charged panhumanism), and others with a more science-oriented emphasis.

Meanwhile, cultural anthropologists and black studies scholars in the academic mainstream, and in some cases the geneticists themselves, have articulated more complex results from the Human Genome Diversity Project and other sources to relatively broad audiences. Thus, while the ethically suspect associations (genetic determinism) retain a strong presence in popular consciousness, there is a growing current that allies this faith in genetics with a science-based antiracism. Another potentially positive association for African Americans is the new heart drug BiDil, which is the first race-specific drug approved by the Food and Drug Administration. After a scandalous history of neglect by medical science, and given the disproportionate number of African American deaths from heart failure, many African Americans (including the Association of African American Cardiologists) have embraced BiDil as a benefit to the African American community, despite its potential implications as support for genetic determinism and reification of race (Hartigan 2008). In 1998 the denial of the existence of children fathered by Thomas Jefferson and born by his slave Sally Hemings was put to rest when the Y chromosome DNA in Jefferson's family line was used to establish a definitive link with the Hemings family.[4]

Thus, the faith in DNA ancestry tracing has multiple foundations and implications, some contradictory: a much welcomed change from science as a foundation for racism to science in the service of the black community (e.g., BiDil, ancestry tracing); an affirmation and reification of the "race" concept, perhaps supporting racism as future genetic claims impinge upon cognition; a contestation of the idea of separate races and affirmation of universal human genetic identity; and a synecdoche for justice and hidden truths ("Just as DNA can uncover the unjustly incarcerated inmate, it can uncover the unjustly hidden ancestry").

SELF-FASHIONING: TECHNOLOGICAL, OBJECTIVE, AND AUTOCHTHONOUS

One way of framing these variations is in terms of "self-fashioning." For example, Orel and Willis-Altamirano (1988) spoke of "technologies of self-fashioning" in the context of product design and consumerism, and Dumit (1997) introduced the phrase "objective self-fashioning" to discuss the ways in which people construct an "objective self" using medical facts, which are then revised and recombined in various ways in relation to other elements of their persona and environment (e.g., someone diagnosed with cancer can view themselves as patient, victim, survivor, and so forth). Much of this has been inspired by Foucault's "technologies of the self." But Foucault's work is primarily a critique; exposing the ways in which a microphysics of power flows through technologies of the self such that individuals are fooled into thinking they are self-governed when they are actually subjects of a dominating "governmentality." Indeed, from a Foucaultian perspective, the story could be read in technophobic terms: African Americans in collusion with genetic ancestry tracing are merely victims of DNA's dominating "biopolitics." But just as "self-fashioned" medical identities should not be dismissed as capitulation to hegemony—for example, they empower individuals for collective action in the social movements born out of geographic "cancer clusters"—such technophobic critiques are also inappropriate summaries for the ways in which the new understandings of DNA analysis can articulate with social justice issues in the black community.

While well-reasoned critiques of technological domination are crucial components of any social analysis, a technophobic analysis is one based in romantic naturalism. Kobena Mercer's "Black Hair/Style Politics" (1988) described the ways in which black hairstyles have often been misinterpreted as having an original naturalistic form, which is then ethically valorized, and contrasted with an artificial form, which is then disparaged. For example, the red "conk" in which African American hair was straightened has been described as a feeble attempt at assimilation (e.g., in *The Autobiography of Malcolm X*). But Mercer points out that the red color was no more a match to white sensibilities than it was a reflection of ethnic naturalism: it was instead hinting at something more independent, what the authors of this essay would term an *autochthonous* self-fashioning. Similarly, he notes that dreadlocks, often assumed to be an "African-roots" style, were found nowhere in Africa previous to its importation from the West (its origins in Jamaica resulting from the influence of laborers from India). And of course post-1960s hip-hop-era hairstyles opened up vast possibilities: "Post-liberated black hair-styling emphasizes a 'pick 'n' mix' approach to aesthetic production, suggesting a different attitude to the past in its reckoning with modernity" (Mercer 1988, 51).

Like Mercer's denaturalized account of black hairstyles, African American naming practices also show a mixture of references to dominant American culture, African heritage, and autochthonous self-fashioning. Lieberson and Mikelson (1995) found that the invention of unique names in the African American community (based on historic records in Illinois and New York)

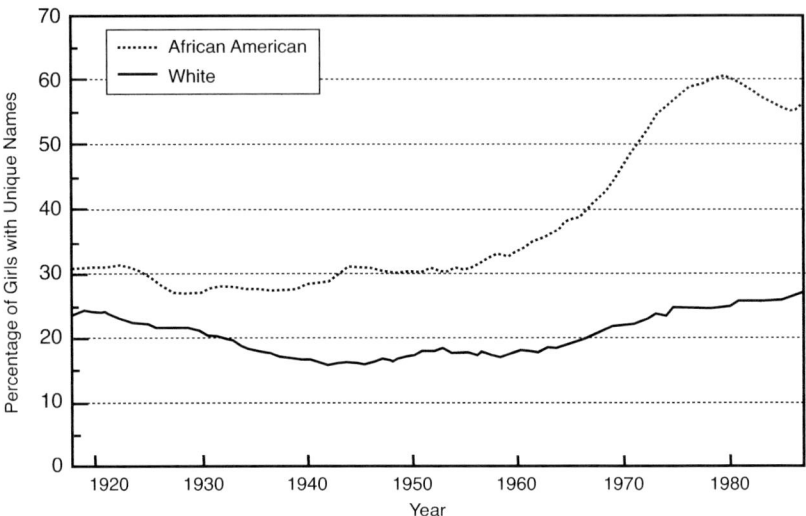

Figure 5.1 Standardized Percentages of Unique Names for Girls Born in Illinois, 1916–1989

dramatically rose after the 1960s, which matches the historical timing Mercer noted for innovations in black hairstyles (see figure 5.1).

Lieberson and Mikelson used a questionnaire completed by 224 subjects (a convenience sample) to investigate the linguistic character of some unique African American names such as Maleka, Shameki, or Shatrye. The authors found that white and black respondents had equal success in guessing the gender of names (about 70 percent), and that specific linguistic conventions for indicating gender (such as the *a* ending for girls) better matched American English conventions than African naming conventions (e.g., in New York 51 percent of African American girls' names ended in *a,* as did 38 percent of white girls' names). Thus, while much of the inspiration for these inventions was clearly from other sources, including Arabic and African, they did include some synthesis with dominant cultural linguistics. What was particularly striking for us, however, was their documentation of the "Frenchification" of the names (such as the *elle* ending). Was this the influence of France's welcoming gestures toward African American expatriates like Josephine Baker, James Baldwin, Sally Hemings, and Richard Wright? Or was it the associations of France with the aesthetics of the upper class? Like the red color of the conk, there are hints here of an autochthonous self-fashioning, a declaration of independence from the restraints of dominant culture and heritage, yet articulating with each in strategic ways.

SELF-FASHIONING AND DNA ANALYSIS

This autochthonous self-fashioning does not end here; one could cite numerous examples throughout African American culture in music, film, performance, dress, and more. The question is, what would it take for black communities to have command of technologies like DNA ancestry—to make them available for projects in technological, objective, or autochthonous self-fashioning—rather than simply capitulate to its authority? There are indications that such agency is already in the making. Social scientist Alondra Nelson reports: "I've spoken with African Americans who have tried four or five different genetic genealogy companies because they weren't satisfied with the results. They received different results each time and kept going until they got a result they were happy with" (quoted in Younge 2006).

While the different results support the skepticism against DNA testing (cf. Palmié 2007), and its ad hoc character violates Popperian norms for scientific discovery, we believe its greater importance lies in the possibilities for an active self-fashioning of diasporic identity. If such uptake is nothing more than selective wish fulfillment—if Oprah Winfrey were to continue to take ancestry testing until she found a result she was happy with—it would be a failure, but so would mere capitulation to what is currently a business-driven version of "science" that also does little to honor Mertonian and Popperian scientific norms (Bolnick et al. 2007). In our vision autochthonous self-fashioning is neither an arbitrary cultural whim nor a predetermined natural law, but rather a mutual coconstruction of the two.

The fantastic fractal shapes of cornrow hairstyles in the African diaspora are one of the best illustrations: the "kinky" or "woolly" quality of African hair is an affordance that inspires stylists to explore what are physically impossible forms for straight hair. Conversely, stylists must prevent damage, maintain grooming, and nurture healthy hair: they coconstruct the very substance that creates the affordance for their styles. Oprah was right to drop her Zulu heritage claims, but there are plenty of cases in which African Americans have been able to use family oral history and historical documents to trace their genealogy; such sources should carry a great deal of weight. Ideally, DNA data and data grounded in such cultural and historical sources should inform each other. It is only when black communities can take the same agency in their command of DNA that they have had in hairstyle, music, and linguistics—an agency that emerges from an intertwining between the physical laws of nature and the infinite creativity of the human spirit—that DNA will take its proper place in the African diaspora. We now shift gears and examine another controversy, illuminating another example of self-fashioning.

OPRAH, 419, AND A CRISIS OF IDENTITY

According to Goffman, "A 'performance' may be defined as all the activity of a given participant on a given occasion which serves to influence in any way any of the other participants" (1959, 15). On April 13, 2007, the *Oprah Winfrey* show presented the episode "What the New Scam Artists Don't Want You to Know." The episode featured Oprah Winfrey and Brian Ross of ABC News discussing "Nigerian scams." The presentation very effectively positioned American citizens as "victims" being scammed and the Nigerians involved as "crooks" doing the scamming. Working off Brian Ross's prior *20/20* investigation of 419 scams, the episode presented a visually engaging exposition of "Nigerian scams" with the stated intent of educating viewers and empowering them to resist these particular scams.

The reaction to this show among what we will tentatively label the "Nigerian diaspora"[5] was overwhelmingly negative and the subject of a great many blog posts and articles. The controversial episode was debated, discussed, dissected, and opined upon in effusive and at times combative conversations across a number of Web sites frequented by communities from or in Nigeria. Indeed, it appears that a number of the participants interpreted the episode personally as a form of criticism, though Oprah took the time to state that the show was aimed not at the "entire country and everybody in the country" but rather at the subset of scammers under discussion. Instead of placating nonscamming Nigerians, Oprah's statement seemed to incite even more ire and contempt, particularly among the 2,490-member Facebook group with the inelegant name of "oprah is an idiot for dissin NIGERIA."[6]

There are diverse ways in which we could approach the study of the controversy that the episode described above engendered. In this paper, however, we will, as Latour (1987) advocated in *Science in Action,* utilize the controversy to examine "identity in the making" by following the discursive trail of the *Oprah* show and the response to the show. We shall posit the episode and the succeeding reactions on the Web as opposing knowledge-generating sites engaged in the process of constructing the identity of ethnicity. The subsequent analysis relies heavily on Erwing Goffman's dramaturgical framework (1959) to situate actions undertaken by the various actors.

Act 1, Scene 1

Oprah begins the show by informing her viewers about a number of American professionals who have been caught up in a terrible scam. This scam, according to Oprah, costs its victims about three billion dollars a year. On cue, Brian Ross expands on Oprah's introduction and explains that the perpetrators are Nigerians. Brian explains that the scams come through e-mail and the U.S. postal system. The volume is staggering, with close to six million scam letters or dud checks from Nigeria alone.

The postal inspector testifies to what Brian has just said, and the discussion quickly moves to establish the bona fides of the "victims" as they are enlisted in the performance as victims/nonfools. Brian then asks (rhetorically), what kind of fool would fall for that? The postal inspector diplomatically corrects Brian, pointing out that these people are not fools; they are victims. Oprah chimes in, buttressing the point that the people being scammed are victims, and should be viewed as such. Responding to his own rhetorical opening, Brian echoes the new role of the scammed individuals. They are "intelligent" and "smart" people who ordinarily would not be susceptible to scams of this nature. However, they have either fallen afoul of greed or been ensnared by appeals to their inner good Samaritans—led to believe they were assisting others in need.

Oprah then turns the conversation to Nigeria and asks why Nigerian scammers in particular are so adept at their trade. Brian explains that Nigeria is a poor country full of "wonderful" and "intelligent people"—but alas, there are not jobs to keep them busy, and thus they have turned to scamming. He even notes that scams of this sort are taught in school. Brian goes on to explain that in Nigeria, no one was arrested for crimes until very recently and details the specifics of the "419" section of the Nigerian code that makes it illegal to obtain money under false pretenses.

A section of his *20/20* report is shown to the audience presenting his trip to Nigeria, including footage of a dramatic apprehension of some suspected scammers. Responding to the footage, Oprah asks about the reason behind the crackdowns. Have they been initiated by pressure from the United States? Brian explains that both the Americans and the British should be credited for pressuring the Nigerian authorities to do what is right; however, the greatest pressure came from the Americans, as Nigerians view them as *mugus* (translated as big fools). This comment draws a visible reaction from Oprah, and Brian reiterates that Americans are seen as "big gullible fools."

As the episode comes to a close, Oprah introduces Sid Kirchheimer, the author of *Scam Proof Your Life,* and they have a conversation about Nigeria and *mugus.* As we shall see, this part of the performance by "Team Oprah" seems to especially agitate "Team Nigeria."[7] In a prescient move, Oprah preempts the coming storm, informing the audience that they do not have to write to let her know that Nigeria is a wonderful place. She is well aware of this fact, and is limiting her comments to the scamming parts of the country. Kirchheimer chimes in, pointing out that Nigeria is not the only source of scams. Scamming originates in Eastern Europe, the Far East, and even from within the United States. Scamming, it appears, is not the exclusive prerogative of Nigerians.

We now switch stages and examine the reaction that occurred in "Team Nigeria." We have chosen to narrow our discussion to three Web sites; a more comprehensive survey would be beyond the scope of this chapter. Furthermore, these particular Web sites—"communities of discourse"—diverge in ways that are productive to the ongoing analysis. The three Web sites are http://www.nigeriaworld.com, http://www.nairaland.com, and http://www.facebook.com.

Act 2, Scene 1: Nigeriaworld.com

Nigeriaworld.com is a news aggregation site. It collates and parses news from a number of Nigerian dailies. In addition, the site possesses a "featured article" section that comprises op-ed-type pieces written by a number of frequent contributors. A look at some site statistics gives us an idea of the audience and reach of Nigeriaworld.com (see table 5.1).

Alexa.com ranks Nigeriaworld.com as the 221st most popular destination for traffic from Nigeria.[8] As table 5.1 indicates, close to 60 percent of the traffic on the site comes from the United States and the United Kingdom combined. Only a paltry 3.7 percent of the site traffic is from Nigeria. We can effectively conclude that the audience of Nigeriaworld.com is located outside of Nigeria, primarily in the United States and the United Kingdom. This is important because identifying the prospective audience enables us to contextualize the content of the discourse.

TABLE 5.1 | ALEXA.COM TRAFFIC DETAILS FOR NIGERIAWORLD.COM

COUNTRY	PERCENTAGE OF USERS
United States	35.8
United Kingdom	23.0
Dominican Republic	5.0
Canada	4.6
Nigeria	3.7
Others (countries with traffic levels below 3.7percent)	27.9

While Nigeriaworld.com has an interactive forum, much of the content relevant to the controversy under examination is in the op-ed articles. A few weeks after "What the New Scam Artists Don't Want You to Know" episode aired, Jude Mbionwu, an author from Atlanta, Georgia, wrote the article titled "Oprah, 419, and Nigeria's Image." In the author's own words:

> I listened with despair as my sister narrated how Oprah ended her show about Nigerian fraudsters by telling everyone to hold on to their letters. She did not want to hear the Nigerians' sides of the story she just told America. A story that further casts Nigeria in much poorer light in the already biased minds of the world and especially the American public: a story about 419 fraudsters. . . . I will write to the millions she just cast in a negative light in front of millions of her fellow Americans. . . . I am, by no means, supporting 419 fraudsters. But just like typical American way of seeing things, she did not want to hear the stories of the people who believe that with $5000.00 paid to a total stranger from the poorest continent in the world, they will be able to receive some deceased person's $1 Million. Oprah should realize that it takes two to tango. Her show should have been about those errant fools who search for the quick millions from Africa. Oh my, what fools. And, oh my, what unrepentant 419 imbeciles for giving Oprah do a show that gives the reason to make people look at me twice with suspicion every time I say I am a Nigerian. It should have been a balanced show, Oprah. For the disrespectful way she presented the show, Nigerians will forever remember her; all the way to ignominy. Unless she uses the same platform to address the true facts. I am not saying that Oprah should not do a show about 419 fraudsters. She should remember to balance your [*sic*] stories.⁹

The central argument is that the episode did not present a balanced point of view. The author calls into question the casting of those being scammed as "victims." It is noteworthy that the author reverses roles from the episode and casts the scammed as "fools." The article is an impassioned affair, and the author appears to have interpreted the episode as relating to all Nigerians, not merely the fraudsters. This same sentiment is echoed in another piece on the site titled "Crime and Punishment: Time for a Real Soul Searching for a Nation." The author of this article, Christopher Odetunde, from Houston, Texas, frames the controversy as an opportunity for "national soul searching":

As Nigeria battles to shed a reputation for corruption, the recent statement, "All Nigerians are Corrupt Regardless of Level" mischievously attributed to Ms. Oprah Winfrey,[10] though sad, is a case in point for soul searching in Nigeria. . . . The accusation by Ms. Winfrey, if false, is a slap in the face for all honest and law abiding Nigerians at home and in Diaspora but, if true, is an opportunity for Nigeria as a viable, law abiding nation to be contrite and embark on real soul searching. . . . In America, for example, an average Nigerian is seen as a thief and Nigeria as being populated by thieves. Some news channels even stated that every Nigerian institution has a department that teaches corruption. . . . The issue is not that Ms. Winfrey made the statement but why on earth will she make a disparagingly blanket statement Nigeria and indirectly black people? There are over 140 million Nigerians, how many of these citizens duped Ms. Winfrey to justify her statement?[11]

A similar position is argued in Okoh Emeka's September 10 article, "Why 'Victims' of 419 Should Not Be Pitied." Okoh, writing from Moscow, reiterates the reversing of roles of victims and scammed, even going so far as to call for a directed media assault on the "victims" of scams:

> The so called Nigerian Scam has also become an issue in the most revered shows in the world run by Oprah Wilfrey [sic], these and many others undoubtedly fall into the grand image tarnishing campaign that most of these media stations who claim independent, free and impartial have carefully orchestrated and they seem ready to carry it out with a force comparable to that of tsunami. . . . There is no denying the fact that 419 scam is a thorn in the flesh of Nigeria as a nation, and the struggle to exterminate it should be an all encompassing fight that not just the anti crime bodies should champion but me and you as individual citizens. The west cannot waste any opportunity to tell the world that Nigeria is the most corrupt, most dangerous, most everything bad, they have virtually succeeded in instigating hate against Nigerians even in our own back yards, pathetically, our citizens and even government have practically fallen prey of this new form of colonization, colonization aimed at dehumanizing us, aimed at reducing us to mere criminals, the type whose focal point is coding the mindset of the world to view every Nigerian in bad light. . . . In matters of world politics and international affairs he who has the power often has the right, and he who is weak can only with difficulty keep from being wrong in the opinion of majority of the world. And

who forms the opinion of majority of the world? The media. That is why we need to change strategy, that is why our diplomatic onslaught should be fierce; we need to be hostile in our counter attack. *Nigeria should use any visible opportunity to tell the west that their citizens are greedy stricken, make it loud and clear at any given chance that only scammers and fraud minded and extremely greedy people fall for 419, our new foreign minister should let them know that falling for 419 is ridiculous and idiotic.* . . . The president should proclaim this even in the UN, yes, it will be a bit scandalous but that is exactly what we need to bring back the issue and allow people ponder it while we quietly work hard to solve the problem at home, even if it means creating a special agency that will handle it.[12]

The author here proposes a radical inversion of the rhetorical strategy employed in the *Oprah* show, calling for a concerted media effort to label "citizens in the West who fall for the fraud" as "scammers and fraud-minded and extremely greedy people."

By design, these Web articles are not open to comment from readers, and we can only speculate as to the effects on the intended audience. While these articles are diverse in tone and scope, there are a few things they share in common:

1. Each article interprets the *Oprah* episode as an attack on all Nigerians, not just the scammers.
2. The authors seem to be simultaneously engaging in two audiences, one of which we can categorize as an *insider* community, and thus most of the articles are written in an inclusive manner. Examples of this include the use of the pronoun "we." In Mbionwu's article his final sentence reads, "They are doing a lot more damage to Nigeria than we realize," and in Okoh's article above, "That is why we need to change strategy."
3. This insider community seems to comprise individuals both within and outside Nigeria, that is, the larger group of Nigerians with whom there is a sense of solidarity—in Goffman's nomenclature (1959), a team.

Act 2, Scene 2: Nairaland.com

Nairaland.com is a forum-driven site. More discussion group than Web site, Nairaland.com hosts about 164,400 members engaged in roughly 89,000 conversations (Topics).[13] The site design is minimalist, with the central

TABLE 5.2 | ALEXA.COM TRAFFIC DETAILS FOR NAIRALAND.COM

COUNTRY	PERCENTAGE OF USERS
United States	19.2
Nigeria	16.8
United Kingdom	14.2
Philippines	4.1
India	2.8
Others (countries with traffic levels below 2.8 percent)	42.9

attraction being the conversations going on in the forum. Members create discussion topics that are of interest to them, and open the floor to others to contribute. Unlike Nigeriaworld.com, the majority of the members utilize aliases (user names) to post. Topically, the discussions cover the entire gamut, from lovemaking to the American presidential election. A large number of the conversations are framed as questions, and can be personal (for instance, asking how to improve one's spoken English) or playful (for example, asking what others are wearing at Christmas). The tagline of the site is "Home of Nigerians and Friends of Nigeria." Established in 2005, it is run by a self-proclaimed capitalist from Ogun State in Nigeria, who appears from his self-image to be in his twenties. Alexa.com statistics give us some insight into the composition of the Nairaland.com community (see table 5.2).[14]

Nairaland.com is ranked eighteenth in Nigerian traffic, indicating that it is highly popular among Internet users there. Interestingly, almost 20 percent of its traffic is generated from the United States. However, unlike Nigeriaworld.com, Nairaland.com has a substantial proportion of contributors and audience from Nigeria.

The controversy we are exploring is represented in a number of conversations, the largest of which is the thread "Nigeria's Image Was Badly Damaged on Oprah Today!"[15] This conversation is one of the largest on the site, constituting twenty-six pages. Although it started out as a discussion about the episode, it morphed into an argument about Nigerian identity politics. The first few posts set the tone for the rest of the twenty-six pages:

Sweet-T (male in Los Angeles, California): Did anybody see Oprah Winfrey's show today? Nigerians and Lagos was paint [*sic*] in dark today. They showed how Nigerian boys are doing all kinds of dating and online scams at the cyber cafes all over Nigeria! And most of these fraudsters are Ibo names. CNN also ran the same kind of program in recent months. Why are Nigerians so damn greedy, especially the Ibo boys!!! They make me sick with all these negative attention to Nigeria. An Ibo boy in

Houston, TX was arrested in his house with $500,000 cold cash in his house. Money he got from doing credit card fraud and internet fraud. Can you imagine??? This is one of the reasons i hate some Nigerians in the western world. They go back to Nigeria and pretend as if they work at the World Bank. Ill-gotten wealth!!!

debosky (male in Onitsha, Anambra, Nigeria): another Igbo basher, una no dey tire(slang for "don't you get tired")??

Sweet-T: @dobsky I'm not an Ibo bashers, if you have access to internet, go to ABC news, CNN, and other Web sites and see for yourself!! They even showed Fred Ajudua's house. Most of the names shown are Ibo boys. The fact is that Ibo boys are too dang greedy and they put Nigeria's image in a terrible shape. I don't care if you are Ibo or not but the fact is that if you know any Nigerian doing illegal and shady business, your duty is to report him to the authorities and let's take our country back NOW!!!!

cheexy (female, Nigeria): It makes me sick too but i kind of frown at the emphasis on Ibo boys. For the fact that few Ibo boys were caught doesn't mean that the Ibos are the only tribe that engage in "yahoo-yahoo." The trend has eaten deep into every tribe and we should all work hand in hand as a nation to restore moral values that have been eaten up by the craze for material wealth. Let us educate the younger generation about the importance of a good name and this country would change for the better.

4Play (male, London, United Kingdom): I agree! These Igbo people are shaming Nigeria with their criminal dealings. Their greed is unparalleled and has led them to become Nigeria's top purveyors of criminality.

Very quickly, the conversation is hijacked and becomes an argument about ethnic identity, with the content of the episode subjected to scrutiny in an attempt to categorize various ethnic groups as more or less corrupt than others. In this thread it appears that Team Nigeria is dissolving, and new coalitions form around other identities—Ibo, Yoruba.

debosky: Rubbish, 'ibo boys like money too much, ibo boys are too greedy' those are tribal stereotypes, and are very wrong. But that is what laces each and every one of your posts. stop maligning the tribe due to a few foolish ones. When they bring out videos of yoruba dudes doing the same like on ABC what did you say then? the issue is that Nigerians are involved, don't make it into an Igbo thing. PS I am not Igbo. @ 4play am I hearing you right?

maxell (female, Nigeria): People lets leave the tribes out of this. The Nigerians on the oprah show could have been from anywhere, this is just side tracking the issue at hand. When are our young men going to stop spoiling our precious Nigerian name? Is it a must that we succumb to anything just to be a millionaire at 22/23 years old? Please lets focus on the main issue here abeg.

4Play: The truth needs to be told sometimes. It is true that Nigeria as a whole has a high rate of delinquency but there is no point pretending its frequency does not vary across the different ethnicities. It is well known that the Igbos are a greedy and money obsessed tribe. This character drives them to commit crime way beyond that of others.

Thus the conversation continues, and soon the original focus is lost in the debate over Nigerian identity politics.

Another thread that attempts to address the controversy is titled "'All Nigerians Are Corrupt,' Says Oprah Winfrey."[16] This conversation is twenty-one pages long, and there is a poll at the start of the thread asking for readers to cast votes. The poll states, "This is a serious matter. Whose side are you on?" Out of 247 votes cast, 85.8 percent were on the side of "Nigeria," and 14.2 percent of respondents were on the side of "Oprah Winfrey." Of particular interest is the manner in which the poll is framed—"Whose side are you on? Nigeria's side or Oprah Winfrey's?"—as an oppositional binary. One cannot be on both sides. The poll is prominently displayed at the beginning of the conversation, like a call to arms—pick your side. Once again, we see a delineating of groups: "Nigerians" at one end, "non-Nigerians" at the other. Team Nigeria is being constructed and performed through oppositional conflict and controversy. The thread begins with an announcement, and immediately there are calls for proof of authenticity:

Fodiyo (male, Kaduna, Nigeria): What do you make of the recent campaign of calumny that Oprah Winfrey was said to have sponsored on the CNN against Nigeria? According to a report i read on the Punch newspaper this morning, she was said to have advised the US govt to sever relationship with Nigeria on ground of corruption. "all Nigerians- regardless of their level of education- are corrupt" she was quoted. The report said her conclusion was because of a Nigerian of Igbo extraction who was said to have stolen 500,000 USD from a gullible foreigner through 'Internet fraud.'

angel101 (female, London United Kingdom): Do u have a link to the story?

MILITIA (female, U.S.): Don't forget that Judge Judy said the same thing on one of her shows last week!!!! Please give us link so we can see for ourselves and have a good debate on this one!!!!

The conversation proceeds with a number of passionate denouncements of Winfrey, drawing more forum members into the debate. The content of the conversation centers on what was or was not said on the *Oprah* show and what would constitute a valid response to the episode. Some forum members discussed the issue as a problem with Nigeria, while others argued that other ethnic groups were involved in the scamming. Some forum member playfully baited others while watching out for the administrator, taking delight in skirting the edge of propriety and forum custom. Twenty-one pages later, and closure over the authenticity of the statement attributed to Winfrey is still unachieved.

Act 2, Scene 3: Facebook.com

Unlike the other two Web sites discussed thus far, Facebook.com does not cater to any particular nationalistic community but rather is utilized by a worldwide audience. Alexa.com reports that it has a worldwide traffic rating of 7, making it the world's seventh most popular site according to Alexa.com's rankings of global Internet traffic. As one of the most popular social networking Web sites, Facebook has a large and diverse community with different and varied interests. We will not delve into a specific description of the Facebook community, as we have done elsewhere. Our interest here lies in a particular group with the inelegant name of "oprah is an idiot for dissin NIGERIA" (i.e., OIAIFDN). However, before we discuss the details of the group, it is necessary to examine the "Group" application on Facebook in order to adequately contextualize the staging of this particular group.

"Group" is a software application that runs on Facebook that allows anyone to stake out a portion of virtual Facebook space and invite others to join in identifying with the aims or direction of the group. Groups not only enable user identification and solidarity but also create a shared space where members of the group can exchange ideas, start discussion forums, share pictures and video, and interact in all the ways that Facebook enables. Reflecting a deliberate design choice on the part of the creators of the site, groups can be formed

TABLE 5.3 | ALEXA.COM TRAFFIC DETAILS FOR FACEBOOK.COM

COUNTRY	PERCENTAGE OF USERS
United States	30.0
United Kingdom	6.1
Italy	5.5
France	5.4
Others (countries with traffic levels below 5 percent)	53.0

with three different levels of access: Open, Closed, or Secret. Open groups are open for anyone to join. All information about the group is visible to the entire Facebook community. Closed groups are more restrictive, new members require administrative approval, and only members can view the group's interactions, though the group is listed in the group directory. Secret groups are the most restrictive. They are not listed, membership is by invitation only, and there is no way to tell who the members are or what they discuss.

OIAIFDN is an open group, and as such anyone can join, all communication is accessible, and the group is clearly listed in the Facebook group search directory. The group's description page is a study in bellicose, belligerent, nearly unintelligible speech:

> oprah winfrey, in her hur long sponsored prgrame on CNN against NIGERIA, just FUCKED with the most populous black nation in the world, you cant spit in the faces of almost 150 million people and expect a happy ending, . . . the moment she opened her mouth to attack the character of all Nigerians, she declared war on each and every one of us, all over the world, and must be made be made to regret this beyond comprehension, for the rest of her life. . . . If u dont understan the gravity of what she just did, let me put it to you this way, CNN IS WATCHED IN OVER 150 COUNTRIES OF THE WORLD, OPRAH IS THE MOST POWERFUL TV PERSONALITY IN THE WORLD, SO WHEN OVER 1 BILLION PEOPLE 'and counting' HEAR YOU ARE A NIGERIAN, YOU'VE LOST THAT BIZNESS DEAL, SHE HAS MADE IT MORE DIFFICULT TO GET INTO SKOOLS ABROAD, SHE HAS MADE IT MORE DIFFICULT TO GET A JOB ABROAD, OUR PEOPLE ABROAD WOULD HAVE TO HIDE THEIR NATIONALITY TO SURVIVE, BECAUSE HAS JUST MADE US THE WORLD ENEMY NO. 1, This are the very few of the numerous effects of her action. NAIJA PEEPS! this woman must pay, and paYback shuld be with everything you've got 'i.e the internet' post on YOU TUBE, on YAHOO, on MSN, MYSPACE, FACEBOOK, on even the HER SITE, . . . What i'm trying to say is hit her hard, shes just one person . . . cuz she just thru a big wrench in the engine of your future.[17]

The description can be read as a call to arms, asking for members of the group to respond to the perceived slight. Once again, the episode on the *Oprah* show is interpreted to be an attack on the entirety of "Team Nigeria."

For our purposes, an analysis of the content generated by the group is not necessary. Content wise, the group differs little from the Nairaland.com conversations, except that it is perhaps a little more vitriolic. Its membership reflects that of the Nigerian-denominated members of Facebook, many of whom are located in different areas of the world. Although it is difficult to tell what percentage is within and what percentage is outside Nigeria, we do not imagine that the distribution will differ much from that of Nairaland.com. Of more interest is the situated nature of the group within the Facebook community.

The administrator of OIAIFDN selected the title of the group, a deliberately provocative name, and left the group as an open group. Currently, the group has approximately twenty-five hundred members who are all individually recognizable and chose to openly associate and identify with this group. As the general custom on Facebook, unlike Nairaland.com, is to use actual names, the implication of the staging of this group is that members are personally identifiable. Thus, taking up membership in the group is discursively identical to announcing to whomever within the Facebook community bothers to investigate that one is diametrically opposed to Winfrey's perceived attack on Nigeria, and that one is sufficiently motivated to do so publicly.

Act 3, Scene 1

An empty theater; empty seats stretch out as far as the eye can see in all directions. Lights come on, focused on a stage to the right of the center. The seats are suddenly filled with a diverse global audience. The *Oprah Winfrey* show episode described above is meticulously performed. Lights fade. After a short period, light comes up on another stage closer to the left. There are three actors onstage from Nigeriaworld.com. They make their impassioned arguments that starkly contrast with the previous performance to the audience that now comprises Nigerians in the diaspora, listening in from multiple continents and nations except that now the stage extends all the way around the theater, encompassing them all. Everyone appears upset and angry as they listen to the rhetoric of "the Three." The lights dim. People are moving around, milling about the place. Some members of the audience move backstage and are joined by others, all young looking. They enter a side room with the label Nairaland.com and begin discussing the performance on the first stage. They are well known to each other, and the discussions take on a tone of familiarity. Multiple nonsynchronous conversations are being carried out simultaneously. Some of them reach consensus, albeit digitally; other conversations seemingly go nowhere; all, however, appear cathartic. One person leaves the room and returns to the global audience. She marks a space and raises a bold flag, mocking and challenging the first performance, calling out for others to join. One by one, they come from all over the world to show their identification and solidarity to the flag.

Although the above description is completely contrived, it serves to amplify the interpretive framework we seek to bring to bear. In Goffman's terms, we can posit that the episode on the *Oprah* show—at least from the point of view of the diaspora Nigerians—calls into question the performance that Nigerians in the diaspora commonly present, that is, that of ethical, honest, productive members of their respective communities. Team Nigeria is therefore a "performance team." Once a member of the team (any individual scammer) is cast as breaking the role and acting in a contrary manner, the credibility of the entire team is called into question. The reaction from teammates is swift.

The Nigeriaworld.com's articles that we analyzed present an interesting case of the blurring between front and backstage, as they simultaneously seek to reassure other team members that the performance has not been discredited, while attempting to reposition the guilty members of the team as belonging to another team of "fools" and thus not adequately representative of the team. This team of "fools," they contend, is one that the scammed and the scammers both belong to. Team Nigeria is shown to be a complex team. Perhaps a more adequate description would be that Team Nigeria is a "metateam" (that is, composed of several different teams). Yet that partitioning is precisely the rhetorical effect that those who label Oprah supporters as fools are trying to achieve; the dissenters themselves would likely prefer to see this as one team with diverse opinions.

The conversations on Nairaland.com present a classic backstage discourse. Team Nigeria struggles to come to grips with the disruption in the performance. Some name-calling takes place, and the metanature of the team quickly becomes apparent, with subteams forming and arguing that the other was responsible for the disruption of the performance. Rhetorically, this is not as strong a move as claiming that the guilty parties should really be seen as a separate team, but perhaps what is lost in rhetorical strength is gained in sustaining open discussion.

From the preceding analysis, the performance in the description of the Facebook group takes place securely on the front stage. By creating an open group that is glaringly negative of Oprah, the Facebook team performs and stages a unified oppositional identity.

FROM OPRAH AND DNA TO AUTOCHTHONOUS SELF-FASHIONING IN THE MAKING

In this paper, we have attempted to trace two controversies orbiting around Oprah, DNA, and 419. Though at first glance they may appear disparate, they operate in a similar fashion in our stories, unearthing controversy and initiating crises of identity. DNA and 419 cause the actors to reconsider the age-old question of "Who am I?" enabling us to examine the processes of what we describe as autochthonous identity in the making: a self-fashioning that appropriates the tools available in a deliberate forging of the self.

NOTES

1. Despite the enormous attention to the results in the popular press, very little attention went to the test itself. Mitochondrial DNA testing examines only the matrilineal heritage, so if Oprah had Zulu ancestry in her patrilineal heritage, it would not have been detected. There are numerous other reasons to be skeptical of the accuracy of this testing. See Bolnick et al. 2007 for an overview.

2. See critiques in Oritz de Montellano 1993, Martel 1994, and Lefkowitz 1996. In contrast to both Asante and his critics, Drake (1984) provides a more balanced portrait, noting that the influence of sub-Saharan Africa on prepharaonic Egypt has solid archaeological grounding and that it is only the claims for black presence during the later pharaonic periods that is controversial.

3. One should give credit to Rowling's conscious opposition to genetic determinism in her fantasy—the characterization of "mudblood" is consistently critiqued as an elitist myth—but it is worth noting that magical ability in the Potter world is primarily inherited, and that the term *mudblood* is critiqued for its disparaging implication and not because the rarity of a magical individual born to a nonmagical family is statistically incorrect.

4. But see Palmié 2007 for a critique of the genealogical testing.

5. The use of the term *diaspora* is in keeping with Edwards's (2001) discussion of the term, retaining the sense that *diaspora* articulates difference.

6. The group size is dynamic. During the six days that we monitored the group, new members joined at an average of eight per day.

7. In our analysis, we are casting "Team Oprah" as the entirety of the various actors in the episode described. The episode is the performance; the set, guests, cameras, and videos are all part of the setting and function as props, enabling the credible performance. "Team Nigeria" will be expanded upon elsewhere in this chapter.

8. http://alexa.com/data/details/traffic_details/nigeriaworld.com.

9. http://nigeriaworld.com/articles/2007/may/033.html.

10. We have been unable to corroborate this statement. It does not appear that Ms. Winfrey ever made such a statement.

11. http://nigeriaworld.com/feature/publication/odetunde/072707.html.

12. http://nigeriaworld.com/articles/2007/sep/101.html; emphasis added.

13. When we visited the site, there were 164,473 members in 88,806 topics: http://www.nairaland.com.

14. http://www.alexa.com/data/details/traffic_details/nairaland.com.

15. http://www.nairaland.com/nigeria/topic-48907.0.html.

16. http://www.nairaland.com/nigeria/topic-68514.0.html.

17. This description is available on the group's Web site within Facebook and is publicly available.

REFERENCES

Anderson, Benedict. 1983. *Imagined communities: Reflections on the origin and spread of nationalism.* Rev. ed. London and New York: Verso.

Asante, M. K., and K. W. Asante. 1985. *African culture: The rhythms of unity.* Westport, Conn.: Greenwood Press.

Bolnick, D. A., D. Fullwiley, T. Duster, R. S. Cooper, J. H. Fujimura, J. Kahn, J. S. Kaufman, J. Marks, A. Morning, A. Nelson, P. Ossorio, J. Reardon, S. M. Reverby, and K. TallBear. 2007. The science and business of genetic ancestry testing. *Science* 318, no. 5849 (October 19): 399–400.

Dent, Gina, ed. 1992. *Black popular culture: A project by Michele Wallace.* Seattle: Bay Press.

Drake, St. Clair. 1987. *Black folk here and there.* Los Angeles: UCLA Center for Afro-American Studies.

Dumit, J. 1997. A digital image of the category of the person: PET scanning and objective self-fashioning. In *Cyborgs and citadels: Anthropological interventions in emerging sciences and technologies,* ed. Gary Lee Downey and Joseph Dumit. Santa Fe: School of American Research Press.

Edwards, Brent Hayes. 2001. The uses of diaspora. *Social Text* 19, no. 1 (Spring): 45–73.

Eglash, Ron. 2007. Multiple objectivity. Available at http://www.ccd.rpi.edu/eglash/papers/mo.doc.

Fisher, Ericka J. 2005. Black student achievement and the oppositional culture model. *Journal of Negro Education* (Summer).

George, Nelson. 2001. *Buppies, b-boys, baps, and bohos: Notes on post-soul black culture.* New York: Da Capo.

Gilroy, P. 1993. *The black Atlantic: Modernity and double consciousness.* London: Verso.

Goffman, Erving. 1959. *The presentation of self in everyday life.* Garden City, N.Y.: Doubleday.

Gould, S. J. 1981. *The mismeasure of man.* New York: W. W. Norton.

Hartigan, John. 2008. Is race still socially constructed? The controversy over race and genetics. *Science as Culture* 16, no. 2.

Latour, Bruno. 1987. *Science in action: How to follow scientists and engineers through society.* Cambridge: Harvard University Press.

Lefkowitz, Mary. 1996. *Not out of Africa: How Afrocentrism became an excuse to teach myth as history.* New York: Basic Books.

Lieberson, Stanley, and Kelly S. Mikelson. 1995. Distinctive African American names: An experimental, historical, and linguistic analysis of innovation. *American Sociological Review* 60, no. 6 (December): 928–46.

Martel, E. 1994. How not to teach ancient history. *American Educator* (Spring): 33–37.

Mercer, Kobena. 1988. Black hair/style politics. *New Formations,* no. 5: 33–54.

Miller, Jon D., Eugenie C. Scott, and Shinji Okamoto. 2006. Public acceptance of evolution. *Science* 313, no. 5788 (August 11): 765.

Ogbu, J. U. 1978. *Minority education and caste: The American system in cross-cultural perspective.* San Diego: Academic Press.

———. 1991. Minority coping responses and school experience. *Journal of Psychohistory* 18: 433–56.

Orel, Tufan, and Susan Willis-Altamirano. 1988. The technologies of self-fashioning: Beyond universality and variance of the industrial product. In Designing the immaterial society. Special issue, *Design Issues* 4, no. 1/2l: 38–51.

Oritz de Montellano, B. 1993. Melanin, Afrocentricity, and pseudoscience. *Yearbook of Physical Anthropology* 36: 33–58.

Palmié, Stephan. 2007. Genomics, divination, "racecraft." *American Ethnologist* 34, no. 2 (May): 205–22.

Perani, Judith M., and Norma Hackleman Wolff. 1999. *Cloth, dress, and art patronage in Africa.* Oxford and New York: Berg.

Younge, Gary. 2006. New Roots. *The Guardian,* February 17.

6 Cyber CVs
Online Conversations on Cape Verdean Diaspora Identities

GINA SÁNCHEZ GIBAU

The question of identity has been a long-standing matter of contention among people who claim Cape Verdean heritage. While the politics of identity formation have played out largely within the national and transnational spaces in which Cape Verdeans reside, these debates are now echoed in cyberspace, through discussion forums, blogs, and Listservs. Cape Verdean usage of information and communication technologies (ICTs) illustrates the ways in which diasporic identity formation has become increasingly facilitated by technological advances, such as the Internet.

In this chapter, I will examine how diasporic Cape Verdeans utilize colonial, postcolonial, and transnationalist discourses in online discussion forums in particular to leverage their positions on the never-ending conundrum that is Cape Verdean identity formation. Specifically, I will focus on how Cape Verdeans, in the islands and in the diaspora, express a variety of ideas about the significance of race, nationality, language, and culture in the construction of "Cape Verdeanness." This examination highlights the complexity of defining Cape Verdeanness across diasporic experiences and how Cape Verdeans have utilized the deterritorialized venue of cyberspace as a tool of identity authentication and legitimization.

My ongoing research focuses on how members of the Cape Verdean diaspora in the United States reconcile their unique cultural identity with their socially ascribed, racialized minority status. I have been most interested in the self-identification practices of Cape Verdeans in terms of "reflexive identity politics," or how they "identify their identifications" (Eriksen 2001, 45). In previous research conducted among Cape Verdeans in Boston, I found that the Cape Verdean diaspora, fragmented into Cape Verdean American and Cape Verdean immigrant segments, offers up competing definitions of Cape Verdeanness and negotiates their cultural and "racialized" identities by enacting situational identities on an everyday basis (Sanchez 1999; Gibau 2005). This process of situational identity enactment takes on added dimensions as it occurs in the larger forum of cyberspace, where virtual social interactionism generates multiple interpretations of Cape Verdeanness.

Examining how Cape Verdeans define Cape Verdeanness for themselves and to others within the context of a virtual community provides insight into how diasporic identities are continually created, negotiated, articulated, and circulated. Likewise, it illustrates how discourses of race, nation, and culture are continuously destabilized in the formation of contemporary diasporic identities.

HISTORICAL CONTEXT

The origins of contemporary debates concerning Cape Verdean identity can be traced back to the initial settlement of the islands.[1] The colonial project was initiated by the Portuguese who first populated the uninhabited islands during the 1460s and then relied on the enslavement of West Africans thereafter.[2] In Cape Verde slavery produced close master-slave relationships, frequent manumissions, tolerance of miscegenation, and a large multiracial population. Over time, these circumstances led the inhabitants to adopt both Portuguese and African cultural practices.

Extensive miscegenation in Cape Verde produced a majority *mestiço* (mixed) or Creole population, who as free individuals occupied interstitial socioeconomic positions undesirable to the colonizers yet unattainable for the enslaved (Meintel 1984). In their service as colonial administrators within Portugal's other West African colonies (e.g., Guinea and Angola), Cape Verdeans became a racially mixed buffer group between the white Portuguese and the black Africans.

Cape Verdeans also relied on social and economic status to determine position within the social hierarchy. For example, according to Richard

Lobban, the term *branco* (white) was used to describe persons who were not only of "apparent European origin" but also "in positions of power" (1995, 54). Moreover, according to Lobban, "higher levels of wealth, power, educational status, and class position . . . lighten[ed] a person's 'racial' classification, while poverty, uncouth behavior, and illiteracy 'darkened' it" (ibid., 57). The resultant social hierarchy in colonial Cape Verde created social distinctions based on a combination of ancestry, phenotype, skin tone, social class status, and island of origin. Given this history, many Cape Verdean migrants entered the United States with preestablished notions of race and colorism, which were subsequently reinforced by the U.S. environment of racial hierarchy.

During the late nineteenth and early twentieth centuries, the ideology of "racial democracy" was promoted in Cape Verde. This ideology, also popular in Brazil at the time, boasted the superiority of racial mixture and proposed that the black segment of the populations would be eventually "whitened out" through successive miscegenation (Bender 1978, 7). While racial democracy exalted race mixture, it simultaneously promoted whiteness as the social and physiological ideal. Paulo Freire and Donaldo Macedo (1987) note how colonization affects the worldview of subjugated people, by defining acceptable behavior and fashioning one's tastes in clothes, food, and other social aspects of life. The internalization of colonial oppression and simultaneous acquisition of upward social mobility led lighter-skinned, privileged Cape Verdeans, especially those who served as mainland administrators, to identify more with the Portuguese than with the Africans.

Through Portugal's Colonial Act of 1930, Cape Verdeans acquired additional rights as "second-class citizens of Portugal" by being given the status of *assimilados* (assimilated). In contrast, the less prestigious status of *indígenas* (wards of the state) was ascribed to the inhabitants of most of Portugal's other West African territories under the same act. These legal classifications further separated Cape Verdeans from mainland Africans; Cape Verdeans were "encouraged to think that they had a greater cultural similarity to the Portuguese and [that] they clearly had little to gain by associating or identifying with Africans" (Lobban 1995, 60).

Although Portuguese colonization of Cape Verde produced a subsequent population that was ranked on a continuum of color and class, European influence was not widespread. The island of Santiago, for example, witnessed the largest influx of enslaved Africans, some of whom were then transported to the New World while others remained. Contemporarily, this historical legacy is understood through popular discourse that constructs Santiago as the most "African" of all the islands. Conversely, Cape Verde's role in the whaling industry during the eighteenth and nineteenth centuries has led to the idea of Brava being the island that experienced the most European influence, through its ongoing contact with American seafarers.

Even through contact with outsiders, each island was relatively isolated from the others. The most travel between the islands was conducted by natives of Brava and Fogo, given their close proximity. As a result, each individual island developed its own insularity. Today, most Cape Verdeans who relocate to the United States tend to refer to themselves and to each other in terms of their islands of origin (e.g., DjaBraba) and, most important, their specific towns (e.g., Nova Sintra) or parishes (e.g., São Lourenço). People have come not only to identify themselves by island of origin but also to understand Cape Verdean nationalism through their particular lived experiences. Unfortunately, this type of insularity has become the breeding ground for stereotypes that persist in the islands and in the diaspora. People from Brava are perceived as the most provincial of all the islands; people from Fogo are believed to be showoffs and braggers; people from São Vicente are perceived as the partygoers. Those from Santiago, for a long time stereotyped negatively as black, use the term *sanpadjuda* to pejoratively describe all of the rest of the islanders.

The ways in which Cape Verdeans express their loyalty to their specific island birthplaces while at the same time differentiating themselves from their fellow Cape Verdeans provide the foundation for contemporary disagreement within the community. Examining how the discourse of race operated historically in colonial Cape Verde, and how this discourse has had a subsequent impact on the politics of contemporary Cape Verdean identity construction on either side of the Atlantic, is necessary for understanding the nature of discord that has been displayed recently in online discussion forums.

IDENTITY CONTESTATION IN CYBERSPACE

Cape Verdeans in the islands and in the diaspora utilize a variety of ICTs; with respect to those who have access to the Internet, most log on to Web sites dedicated to social networking (e.g., ClubNino.com), disseminating community and entertainment news (e.g., FORCV.com), and commerce (e.g., Caboverdeonline.com). Many of the Web sites are hybrid in form, containing elements of all of the above. In addition, there are other ways in which Cape Verdeans are engaging in these technologies, namely, the development of individual Web pages and pages for academic use (e.g., Cape Verdean college student organizations). Discussion forums in particular provide a platform for a variety of topics, including immigration, racial and cultural identity, language, politics, tourism, religion, education, business, and entertainment.

Back in November 2005, a flurry of chatter was generated on "capeverdeFORUM," a Yahoo discussion group of 171 people interested in topics related to the Republic of Cape Verde and the Cape Verdean diaspora. Many, though not all, of the members of this online collective are of Cape Verdean ancestry.[3] The chatter, focused on the topic of Cape Verdean identity, began in response to a posting by the group's moderator of a lecture delivered by the prime minister of Cape Verde, Jose Maria Pereira Neves, at Brown University in October 2003.[4] The title of his lecture was "New Horizons in African Leadership in a Globalizing World." The lecture, though not necessarily controversial in tone, caused a stir in this online community, as was evident in the sheer volume of postings that resulted. The following excerpts provide some indication of the catalyst behind the online discussion:

> Ladies and Gentlemen, I believe in the future of my continent, Africa. It is crucial for Africans and African leaders to believe in the future of the continent. We need to be optimistic, but above all we need to be proactive. We have to be able to ensure Africa's transformation. To do so it is crucial to maintain a sense of African ownership over the process of transformation. At the same time, it is equally important for us to build a culture of efficiency and of results in Africa. . . . We were the first country in Africa to initiate a democratic transition process, back in 1990. Many observers are of the opinion that ours was the most democratic of these transitions.[5]

The discussion that ensued, lasting from November 11 to December 8, was generated by an inquiry from a member as to why Prime Minister Neves kept referring to Cape Verde as an African nation. The discussion that followed revealed a marked split between those members who were in agreement with the prime minister's conceptualization and others who offered correctives emphasizing Cape Verde's unique history and culture in contradistinction to the continent. At the heart of the diasporic standoff that occurred on capeverdeFORUM was the issue of whether Cape Verde could or should be described as an African country. To many people, this could actually be interpreted as a nonissue; many people are aware of the fact that Cape Verde is a member of the African Union, for example. But for certain members of capeverdeFORUM, the question led to more extensive conversations related to ideas of race, culture, and identity pertaining to the islands and the diaspora.

The number of postings responding to the initial inquiry into Cape Verde's status as an African nation was stunning. Of the 103 messages posted in November, 71 pursued this thread related to Cape Verdean identity. What became interesting to me was the development of this conversation, and how specifically the idea of Cape Verdean identity became multiply articulated through discourses of race, culture, color, nation, politics, history, gender, and language. What was equally fascinating about this exchange was the discursive impact that the underlying social context had on these articulations; although the conversation was occurring in what is considered an unbound, transgressive space (the Internet), it was ultimately shaped by the physical context of *where* people were posting their messages (primarily in the United States). In other words, the postings illustrated how the debate on Cape Verdean identity was affected by the location of the persons posting the messages, as "diasporic citizens" (Laguerre 1998) residing in a racialized society. Language use was also indicative of one's social and political positionality; although English was used predominantly, use of Kriolu by both Cape Verdean immigrant and Cape Verdean American participants often informed and enhanced the virtual performance of ethnicity.[6]

Much of the observed discussion focused on the distinctions between two primary conceptualizations of "Cape Verdean": one group of individuals argued that Cape Verdeans, though having African heritage, were not African but were indeed a unique group formed out of a mixture of cultures; the other group comprised individuals who countered that Cape Verdeans were African by virtue of their heritage, their culture, and the islands' geographical and political affinities with the continent. Though there were some concessions offered between the two groups (e.g., "Fortunately, we can all agree that we are Cape Verdean"), the justifications for each line of argument proved to be emblematic of the historical and political conditions that have shaped Luso-African postcolonial realities in diaspora.

What is obvious from the outset is the conflation of nation with identity. Of the 71 postings,[7] 15 demonstrated the belief that Cape Verde was an African nation, and thus Cape Verdeans were Africans. Seven additional postings qualified this belief with another conflation, this time of race with nation: because Cape Verdeans were defined as African, they were also defined as black people. The one person who offered the most postings subscribing to this viewpoint went so far as to state that "unequivocally, Cape Verdeans are Africans, even if they are 'white' and the majority are not such."

Much of the discussion of Cape Verdeans as African hinged on essentialized notions of identity as "rooted" in a particular geographical homeland. For many diasporic populations, Cape Verdeans included, a sense of community is often predicated on the beliefs in common (biological) origin, homeland, language, and cultural traditions that define the ties that bind. In this vein, the diasporic Cape Verdeans postulating the "Cape Verdean as African" thesis utilized essentialist depictions of Africa as the primary site of identity formation in their conceptualizations of Cape Verde as geographically, culturally, and politically part of continental Africa. Some of the more prevalent indications of this depiction included political gesturing to Amilcar Cabral, the well-known freedom fighter who toiled for the liberation of both Cape Verde and Guinea-Bissau. Many members referenced Cabral's re-Africanization platform and utilized his oft-quoted "return to the source" to justify their positions.

This sentiment was observed more recently, in February 2007, when the discourse on Cape Verdean identity made its way into another forum, this time one with a more general African membership, at Africanpath.com. The moderator of this forum posted an e-mail that he had received from a Cape Verdean American who identifies as black:

> Do y'all post news on Cape Verde? A lot of people forget about this island nation off the west coast of Africa and its frustrating and insulting! There's a debate about the racial identity of Cape Verdeans. Some even choose to identify themselves as Portuguese as opposed to African and that's an insult that they would even want to identify themselves with the people that colonized and enslaved us! I know I am African! Cape Verdean Americans are the only group of African Descendants that can readily race their roots back to the motherland due to the fact that we are the only African descendants to come to America voluntarily! How do you feel about Cape Verdeans and the whole racial identity issue? Signed—Northern Light.

The moderator posted this e-mail as a means of creating discussion. In his initial response, the moderator discussed Cape Verdean identity and that of people from Mauritius and Egypt in terms of geography, meaning that any country or group or people identified with the African continent or marked off as an African territory is African, regardless of race (e.g., white farmers in Zimbabwe) or culture (e.g., "Middle Eastern" Egyptians). He further responded that one does not have to be black to be African. The seven Africanpath.com responders to this posting (three were of Cape Verdean ancestry) cited without question their belief that Cape Verdeans were Africans ("African islands"); one person offered the proof of Cape Verdeans' being covered in African media (i.e., a newspaper article).

Other members of capeverdeFORUM stated their belief that Cape Verdeans are African because they have African roots as part of their historical and cultural makeup, roots that should be acknowledged and remembered. Indeed, one proponent of this discourse asserted that the diaspora youth of her generation (e.g., those in their twenties) are "re-awakening to their African roots." Many of these individuals also made a subsequent conceptual leap to blackness, stressing their affinity with African Americans as, again, "rooted" in common experiences of racism and discrimination. What is important to mention, however, is that the conversation pertaining to remembering one's African roots was initiated and continued in response to the postings of those who postulated the thesis that Cape Verdeans were "just Cape Verdeans."

The latter viewpoint was supported by the discourse of race mixture, whereby Cape Verdeans were described as being "not just African" but rather a mixture of many "races" that created a unique cultural group. The members who offered this viewpoint were often quick to acknowledge their African heritage, even in the form of the essentialized "root," but they also strongly identified Cape Verdeans as having a distinct culture apart from continental Africans. Interestingly, many of these members, as the conversation ensued, then abandoned their acknowledgment of an African contribution by stating that Africa had nothing to do with Cape Verdeans, or, more specifically, that Cape Verdean culture bore absolutely no resemblance to African culture. In justifying this position, some members offered the contributions made by the Portuguese and other European cultural groups to Cape Verdean culture. The final gesture solidifying the dismissal of an African contribution, however, was the postings of the primary proponent of this thesis, posing the question, "Why [is there] this constant need to look back and relate with our past?" His position appeared to be a de-essentialized, "uprooted" one on the surface, but ultimately postulated an amalgamated Cape Verdean identity, one devoid of its foundational cultural content.

Turning again to the more recent example from 2007, the day after the Africanpath.com posting, the moderator of another Cape Verde group, FORCV.com, in turn posted just the Africanpath.com moderator's comments on the Web site but not the attendant responses. To this, the Cape Verdeans on FORCV.com responded that the posting was "propaganda and manipulation." For the most part, the majority of these posters offered perspectives that highlighted the multicultural and multiracial aspects of Cape Verdeans. However, one person in particular focused on the idea that people who identify with Africa are preventing "white" Cape Verdeans from identifying with their European and other non-African roots. This opinion sparked a heated dialogue between two individuals in particular, which quickly spiraled out of control, each hurling insults meant to dismantle the other's intellectual and moral credibility.

Insidious distinctions made between Cape Verdeans and Africans indicate the persistence of global stereotypes and racist perspectives nurtured in the colonial and postcolonial contexts of the islands and the diaspora. Such stereotypes and perspectives reinforce the historical expressions of Cape Verdean elitism, bigotry, and colorism. On capeverdeFORUM, one member expressed dismay at the influx of Senegalese to Cape Verde; she was upset because she was unable to buy authentic Cape Verdean souvenirs. The nature of the global marketplace dictates patterns of tourist consumption whereby Senegalese merchants are now found in Cape Verde, selling a variety of products, including fabrics and animal sculptures, for example. Unfortunately, the interaction between this member and a Senegalese vendor in Cape Verde reinforced global stereotypes of Africa that further confirmed for her that Cape Verdeans were nothing like Africans ("When did we have lions, tigers, etc. in Cabo Verde . . . It's ridiculous"). She likened the demographic changes in Cape Verde to an imposed "Africanism." This same person provided comments that expressed her desire to go back to the "good ole days," for her the days of privileged and nonoppressed class position during the colonial period. Perhaps the most disturbing of all of the postings came from a member who, after arguing his "Cape Verdean only" position, stated: "My roots are in Cabo Verde, not Africa. My language is Crioulu or Portuguese, not some African tribal dialect. My music is the morna and coladera with a full orchestra, not drums. My dwelling is a traditional house with painted walls and a tiled roof, not a mud hut."

Toward the end of the month, the moderator stepped in to quell the exchange of insults by suggesting that "care and caution" be exercised so that the discussion could progress and not in effect combust. He or she acted in response to postings that expressed ageism (e.g., "the elders need to step aside") and animosity between diaspora- and island-born Cape Verdeans (e.g., "implicit in [her] remark is this authenticity badge"). Unfortunately, this redirection could not be sustained; some members engaged in self-censorship, while others asked to be removed from the group because of their "unpopular" perspectives.

In the more recent exchange on FORCV.com, the moderator did not step in, and thus the conversation rife with verbal assaults and unsubstantiated posturing unfortunately continued unabated, with both sides trying to position Cape Verdeans as being either more European or more African. In the end, this exchange effectively ended with one member making an assessment of the person deemed a "wannabe" white Cape Verdean (who lives in France), based on this person's own disclosure of personal experience:

> You say that people can describe you as either a black, an arab or a portuguese, depending on what you tell them. Well that settles the debate as far as I'm concerned, because if you really weren't black, as you so desperately want us to believe, you wouldn't have been mistaken for or considered to be one, no matter what you say to people. But yet people do consider you to be a black, a noir. Why? Because you are one. In short, to make my case, consider this analogy: you share a culture and geographical origin with the late Eugenio Tavares, but he, unlike you, I would bet dollars to donuts, wouldn't have been taken for being black, regardless of whatever he had told people. Why? Because he wasn't phenotypically black. You are, thus you are considered one. End of debate.

Indeed, the accused person did not continue the line of debate, but resurfaced again to respond to a different topic later in the month. Ultimately, these conversations continually resurface, for as Mitra rightly observes, "there is no opportunity to have the 'final word' in cyberspace" (2001, 38). Therefore, there is a need to understand Cape Verdean diasporic identity as in process and perhaps always "under construction."

CONCLUSION

In the mid-1990s, as the Internet became a regular facet of our everyday lives, one of the ways it was envisioned was as a powerful communicative tool for community building (Ebeling 2003, 98; Everett 2002, 127; Nakamura 2008, 4). However, the Internet now not only functions as an arena for the creation of global connections and subjectivities but can also provide a space that allows for political contestation (Bernal 2005; Everett 2002) and conflict (Axel 2004) that could not happen otherwise in actual nationalist contexts.

The discussions surrounding Cape Verdean diasporic identity formation occurring in cyberspace illustrate a similar way in which the Internet is functioning as a vehicle through which unconstrained, "difficult dialogues" can occur. Although on the surface Internet forums can appear to illicit such openness due to an implicit anonymity on the part of participants, the Cape Verdean example illustrates how contestations over identity can occur even when there is a hint of identity disclosure (e.g., last name or place of residence).

The concept of diaspora has become increasingly understood as a "globally mobile category of identification" (Axel 2004, 27) for communities with access to ICTs. Revealing Cape Verdean understandings of their own individual and collective identities provides insight into the complex and often conflicting manner by which Cape Verdeans construct, assert, and negotiate their diasporic identities in a global context. Moreover, the process of Cape Verdean identity formation in cyberspace demonstrates that "what connects people also fragments them" (Bernal 2005, 662). Finally, these online conversations among Cape Verdeans, as contentious as they can become, are still a means through which the collective "voice" of a relatively marginal community can be spoken and potentially heard by a wider audience (Mitra 2001). What remains important to consider when examining the proliferation of digital communities are the relations of power that inform them; beyond the obvious issue of access, questions of ownership and maintenance of sites as well as the dominant language of communicative exchange on these sites must also be considered.

NOTES

1. The Sotavento (Windward) Islands include Brava, Fogo, Santiago (housing the capital, Praia), and Maio. The Barlavento (Leeward) Islands consist of Boa Vista, Sal, Santa Luzia (uninhabited), São Nicolau, São Vicente, and Santa Antão. There are also eight *ilheus,* or islets.

2. The institution of slavery in Cape Verde lasted from 1462 to 1869 (and illegally until 1878).

3. "Cape Verdeans" are people who trace their ancestry or that of their relatives to one of the nine inhabited islands constituting the Republic of Cape Verde, located off the west coast of the African continent.

4. The lecture is available at http://brown.edu/Administration/News_Bureau/2003-04/03-029t.html.

5. The speaker was indicating the democratic process of transitional government that has occurred during Africa's second postcolonial era.

6. Kriolu is the national yet unofficial language of Cape Verde. It is derived from fifteenth-century Portuguese and various West African languages. Kriolu is a distinct language that is often mistaken for a dialect of Portuguese. Efforts continue to be made in both the islands and in the United States to standardize the orthography of this orally transmitted language through the Alfabeto Unificado para a Escrita do Cabo-Verdiano, or ALUPEC, system.

7. Many postings were multiple, from a single individual.

REFERENCES

Axel, Brian Keith. 2004. The context of diaspora. *Cultural Anthropology* 19, no. 1: 26–60.

Bender, Gerald. 1978. *Angola under the Portuguese: The myth and the reality.* Berkeley and Los Angeles: University of California Press.

Bernal, Victoria. 2005. Eritrea online: Diaspora, cyberspace, and the public sphere. *American Ethnologist* 32, no. 4: 660–75.

Ebeling, Mary F. E. 2003. The new dawn: Black agency in cyberspace. *Radical History Review* 87: 96–108.

Eriksen, Thomas Hylland. 2001. Ethnic identity, national identity, and intergroup conflict: The significance of personal experiences. In *Social identity, intergroup conflict, and conflict reduction,* ed. Richard D. Ashmore, Lee Jussim, and David Wilder. New York: Oxford University Press.

Everett, Anna. 2002. The revolution will be digitized: Afrocentricity and the digital public sphere. *Social Text* 20, no. 2: 125–46.

Freire, Paulo, and Donaldo Macedo. 1987. *Literacy: Reading the word and the world.* South Hadley, Mass.: Bergin and Garvey.

Gibau, Gina Sánchez. 2005. Contested identities: Narratives of race and ethnicity in the Cape Verdean diaspora. *Identities: Global Studies in Culture and Power* 12, no. 3: 405–38.

Laguerre, Michel S. 1998. *Diasporic citizenship: Haitian Americans in transnational America.* New York: St. Martin's Press.

Lobban, Richard. 1995. *Cape Verde: Crioulo colony to independent nation.* Boulder: Westview Press.

Meintel, Deirdre. 1984. *Race, culture, and Portuguese colonialism in Cabo Verde.* New York: Maxwell School of Citizenship and Public Affairs, Syracuse University.

Mitra, Ananda. 2001. Marginal voices in cyberspace. *New Media & Society* 3: 29–48.

Nakamura, Lisa. 2008. *Digitizing race: Visual cultures of the Internet.* Minneapolis: University of Minnesota Press.

Sanchez, Gina Elizabeth. 1999. Diasporic (trans)formations: Race, culture, and the politics of Cape Verdean identity. Ph.D. diss., University of Texas.

7 Nationalist Networks
The Eritrean Diaspora Online

VICTORIA BERNAL

The ways that Eritreans in diaspora are engaged with cyberspace reveal some of the ways that transnational migration, coupled with new technologies of communication, is transforming political participation. The Internet may be the quintessential media for diasporas because it so easily bridges distance and dispersal. But it would be wrong to see the new technology as simply facilitating or speeding up communication across social networks that are already there. New media, especially the Internet, are making possible new kinds of communicative spaces and practices. New discursive communities are emerging that, while they may, as in the case of Eritreans, build upon existing social networks on the ground, bring them together and extend their membership, purpose, and significance in novel ways.

Although new information technologies make possible such things as cheap and immediate communication among dispersed interlocutors across long distances, technological potential itself does not determine and cannot by itself explain the uses to which it is put. Despite their global scale and their lack of a physical location, Internet communications are not independent of cultural context. The ways in which they are given form, character, and meaning relate to social systems and political structures on the ground (see, for example, Van Den Boss and Nell 2006; Miller and Slater 2006;

and Whitaker 2006). In the Eritrean case, the history of nationalist organizing and the state's interest in mobilizing the diaspora has led to an Eritrean online public sphere with distinct characteristics whose individual posts and overall patterns of posting can be understood only in relation to the transnational field of Eritrean politics of which it is a part (see also Bernal 2004).

DIASPORA AS AN ERITREAN WORLD WIDE WEB

During the course of the thirty-year struggle for independence from Ethiopia, an estimated one million Eritreans fled their country (UNICEF 1994). Thus, at the time of independence nearly one out of every three Eritreans was living outside the country. Although physically located outside Eritrea's boundaries, Eritreans in diaspora were actively involved in the struggle for independence. The primary nationalist movement, the Eritrean People's Liberation Front (EPLF) was organized transnationally with cells and affiliated organizations in many countries. These political organizations connected Eritreans in diaspora to the struggle's leadership within Eritrea (Hepner 2005). The EPLF was extremely successful in mobilizing the diaspora and harnessing its energy and resources for the nationalist cause. While solidly based within Eritrea, the EPLF extended far beyond Eritrea's borders and maintained communications with Eritreans abroad (Al-Ali, Black, and Koser 2001).

During the course of three decades of struggle Eritrean organizations and individuals in diaspora organized political support, staged public events, held demonstrations, and engaged in public relations efforts on behalf of the Eritrean cause. They funneled resources and supplies to the Front. In some cases Eritreans even left the safety of their new homes to join the ranks of EPLF fighters on the battlefront in Eritrea. For Eritreans in diaspora the nationalist movement served as an organizing force in their experiences of exile and forced migration, connecting Eritreans to each other, to their homeland, and, not incidentally, to visions of a better future.

At the same time, the experiences of escape, displacement, and survival in new lands produced new forms of community and new bases for the connections among Eritreans living in diaspora (Woldemikael 1996). The isolation, discrimination, and disenfranchisement experienced by Eritreans in diaspora contribute to their continued identification as Eritreans, despite holding citizenship of other nations. Shared personal histories of lives disrupted, loved ones killed, and families separated by war provide an impetus for Eritreans in diaspora to keep in contact with fellow Eritreans who understand firsthand what they have been through. The histories of Ethiopian oppression, war, and displacement that members of the diaspora share thus contribute to the bonds of community and connectedness even among Eritreans who had no prior relationship with each other back in Eritrea. In addition, because so many Eritreans fled the country, many Eritreans in diaspora are in contact with former neighbors and classmates as well as relatives in diaspora, creating a complex web of interconnections. Part of what cyberspace offers Eritreans in the diaspora is a mending of ruptures in the social body and in individual subjectivity, through the ability of the Internet to bridge distance or at least render it invisible, making physical location irrelevant.

Eritrea's achievement of independence in 1993 coincided with the development of the Internet. Eritreans in diaspora who were already politically linked to the EPLF and socially connected to each other throughout the diaspora were quick to perceive the benefits of digital communications. Dehai, now http://www.dehai.org, was the first and foremost Eritrean Web site that created an online community of posters and readers, and any discussion of Eritreans in cyberspace must start there. Dehai actually predates the Internet. It was founded in 1992 by a group of techno-savvy Eritrean professionals and computer science students in the Greater Washington, D.C., area as an earlier form of computer-mediated network. Thus, Eritreans began their efforts in digital communications at a time when many Americans, such as myself, had no access to or interest in computerized communications. Dehai continued to evolve as new information technologies developed and became a Web site.

The Web site facilitated the maintenance and coordination of Eritrean networks across geographic distances and the mobilization of Eritreans around the world for nationalist purposes. Dehai served as well to facilitate the involvement of Eritreans outside of Eritrea in the economic and political life of the new nation. The openness of cyberspace, where users create content and costs are extremely low, made it possible for Eritreans to create Web sites that serve their particular needs. This was important for Eritreans in diaspora because there are no media outlets in the countries where they settled that cater to their interest in Eritrean current events. Eritreans often find mainstream media coverage of Eritrea either lacking altogether or misinformed. In the words of one Eritrean poster describing the hunger for news and politics during the liberation struggle:

> Before the momentous liberation of Eritrea in 1991, just about every Eritrean was a consummate politician digesting, debating and predicting every news item in any way related to Eritrea. "Dehai adi intai alo" ["what news is there of our home country?"] was the usual introduc-

tion to any conversation between Eritreans, and there was plenty to talk about starting with the ever reliable news from the EPLF and usually misinformed or unsophisticated world press reporting on Eritrean affairs (usually quickly dismissed after a brief analysis with "iziom intay aflititwom"—"what do they know"). (Posted on Dehai in March 1998; bracketed translation of transliterated Tigrinya added; translation in parentheses is original)

Eritreans in diaspora harnessed the power of the Internet to pool their knowledge and share news and analysis. Dehai's subtitle, "Eritrea Online," suggests that it is meant to be a kind of virtual Eritrea and to serve as a "national" space within cyberspace. Dehai remained for many years the predominant Internet link for Eritreans around the world, allowing them to connect to each other throughout the diaspora, spread news of Eritrea, and debate the Eritrean political issues of the day. Messages are written in English with occasional transliterations of Tigrinya (the dominant language of Eritrea). Dehai is a transliterated Tigrinya word that means both "voice" and "news."

Dehai has two main components. One is a news list where people can post published news or links to published news related to Eritrea. The other is a discussion list or message board devoted to Eritrea-related issues. Both of these are archived on Dehai. The Dehai charter posted in 1995 states, "The main objective is to provide a forum for interested Eritreans and non-Eritreans to engage in solving Eritrea's problems by sharing information, discussing issues, publicizing and participating in existing projects and proposing ideas for future projects."

Once it became a Web site, Dehai could be accessed by anyone; however, only members can submit postings and gain access to Dehai archives. Dehai has served as a means of mobilizing Eritreans in diaspora and soliciting resources for nation-building projects but also (as will be discussed later) for waging the 1998–2000 war with Ethiopia. For many years Dehai had an annual fund-raising drive to benefit Eritrean orphans, and Dehai also raised funds for numerous other causes and projects, including repatriation and emergency assistance for refugees and the purchase of a professional camera to send to Eri-TV, the government television station, among many others. Typically, the list of donors by name and with the specific amount paid or pledged is circulated on Dehai. Needless to say, this works as peer pressure to increase participation as well as a means of giving recognition to those who give. Indicating the level of trust within the Eritrean community, in some cases the donors are told simply to make out their checks in the name of the particular Eritrean who is leading the fund drive.

Over the past two decades, cyberspace has facilitated the participation of the diaspora in the formation of Eritrea's national institutions and political culture in a number of ways. The formulation of Eritrea's constitution was discussed at length, as has been the requirement that all youth in Eritrea receive military training and perform national service, among other major policies. Eritreans in diaspora have created a Web-based public sphere for debating national issues that has become especially significant given the lack of press freedom and civil society within Eritrean territory. Dehai's charter states, for example, "The purpose of this Charter is to promote freedom of expression, and to facilitate an open environment in which members may express a diversity of opinion that is to be both welcomed and respected. All subscribers are duty bound to abide by and preserve the spirit of the Charter."

Dehai has no commercial content or links, and it has been created, maintained, and developed by its core founders and by its users, particularly its posters, many of whom have voluntarily devoted large amounts of time to Dehai for years. Those who created it and administer it do so as volunteers, as are those who send in news items and those who contribute to discussions and debates. For many of them, this is not a small amount of time. As one poster wrote in response to another on Dehai in December 1996, "I did join Dehai not 'to keep me from boredom,' as you have testified. I might be the last person who has spare time. My self-exposure would help you to understand my situation & what derives me to have a perspective different from yours. I am a 41 year old, a father of four who drives a taxi during the day & goes to school during the night on part-time basis (to make-up my lost school years) & only I am the only one working because of my wife's health reasons. This unnecessary self-revelation is to show the time I sacrifice in Dehai." Eritreans would understand his mention of "lost school years" as a reference to the disruptions caused by the liberation struggle, and his post, in fact, goes on to mention that he was a guerrilla fighter.

Although Dehai's founders and some of the earliest participants were self-described "techies" employed or studying in computer-related fields, Dehai's appeal was wider, and the success of new Eritrean sites shows the continued appetite for Eritrean-related content, for debates, and even for passive (i.e., silent readers') communication with fellow Eritreans among the diaspora. Some of the other Web sites allow commercial links, but even so the Web sites are not commercial enterprises or sources of profit for their creators and managers. They are essentially labors of love and perhaps obsession on the part of those who maintain them and on the part of the devoted posters who produce a steady stream of content that draws readers. Asmarino.com, for example, recently posted desperate appeals soliciting donations from its readers and posters to help defray the site's expenses.

What is particularly significant is that these electronic networks of Eritreans, unlike the earlier nationalist networks, were initiated independently of the Eritrean leadership and were not and are not centrally coordinated by the EPLF, the People's Front for Democracy and Justice (PFDJ), which is the postindependence form of the EPLF that became a ruling party, or the Eritrean state. Dehai was outspokenly nationalist in purpose and orientation, but it was nonetheless an independent initiative and has remained so. Since the end of the 1990s, Eritreans have developed a number of other Web sites, such as http://www.asmarino.com, http;//www.meskerem.com, http://www.hafash.com, and http://www.awate.com, among others. All of the sites can be considered nationalist in their devotion to Eritrean nationhood, but sites such as Awate.com and Asmarino.com emerged in part as outlets for critical views of the PFDJ and President Isaias Afewerki.

The stated goals of these other major Web sites are similar to Dehai's. For example, on June 15, 2004, Asmarino.com posted "an important notice to our readers and contributors" that began, "Asmarino.com is not only an information center, but also a meeting point. We believe that many minds, different perspectives, and civil debate are necessary . . . Asmarino.com . . . provides an open forum for the presentation and refinement of ideas relevant to Eritrean communities." One can see Eritreans as modeling online the kind of open public sphere they wish to see within Eritrea itself.

NEW MEDIA AND THE POLITICS OF INFORMATION

In recent years, Eritreans have come to question "the reliability" of government sources. Moreover, the availability of satellite television has given Eritreans abroad another avenue of access to news of Eritrea, and many watch Eri-TV in someone's home or in places where Eritreans gather, such as Eritrean restaurants. However, access to Eri-TV via satellite television has not caused a decline in Web traffic, since Eri-TV is government controlled and is perceived as providing a limited view on national affairs. In contrast to official programming, the content produced by posters is independent and the perspectives varied, including irreverent and, in recent years, sharply critical perspectives. Moreover, the Web as a medium, unlike television, has a social and interactive dimension. This is particularly true of the Eritrean online public sphere, which is characterized by exchange and debate, and where posters often address each other by name as they respond to each other's statements.

While the Eritrean state controls television and other media based in Eritrea, it cannot control the access of Eritreans abroad to information or restrict their political expression. As a communicative space, cyberspace offers considerable freedom from censorship as well as freedom from politically motivated violence by state actors or other parties. In fact, I suggest that for Eritreans the possibility of exploring ideas without great fear of official reprisal or violence is another one of the underlying attractions of Internet communications.

Living outside of Eritrea has exposed members of the diaspora to alternative politics (for example, Eritreans in North America and Europe have had at least informal ties to progressive organizations of various kinds through their involvement in the liberation struggle), and they have been exposed to media and educational environments where ideals of freedom of expression are promoted. Thus, the development of spaces for free expression and analysis is also grounded in the experience of diaspora. Eritreans in North America and Europe in particular have been influenced by their experiences of the more open public spheres in their new countries of residence.

Although debates can be heated, and it is not out of the ordinary for posts to rail on for several single-spaced pages, neither is it unusual for posters explicitly to cite the values of the public sphere and to call for respectful exchanges, often with reminders of the larger national interests at stake. For example, in a July 2003 post, an author remarked, "After all, it's not about me or you, its about major issues that affect the country."

For Eritreans, the Internet has reduced the constraints of space and time in communication across the distances and dispersal of diaspora. More significantly, the Internet has also brought into conversation an unprecedented range of social personalities across divisions of ethnicity, religion, class, age, education, and, perhaps to a lesser degree, gender. A Dehai post from December 1996 raised the issue of generational status in relation to discussing Eritrea this way: "You reminded us many times about your seniority & life time experience. I give greater respect to respectful people; especially those who are older than I am. . . . [A]t the same time, older doesn't mean wiser; older doesn't mean that you have greater love for our country comparing to the new generation."

Online, Eritrean Christians and Muslims contribute to the same debates, and professionals engage interlocutors who possess only an elementary school education. This indicates the potential of the Internet to change politics, particularly through giving voice to the views of people who are not in authoritative positions. The Internet can also serve to mask or reduce the effects of status inequalities among participants in the public sphere. Some of Dehai's most highly regarded and prolific posters work as parking lot attendants and taxi drivers, as do many Eritreans. Their status as media personalities on Dehai is not based on their established social status in the United States or in Eritrea. Therefore, it is important to note that the Internet did not simply speed up or make more convenient the social and political links that already existed; it brought into being something new and distinctive in Eritrean political culture: a relatively open public sphere.

Lack of access to computers and literacy clearly are barriers to using the Internet, though it is important to remember that cyberspace exists not in a vacuum but in a larger field of communication flows via word of mouth, telephone, fax, and other means (González-Quijano 2003). Therefore, the reach and influence of the Internet extend to those who may not go online themselves. This may be particularly true for highly socially networked cultures like Eritrea's, where ties based on ancestral villages, neighborhoods, and extended kinship relations link people to many others in complex chains of relationships.

NEW MEDIA AND NEW FORMS OF CITIZENSHIP

The political activities of Eritreans on the Web suggest that cyberspace has implications for new formulations of citizenship and sovereignty and the ways the nation is imagined as community. Eritreans in diaspora in North America and Europe are for the most part legal citizens of the countries where they reside and earn their living, but they are emotionally pushed and pulled into Eritrea's national politics (Conrad 2005; Tezare et al. 2006). Their sense of themselves is intertwined with Eritrea's destiny as a nation, even if they have no intention of returning "home" permanently.

The Eritrean state, for its part, has acted to include the diaspora as a component of Eritrea, extending recognition to them and to their offspring born and raised abroad through issuing them national identity cards, subjecting them to taxation, and continuing to court their political support and solicit their financial support on an ongoing basis with a steady stream of official visits and state-sponsored events and outreach activities. While dual citizenship and citizenship through descent rather than birthplace are not unique to Eritrea, what stands out in the Eritrean case is the size of the diaspora relative to homegrown Eritreans, the relative economic power of the diaspora, and the political sway they hold.

The Eritrean diaspora has a special relationship to the state, which is dominated by former EPLF leaders who transformed the Front into the ruling party, the PFDJ. The relationship is rooted in the transnational history of the struggle for independence. The ambiguity of boundaries in cyberspace is also congruent with the transnational reach of Eritrean nationalism, which flows not only from the diaspora toward Eritrea but from the Eritrean leadership in Eritrea out to the diaspora.

Most Eritreans within Eritrea did not have access to the Internet until the year 2000. Significantly, members of the government did have Internet access, however. Thus, while Eritrean activities in cyberspace have been largely confined to the diaspora until recently, in practice this meant that the diaspora had a special conduit of communication to the Eritrean leadership. The government is widely believed to monitor Dehai and other sites, and government officials, including President Isaias Afewerki himself, are believed to have participated in online debates using assumed names. It is also widely assumed that some posters are government plants or mouthpieces used to transmit and defend the party line online. As I write, http://www.awate.com is filled with posts about slanderous statements made by one poster, believed to be acting for the government, defaming a member of the "awate team," as the site's core management calls itself.

WAR AND THE WEB, 1998–2000

Perhaps the most significant impact of Eritrean activities in cyberspace relates to war. The outbreak of war with Ethiopia in May 1998 moved Eritreans in diaspora to mobilize resources, wage public relations campaigns, and keep in touch with unfolding events in Eritrea. Dehai was central to all of these efforts. Beginning in the spring of 1998, Dehai was largely taken over by postings related to the unfolding conflict. The war appeared to threaten the very survival of Eritrea as a nation and was regarded by Eritreans in diaspora as a crisis that motivated them beyond discussion to action. Both Ethiopia and Eritrea waged public relations campaigns in the international media that served as a second battlefront. One of the noteworthy aspects of Dehai postings during the war was the vehemence and hatred expressed by Dehaiers toward Ethiopia and the passion with which posters supported Eritrea's war effort in their outpourings of emotion as well as their contributions of financial resources to support Eritrea's war effort.

The war rallied Eritreans who might have disagreed with each other on various issues and over the degree of their support for EPLF/PFDJ. Together they focused on Eritrea's survival as a nation. On June 16, 2000, Eritreans organized a demonstration in front of the UN headquarters in New York to protest the United Nations' "silence and inaction" on behalf of Eritrea. The call to participate was posted on Dehai on June 14. Telephone and word of mouth were also used to spread the call to demonstrate, but the speed and reach of Dehai surely played a role. A June 15 posting on Dehai explains, "I and other volunteers went through a list of Eritreans in the Tri-state area (NY, NJ, CN) [New York, New Jersey, Connecticut] and invited everybody to come to the demonstration at the UN this Friday. While many of those we called were encouraging and pledged to come, some excused themselves for a number of 'convincing reasons.' Unfortunately, we did not have a chance to respond to every pessimistic, but to collectively reach them through the Dehai medium." The post continues by listing four common "excuses" and responding to each one with rationales for participation.

Eritreans in diaspora used Dehai and other Web sites not simply to express themselves during this crisis but also to organize and act to affect the handling and the outcome of the crisis. They publicized Eritrean (as opposed to Ethiopian) perspectives on the border war, mobilized fund-raising efforts, and coordinated actions, such as the demonstration in front of the UN building in New York. Responding to these efforts, the Eritrean government established a national defense bank account to receive the donations from abroad. It is worth noting, moreover, that these donations were not earmarked as humanitarian aid to alleviate the suffering caused by the war but were aimed explicitly at bolstering the Eritrean state's capacity to wage war. Messages with subject lines like "Let's support our troops!!!!" and sign-offs saying "Proud to be Eritrean!!!!!!" and "Awet n hizbi Eritrea Zikre nswat AHwatnan AHatnan" (Victory to the people of Eritrea, Remember our martyred brothers and sisters) capture the flavor of these wartime posts. One mentions that pledges totaling fifty-five thousand dollars were raised in two hours, with fifteen thousand dollars paid on the spot (posted on Dehai, June 16, 1998). The monetary contributions from the diaspora were officially acknowledged by the governor of the Bank of Eritrea, who described them as "beyond anybody's imagination" (Voice of America, June 24, 1998).

The response of the diaspora to the outbreak of war with Ethiopia in 1998 demonstrates among other things the key role of the diaspora in Eritrean politics and the ways in which they engaged with the Internet not just to discuss events or keep up to date on the breaking news of the conflict but to try to affect events on the ground. Dehai reached its peak during the 1998–2000 border war between Eritrea and Ethiopia when it became the source for breaking war news, for fund-raising and organizing support, and for analysis and discussion of the war, the media coverage of the war, and all of the implications of these.

CYBERSPACE, CIVIL SOCIETY, AND DISSENT

Even in the midst of supporting the war effort, Eritreans used Dehai to express their critical views to the Eritrean government. For example, in the fall of 1998 a petition addressed to President Isaias asking for due process and amnesty for an Eritrean journalist thrown in prison was circulated on Dehai to collect signatures. It stated in part, "We the undersigned Eritreans living in the Diaspora humbly bring to your attention an issue that has been of great concern to many other Eritreans," and then, after petitioning on behalf of the journalist, continued, stating, "We would like to make it clear to your Excellency and to the whole world that even those amongst us who hold differing political outlook [sic] to those of your government have also decided to stand united with our people, our army and our government for the defence of Eritrea's hard-won sovereignty and territorial integrity" (posted on Dehai, October 20, 1998).

The end of the border war in December 2000 and the shift out of crisis mode to more routine political issues and debates led to decreased activity on Dehai as people's attention drifted. But that may not be as significant as the rapid rise of a handful of other Eritrean Web sites competing for the same readers and writers as Dehai. Political rifts emerged within the ruling circles of Eritrea. The war and the government's handling of it strained people's loyalties. The government's response to criticism within Eritrea was to crack down on any suspected critics. The independent press was closed down, and journalists were arrested and jailed without trial. A new political era dawned within Eritrea and in the diaspora. The diaspora was able to express itself with a degree of openness that made the cracks in the facade of Eritrean unity more visible as the online public sphere that had centered on Dehai fragmented.

The fragmentation of Eritrean cyberspace into multiple sites or spheres also reflects the centrifugal force toward plurality that some scholars see as characteristic of the Internet as a medium (Becker and Wehner 2001). However, despite such tendencies in the technology, Dehai was able to hold Eritreans together for years despite (or, as I have argued elsewhere, because of) political conflict on- and off-line (Bernal 2005). Perhaps the eventual fragmentation of Eritrea's virtual public sphere should cause us to consider more seriously the ways in which the Internet facilitates social divisions as well as connections. But these divisions arose not from the technology itself but from the wider field of politics and war beyond it. In the case of Eritrean Web sites, the struggles and ruptures are not simply those of everyday life reproduced online, however. The public sphere in cyberspace made possible certain kinds of discussion and critique that could not be expressed elsewhere, and widened the participation in debates by linking interlocutors who were unknown to one another and geographically dispersed, thus transforming politics and the relationship among citizens, states, and territory.

With the end of the border war and the government's heavy-handed repression of independent media and civil society within Eritrea, the diaspora began to express criticism of President Isaias and the PFDJ more openly, particularly through a range of newer Web sites whose posters were more receptive to dissenting views. The diversity of Web sites grew, with some reflecting distinctive ethnic or religious points of view on Eritrean politics.

The fluorescence of Eritrean-diaspora Web sites since 2000 has decentralized Eritrean online culture so that Dehai no longer represents the public sphere but is part of a range of sites that together can be considered an online public sphere. Several successful competitors to Dehai were launched; most notably http://www.asmarino.com appears to have overtaken Dehai in popularity.

The postwar situation is also different in that ordinary Eritreans within Eritrea obtained the ability to go online in publicly and privately owned cybercafes in 2000. The content of Eritrean cyberspace remains generated largely by the diaspora despite these developments. This may in part be due to the predominant use of English, although English is used as the language of instruction in Eritrea above primary school.

In recent years, new technologies for producing Ge'ez script, the system that is used for writing Tigrinya, have made it possible for some people to post in Tigrinya. On Asmarino.com, for example, English predominates, but there are many posts in Tigrinya and neither language is translated, so full participation assumes competency in both languages, which for those who have dominated content production, Eritreans in diaspora in North America, is more or less a given. This may increasingly be true of new generations of Eritreans in Eritrea due to the use of English as the language of instruction, and the de facto if not de jure use of Tigrinya as the national language.

The suppression of free expression within Eritrea and state control of the media have given a heightened significance to the online public sphere created by Eritreans in diaspora. The spaces of diaspora and cyberspace could be seen as constituting a kind of offshore civil society and virtual public sphere for the nation of Eritrea that cannot at present take shape within national territory.

CONCLUSION

To realize the promises of Eritrean nationalism, issues of development and democracy had to be pursued with the same zeal as independence. Eritreans used cyberspace to overcome their dispersal and to participate in the new nation. As the initial euphoria of independence grew to be replaced by disappointment with the Eritrean leadership, cyberspace became important in new ways, offering Eritreans another kind of independence, independence of thought and political expression, not controllable or punishable by the Eritrean state. Moreover, in this receding horizon of political desires where objects remain out of reach, cyberspace could serve as a repository for the yearnings and visions of Eritreans in diaspora for the best of all possible Eritreas, even as it served as a means for individuals to connect with other Eritreans and even to their own "Eritreanness."

The ways in which Eritreans have used information technologies is worth documenting and investigating in its own right, and the study of Dehai and other sites can be used to reflect on the meanings and potentials of online life and cyberspace for politics. In considering Eritrean activities in cyberspace, we must understand them in the context of the history of strong transnational networks throughout the diaspora that were largely centralized under the management of the EPLF as a protogovernment, and that subsequently the PFDJ has tried to keep under its control or at least under its influence. These established relationships linking the diaspora to nationalist fronts and eventually to the Eritrean state have allowed for translations between online and off-line social and political worlds that are not available to all diasporas or migrant populations. In fact, the great degree to which Eritrean transnationalism has been promoted and orchestrated by nationalist leaders and the state is unusual.

Both the new media of cyberspace and the new spaces occupied by transnational migrants can be understood as spaces beyond the reach of government to some extent. As such, these spaces allow the possibility of political experimentation. Through cyberspace the Eritrean diaspora has not only participated in Eritrean politics but also experimented with new ways of practicing Eritrean politics, creating a public sphere that has no counterpart within Eritrea.

REFERENCES

Al-Ali, Nadje, Richard Black, and Khalid Koser. 2001. Refugees and transnationalism: The experience of Bosnians and Eritreans in Europe. *Journal of Ethnic and Migration Studies* 27, no. 4: 615–34.

Becker, Barbara, and Josef Wehner. 2001. Electronic networks and civil society: Reflections on structural changes in the public sphere. In *Culture, technology, communication,* ed. Charles Ess. Albany: State University of New York Press.

Bernal, Victoria. 2004. Eritrea goes global: Reflections on nationalism in a transnational age. *Cultural Anthropology* 19, no. 1: 3–25.

———. 2005. Eritrea on-line: Diaspora, cyberspace, and the public sphere. *American Ethnologist* 34, no. 4: 661–76.

Conrad, Bettina. 2005. We are the prisoners of our dreams: Exit, voice, and loyalty in the Eritrean diaspora in Germany. *Eritrean Studies Review* 4, no. 2: 211–61.

González-Quijano, Yves. 2003. The birth of a media ecosystem: Lebanon in the Internet age. In *New media in the Muslim world,* ed. Dale Eickelman and Jon Anderson. Bloomington: Indiana University Press.

Hepner, Tricia. 2005. Transnational Tegadelti: Eritreans for liberation in North America and the Eplf. *Eritrean Studies Review* 4, no. 2: 37–84.

Miller, Daniel, and Don Slater. 2006. Being Trini and representing Trinidad. In *Cybercultures: Critical concepts in media and cultural studies,* ed. David Bell. London and New York: Routledge.

Tezare, Kisanet, Tsehay Said, Daniel Baheta, Helen Tewolde, and Amanuel Melles. 2006. *The role of the Eritrean diaspora in peacebuilding and development: Challenges and opportunities.* Toronto: Selam Peacebuilding Network.

UNICEF. 1994. *Children and women in Eritrea.* Situation report, UNICEF.

Van Den Boss, Matthijs, and Liza Nell. 2006. Territorial bounds to virtual space: Transnational online and offline networks of Iranian and Turkish-Kurdish immigrants in the Netherlands. *Global Networks* 6, no. 2: 201–20.

Voice of America. 1998. Carol Pineau reporting from Asmara. June 24. Available at http://www.voa.gov.

Whitaker, Mark. 2006. Internet counter counter-insurgency: TamilNet.com and ethnic conflict in Sri Lanka. In *Native on the Net: Indigenous and diasporic peoples in the virtual age,* ed. Kyra Landzelius. London and New York: Routledge.

Woldemikael, Tekle. 1996. Ethiopians and Eritreans. In *Case studies in diversity: Refugees in America in the 1990s,* ed. David Haines. Westport, Conn.: Praeger.

8 Keeping the Link
ICTs and Jamaican Migration

HEATHER A. HORST

In the summer of 2007 news and weather agencies began tracking the progress of Tropical Storm Dean in the mid-Atlantic. A week later, the storm moved into the Caribbean, steadily growing in intensity. By August 17, Dean was reclassified as a Category 4 hurricane, and the island nation of Jamaica was placed on a hurricane watch. The same day, Jamaican prime minister Portia Simpson-Miller called together the National Disaster Preparedness Council to discuss evacuations and the transformation of the national arena in Kingston into a shelter.

Jamaicans as a general rule believe that God can and does counteract the vagaries of Mother Nature in Jamaica, a country that is known in the *Guinness Book of World Records* for possessing the most churches per square mile.[1] Indeed, Jamaicans routinely attribute the historical tendency of hurricanes to veer south and thus spare the island to the collective power of prayer and faith. Yet even with God on their side, the prospect of Hurricane Dean incited worry and concern among Jamaicans dispersed across the globe. Those who remembered the devastation of Hurricane Gilbert in 1988 and Hurricane Ivan in 2004 were all too aware of the difficulties of coping with the aftermath of the storms in a cash-strapped country.

To compound matters, the national election between (then) Prime Minister Portia Simpson-Miller of the People's National Party and (now) Prime Minister Bruce Golding of the Jamaican Labour Party was slated for August 27, 2007, seven days following the hurricane's anticipated arrival. Already quite a violent and virulent race, the potential for looting and other post-hurricane violence weighed heavily on people's minds. Finally, and especially for Jamaicans living in the United States, the Hurricane Katrina disaster in 2006 stood as a poignant reminder of the ways in which racially inflected discrimination continues to shape relief efforts, the slow response to the needs of African Americans in New Orleans often viewed as a microcosm of the ambivalence of the world to the ails of black people in the Caribbean and other parts of the African diaspora. Jamaicans readily acknowledged that their friends and family would probably be the most viable form of state or foreign aid in the storm's wake, and a small number of Jamaican migrants even flew home to wait out the storm with their families in an effort to avoid the feeling of helplessness.

On the night of August 19, 2007, Hurricane Dean finally reached the island of Jamaica. Leading up to the storm, the Jamaican diaspora monitored the progress of the storm on CNN, the Weather Channel, and other news agencies. As the storm gained momentum, the nation's telecommunications system suffered. Cell phone tower masts went down, as did landlines throughout the island. Even in areas with continued service, the frequency of calls made tied up the phone lines, and it became particularly difficult to get through to the island's rural towns and remote districts. In place of the telephone, Jamaicans began tuning into Power 106 FM, a local talk radio program airing 24/7 coverage of Hurricane Dean before, during, and after the storm. Jamaicans off the island also started logging onto Power 106 FM over the Internet through Go-Jamaica (http://www.go-jamaica.com), a Web portal that hosts the online editions of the *Jamaica Gleaner, The Star,* as well as *The Voice* out of the United Kingdom.

Moreover, and with the collaboration of the Jamaica National Building Society and People's Telecom that provided local and toll-free numbers to Power 106 callers, Jamaicans living in the UK, United States, Canada, and elsewhere could not only listen to the live streaming of the Hurricane Dean coverage but also call in to send messages to their loved ones in Jamaica. Alongside official reports from the Office of Disaster Preparedness and Emergency Management, throughout the night messages of love and concern went out to people by Jamaicans calling from the United States,

Canada, and the United Kingdom to check in on their families. At 2:40 in the morning a man from the United States called to send a message to his family in St. Thomas, a parish in eastern Jamaica, which was hit hard during the storm to convey the following message, "I am just calling to wish all my family, mum, sister, brother, Karin, Kimoy, Kimberley and friends just want to wish all of them, hope them safe and we'll see them soon" (Power 106 FM, August 20, 2007).

In addition, Jamaicans who were listening called in to share their own eyewitness accounts of the storm, providing information concerning the intensity and aftereffects of the hurricane as it traveled from Morant Point to Negril. For instance, a young woman in Jamaica called in from her family's leaking house in the parish of Manchester to share her situation: "We are just sitting here listening to you, and we wanted to make sure that if my sisters and my other family members if they're hearing us that the majority of the family is alright, but the roof is gone" (Power 106 FM, August 20, 2007, 2:32 AM). For me and others who stayed up through the night worrying about how our families and friends would manage to weather the storm, Power 106 became a lifeline, and we felt eternally grateful for any and all information we could glean from afar.

The Hurricane Dean crisis and radio-Internet coverage of the event represent a very poignant reminder of the powerful ways in which digital media and technology can be mobilized to facilitate transnational communication across space and time. Borrowing from Paul Gilroy's (1993) conceptualization of *roots* and *routes* to describe the relationship between displacement, movement, and power within the African Atlantic diaspora, this chapter outlines the media and technological landscapes of Jamaican migrants in terms of these two distinctive but increasingly interlocking processes that motivate Jamaican migrants' media practices.[2] I employ the term *rooting* to describe practices that reflect the desire of Jamaican migrants' use of digital media and technology to understand their heritage and identity. More specifically, rooting often (though not always) involves a symbolic engagement with the notion of "Jamaica" and "Jamaican culture" among Jamaican migrants as they are mapped, read, and negotiated in digital worlds. By contrast, I characterize *routing* as the more mundane, or routine, practices of keeping in touch with specific individuals, including family and friends using digital media and technologies. Throughout the chapter, I illustrate Jamaican migrants' varied engagements with new information and communication technologies and the ways in which these may be changing in the era of convergence culture (Jenkins 2006).

CREATING AND MAINTAINING ROOTS THROUGH DIGITAL MEDIA

As has been well established in the literature on transnational migration and diasporas, migrants develop an orientation that incorporates their country (or countries) of migration as well as their homeland(s) (Basch, Glick-Schiller, and Szanton-Blanc 1994; Smith and Guarnizo 1998). Historically, a variety of media and other forms of communication technologies have played an important role in the maintenance and shaping of these transnational orientations. For example, in the areas of New York, London, and South Florida, which are heavily populated with Jamaicans, a number of shop vendors sell copies of local Caribbean-focused newspapers as well as the *Jamaica Observer, The Star,* and the *Jamaica Gleaner,* the island's leading newspapers. In Washington, D.C., the radio station Caribbean Vibes Radio not only plays Jamaican music but also disseminates information about politics, cricket and football matches, and other current events of interest to Jamaicans (and other Caribbean people). They also occasionally host Jamaican DJs from other disparate parts of the diaspora. Other stations cater to the prominent Jamaican and Caribbean Christians in the diaspora by playing Caribbean-inspired gospel music.

Although middle-aged Jamaican migrants exhibit a notable presence in the Jamaican digital diaspora online, the Internet represents a particularly important location for Jamaican diasporic youth trying to understand themselves and their place within the wider North American and European society.[3] In an article focusing on one-and-a-half-generation Jamaicans in Canada, Plaza (2006) argues that for youth, and particularly one-and-a-half-generation Jamaicans in Canada, questions of identity and the formation of identity consciousness commonly occur when they attend college. From joining Caribbean Association student organizations to the development of interests surrounding their Jamaican, Caribbean, or West Indian heritage, Jamaican university students emphasize that young adulthood remains a crucial time in their lives to define and grapple with their awareness of being Jamaicans and, in turn, reach out to make intellectual and emotional links to their country or parents' country of birth. In addition to a pilgrimage to Jamaica to learn more about Jamaican culture, young Jamaicans in Canada and North America often join projects that focus on alleviating the conditions of poverty and violence in Jamaica (see also Simmons and Plaza 2006). These engagements typically involve searching for information about Jamaica, their Jamaican families, and Jamaican history in online venues and, in turn, reinscribing the Jamaican presence in digital spaces through their own Web sites, pages, videos, and music.

Alongside traditional media channels, video-sharing venues, social network sites, and other Jamaican-oriented Web sites have assumed a prominent role in the sharing and shaping of common identities, ideologies, and localized interests among different generations of Jamaican migrants. For example, youth in the Jamaican diaspora have increasingly become interested in social network sites like MySpace, Facebook, Hi5, Orkut, and a range of others. Many college students create and maintain connections via Facebook with other college students of Jamaican descent in North America and England. In some instances, these connections are made with family members who are attending Jamaican universities, such as the University of Technology, Northern Caribbean University, and University of the West Indies. In these spaces, members keep abreast of local music and entertainment events and may join discussion groups on topics ranging from "Why do successful Jamaican men feel they have a right to keep more than one woman?" to "How many Jamaicans do you know in diaspora?" and "The Jamaican Language," where people write in to exchange ideas about the value of speaking, writing, learning, and using "patois" (Jamaican Creole). As of February 2008, there were 38,042 members in a Facebook network entitled "Jamaica," with members spread across North America, Jamaica, and (to a lesser extent) the Caribbean and Europe. By way of comparison, there are 57,552 members in the Berkeley network and 1,130,548 in the Toronto network.

Another increasingly popular venue for the Jamaican diaspora online includes *vlogging* sites like YouTube (http://www.youtube.com>), where individuals view, upload, and share short videos.[4] These video clips range from comedy shows, commercials, music videos, and sports matches, which originally aired on traditional media, to remixes of mainstream media and informal videos produced by amateurs (sometimes referred to as DIY, or do-it-yourself, videos). They range in content from Dancehall DJ Beenie Man and D'Angel's wedding ceremony in 2006 to videos of weekend dance-hall parties and individuals performing the Dutty Wine, a popular Jamaican dance in 2005 and 2006 that involved rapid rotations of one's neck and behind while moving one's legs like a butterfly. Although most of these videos tend to originate outside of Jamaica where the tools of production are more readily available, Jamaicans on the island increasingly produce and upload videos to sites like YouTube. Many firsthand accounts of Hurricane Dean were created by Jamaicans who turned their cameras on the blowing trees, darkening skies, and (later) the aftermath of the storm to give those "in foreign" the opportunity to virtually experience the hurricane (2kharrier 2007).

Alongside circulating information and images of Jamaican popular culture, sites like YouTube become spaces for the production and reproduction of Jamaican culture, broadly defined. For instance, this response to a video titled "dutty wine white" reaffirmed the traits and behaviors that are typically attributed to Jamaican (and Caribbean) culture through the assertion of difference between Afro-Caribbean and white culture: "God its true when they say white boys cant dance like dem from the carribean . . . unless they take classes but still that shit doenst come naturally" (99calleri99 2006). Even when there are positive comments, as in the responses to "white boy wines to dancehall," the origin of the white boy's ability to dance is attributed to skills derived from others: "Dis white bwoy must fi have a black girlfriend. Who teach im fi wine so?" (ibid.). While some women (who imply in their comments and profiles that they are Jamaican) may express amazement at a particular (white) man's ability to wine, for men the white boy's prowess on the dance floor while doing the Dutty Wine reinforces many Jamaican well-established anxieties over sexuality and nonnormative sexuality, which is pervasive in dance-hall culture (Cooper 2004). For example, a man responds to "white boy wines to dancehall" as follows: "Him gd, rele gd, mi nah call him a batti bwoy cah mi nuh wha fi offend him. keep up wid di dancin my yoot and do mi a fyava and teach some a dem white girls fi dance and keep wid di I dem surely need a lesson from you" (Him good, really good, I'm not calling him a batty boy [a gay man] because I don't want to offend him. Keep up with the dancing my youth and do me a favor and teach some of those white girls to dance and keep with the rhythm they surely need a lesson from you) (Soccersocks23 2006). Given the racial hierarchies that continue to be pervasive within American society (see Vickerman 1999), these spaces become locations where Jamaican migrants affirm, negotiate, and redefine Jamaican culture through popular culture.

In addition to asserting and affirming Jamaican values, speaking or writing in Jamaican patois also remains an important part of Jamaican diasporic identity formation and negotiation in digital spaces. Although patois is generally used for oral communication, in online spaces Jamaicans use formal and informal linguistic devices to communicate patois, or Jamaican Creole. Lars Hinrichs (2006), who studied code switching and the use of Jamaican patois and English in university students' e-mail messages, argues that in its emergent, written forms, Jamaican Creole is increasingly being utilized symbolically. Hinrichs further suggests that these new forms of Jamaican Creole that continue to emerge in digital formats represent a shift away from code switching (or switching back and forth between standard Jamaican English and Jamaican Creole). Youths' acceptance of Jamaican Creole as a legitimate form of expression in written communication represents a divergence from the stigma that influenced its use (or lack of use) among older individuals. These changes not only suggest that "the reproduction of legitimate culture in enclave social spaces—such as Montreal, Toronto, or Winnipeg, where events such as Carifesta, Caribana, or Jamaica day occur annually" (Plaza 2006, 8)—plays an important role in defining Jamaican culture but also recognize that digital spaces have become important locations for creating and transforming Jamaican culture.

ROUTING AND REROUTING JAMAICAN FAMILY CULTURE

Unlike rooting practices that involve the use of information and communication technologies in order to create, maintain, and negotiate symbolic relationships with the island of Jamaica and Jamaican culture, routing reflects communication practices that are formed with the aim of solidifying, navigating, and maintaining tangible connections with individuals, friends, and families within and between nodes, points, and spaces of the Jamaican world. Although Jamaicans possess a long and varied migration tradition dominated by movement to and from the United Kingdom, United States, Canada, and elsewhere in the Caribbean, communication between migrants and their friends and families in Jamaica remained sporadic throughout the twentieth century because mail communication and telegrams often remained expensive and unreliable. Geographer Elizabeth Thomas-Hope (1992, 131–32) reports that in the late 1980s, 37.52 percent of the populations she surveyed never or rarely received letters from a Jamaican migrant living abroad. Just over 28 percent received letters from abroad occasionally, and 33.91 percent of the individuals surveyed received letters frequently, with rural areas receiving the lowest levels of correspondence. Almost 7 percent of respondents had no foreign contacts at all. Correspondence via telephone was even less reliable, particularly between rural Jamaicans and Jamaicans living abroad.

To understand the dynamics of "routing" in shaping the dimensions of Jamaicans in the digital world, a picture of the media and technology infrastructure in Jamaica should be understood.[5] In 2003 there were only seventy thousand Internet connections in Jamaica. Half of these connections were with Cable and Wireless, the main incumbent telecommunications company in Jamaica and most of the Caribbean (the rest of the subscriptions are registered with a half-dozen smaller companies). An Allen Consulting Group

(2002) report, which attempted to assess the potential for e-commerce within Jamaica, suggested that only 9 percent of companies had Internet access, and less than 2 percent possessed Web sites, which implied that commercial access is not much greater than private access during this same time period. Not surprisingly, the most developed sector at the beginning of the decade was the tourist industry with its attention to accommodation, travel, transportation, and online booking. However, outside of tourist areas, very few Internet cafés exist (Anderson Consulting Group 2002). By 2004 there were six major cafés in the main areas of Kingston, with a population of eight hundred thousand. In Portmore (population of two hundred thousand), a dormitory suburb of Kingston with few international visitors, there were three commercial and one nongovernmental organization–based Internet cafés.

However, there is evidence that this situation is changing. Whereas at the turn of the millennium, Internet penetration rates were estimated at 3 percent, penetration rates increased to 24.7 percent of the population by 2005; at the same time, Canada's Internet penetration rate lingered around 67.5 percent (Internet World Stats 2007). Recent reports by the ITU (2008) suggest that more than 53.8 percent of the population in Jamaica now use the Internet, and actual Internet subscription in 2007 was estimated at eighty-five thousand subscriptions. In effect, we are seeing a moment when the interactions between Jamaicans living outside Jamaica and those in Jamaica have the potential to intensify, sometimes in unexpected ways. For example, to keep in touch with relatives who live abroad, one individual often assumes the management of transnational family communication via the Internet. Students and working- and middle-class Jamaicans with jobs at the university and in the private sector use their access to the Internet at work and school to maintain contact with their families and friends living abroad.

These mundane communications can range from sharing key news about the family, including the coordination of visits or the birth of new children. Most Jamaicans I have interviewed stress that communication by e-mail is uncertain and thereby not the most reliable form of communicating about synchronous, or time-sensitive, issues. Instead, the most frequent messages tend to involve the circulation of jokes, poems, "positivity" messages, and other chain mail, which are cut and pasted or forwarded, rather than original content. Many of these e-mails fall into the category of "link-up" e-mails, or messages that are intended to keep in touch for the purpose of maintaining connection in and of itself (Horst and Miller 2005). One example of this type of e-mail is "God has amazing things in store for you" (originally written in red fourteen-point font):

GOD HAS AMAZING THINGS IN STORE 4 U! (This is a heavy prayer), Stars do not struggle to shine, rivers do not struggle to flow, and you will never struggle to excel in life, because you deserve God's BEST. Hold on to your dream and it shall be well with you . . . Amen. Good Morning, your dream will not die, your plans will not fail, your destiny will not be aborted, and the desire of your heart will be granted in Jesus' name. The eyes beholding this message shall not behold evil, the hand that will send this message to others shall not labor in vain, the mouth saying Amen to this prayer shall laugh forever, remain in God's love. Money will know your name and address before the end of this month. If you believe, send it back to all your friends. None goes to the river early in the morning and brings dirty water. As you are up this morning, may your life be clean, calm and clear like the early morning water. May the Grace of the Almighty support, sustain and supply all your needs according to His riches in glory. Amen. Have a wonderful day in Jesus' name. The will of God will never take you where the grace of God will not protect you. Love the Lord. See something good happening to you today, something that you have been waiting to experience. This is not a joke; you are going to receive a Divine visitation that will move your life forward mightily by the outworking of God's unstoppable Healing & Delivering Power. Do not break this prayer; Say a big Amen, and if you believe it send it to a minimum of seven people. You are blessed in Jesus' name! A teacher gives, a teacher shares; but most of all, a teacher cares!! (ibid.)

Other messages reflect Pentecostal teachings and are often a critique of Jamaican society, as in "Don't let the devil leave his bags," which was circulated in the summer of 2007:

YOU PUT THE DEVIL OUT, BUT DID YOU LET HIM LEAVE HIS BAGS? This is powerful! You got out of a bad relationship because it was bad, but you are still resentful and angry (you let the devil leave his bags) You got out of financial debt, but you still can't control the desire to spend on frivolous things (you let the devil leave his bags) You got out of a bad habit or addiction, but you still long to try it just one more time (you let the devil leave his bags) You said, I forgive you, but you can't seem to forget and have peace with that person (you let the devil leave his bags) You told your unequally yoked mate that it was over, but you still continue to call (you let the devil leave his bags) You got out of that horribly oppressive job, but you are still trying to sabotage the company after you've left (you let the devil leave his bags) You cut off the affair with that married man/woman, but you still lust after him/her (you let the devil leave his bags) You broke off your relationship with that hurtful, abusive person, but you are suspicious and distrusting of every new person you meet (you let the devil leave his bags) You decided to let go of the past hurts from growing up in an unstable environment, yet you believe you are unworthy of love from others and you refuse to get attached to anyone (you let the devil leave his bags) When you put the devil out, please make sure he takes his bags! HAPPINESS KEEPS YOU SWEET TRIALS KEEP YOU STRONG AND SORROWS KEEP YOU HUMAN FAILURES KEEP YOU HUMBLE SUCCESS KEEPS YOU GLOWING BUT ONLY GOD KEEPS YOU GOING! in The Remainder of 2007, Let the devil Take his bags with him! Be Blessed, Healthy and Happy.

These messages that circulate in e-mails and other online venues are often printed, photocopied, and affixed to bedroom walls, doors, and workspaces, although the particular jokes, information, and communication that make it into other contexts remain at the discretion of the primary communicator, or information gatekeeper.

Although the Internet certainly is beginning to have an impact on Jamaican transnational communication practices, the growth of the cell phone industry over the past two decades represents one of the most transformative new information and communication technologies for Jamaicans, at home and abroad (Horst 2006; Horst and Miller 2006). Largely due to the pricing structures of the American and Canadian telecommunications market, Jamaicans living abroad continue to rely on a combination of phone cards and landlines to initiate communication with their friends and families throughout Jamaica. However, the prevalence of mobile phones throughout Jamaica (2.7 million mobile phones in a country of 2.7 million people) has enabled more direct forms of communication, often without the gatekeepers who have dominated both phone and Internet communication historically.

Mobile phones enable near-synchronous communication and thereby enable the extensive microcoordination of the exchange of remittances. Others use the mobile phone to coordinate visits as well as to organize barrels of food and other provisioning gifts for the transnational families. It is used to keep one up-to-date on the health and welfare of aging parents and grandparents as well as children and grandchildren (Horst 2006). The personal and pervasive incorporation of mobile phones in Jamaica represents one of the key changes in communication as the Jamaican diaspora goes digital.

Although it is unlikely that the Internet will ever replace phone communication for Jamaican migrants with families who live in Jamaica's rural hinterlands, there are spaces where the Internet and digital media generally continue to enhance the experiences of family members separated by time and distance. One of the most important contributions of the Internet, particularly for middle-class Jamaican families, revolves around the sending and sharing of digital pictures. During my fieldwork in Jamaica in 2004, a young woman named Simone gave birth to a child. Her babyfather's mother (i.e., the child's grandmother) lived in Florida and had not yet had the opportunity to visit Jamaica and see her new granddaughter.[6] When the baby was a month old, one of her aunts purchased a digital camera. Simone and her babyfather dressed the baby and took pictures of her and later downloaded them at a friend's house and then e-mailed them through a Yahoo account, which they checked intermittently at friends' homes. The pictures were sent to a sister who lived near the baby's aunt in Florida, and the sister then downloaded the images so that she could bring them to the grandmother's home. Simone's baby's grandmother was delighted to see the baby and immediately called her son to fawn over the child and how much the baby "favored" her father. Many of the young women I knew with access to the Internet also began using social network sites like Hi5 to post pictures for their Jamaican friends and family. Rather than registering with official album sites, such as Kodak Gallery, Shutterfly, and Flickr, many Jamaicans living in the United States, Canada, and Jamaica set up accounts on social network sites and upload pictures of their children, nieces, and nephews (rather than themselves, as is the normative practice in North America). Much like the video cassettes that circulate through Jamaican transnational families after weddings, funerals, graduations, and other rites of passages, these digital routing practices, or acts, which keep family connected across time and space through digital media and technology, continue to transform the experience of family in the Jamaican diaspora.

THE CONVERGENCE OF ROOTS AND ROUTES?

While the somewhat ambiguous notion of "the Internet" tends to dominate discussions of diasporas and transnational communication, throughout this chapter I have tried to illustrate that a variety of digital media and information and communication technologies now shape the contours of Jamaican migrants' communication within and between the diaspora. My focus on framing these practices in terms of rooting and routing is an attempt to acknowledge the motivations and strategies that underpin the different practices in the digital diaspora. In this final section I want to consider these processes and the extent to which, in the words of Henry Jenkins when discussing convergence culture in the United States, the "technological, industrial, cultural and social changes in the ways media circulates within our culture" may also be transforming the experience of Jamaican migration and transnationalism (2006, 282).

In examining the Jamaican presence in the digital world, it becomes evident that most Web sites that focus on Jamaica are hosted outside of Jamaica. As a result, these sites are often geared toward rooting. One of the landmark Jamaican Web sites, Top5Jamaica, was developed by Sandor Panton, who left Jamaica in 2001 to work for a small Internet marketing company in Canada. Panton began the site as part of a master of science program at the University of the West Indies in 1997 and decided to post it live on the Internet at the end of 1998. By mid-1999 the Web site was generating such serious traffic that Panton decided to register it and develop the site further. Much of this enterprise involved searching for sites on Panton's own time and money. By 2004 the site wielded approximately eight thousand unique visitors per day, although most of the traffic was derived from Jamaicans living abroad and foreigners seeking information about Jamaica. Approximately 70 percent of Top5's visitors are from North America (the United States and Canada) and the United Kingdom, and Top5 now attracts enough advertising to pay for itself. Similarly, Jamaicans.com, a Web site devoted to consolidating information about Jamaica that went online in 1996, is owned by a Jamaican living in Florida. In effect, the Jamaican Web is Jamaica through the eyes of the Jamaican diaspora.

The dominance of the Jamaican does seem to be shifting as Jamaica's middle class comes online. An increasing number of blogs are being produced, ranging from profiles on social network sites geared toward friendships to dissemination of information about religious youth groups and churches. Original content, such as short videos of Hurricane Dean uploaded by Jamaicans, is now surfacing, garnering comments from people on and off the island. These personal but one-to-many forms of communication by Jamaicans alter the ways in which Jamaica (and rooting more generally) is imagined in digital spaces. These shifts in the power to define, describe, and assert "Jamaican culture" are also taking place in the more routing-oriented practices of Jamaicans in the diaspora.

For example, mobile phone ownership in Jamaica creates an opportunity to initiate the process of communication. Although readily acknowledging the tenuous and uncertain economic status of many Jamaicans living abroad and the sacrifices many individuals make to support their families in Jamaica (Vickerman 1999), placing the power to call (and control over the duration of the call) in the hands of Jamaicans living in Jamaica destabilizes the traditional power relations, particularly the sense of powerlessness, frustration, and indebtedness many Jamaicans may feel toward their friends and family members abroad. This sense of inequity is tied to the historical importance attributed to travel as well as the escape from the everyday realities of life in Jamaica. In addition, going abroad provides opportunities for material advancement, information, and resources that migrating abroad is perceived to afford (Thomas-Hope 1992; Olwig 2007).[7]

While "routing"-motivated participation is often particular and pragmatic in nature and "rooting"-motivated participation in the digital diaspora may be more symbolically oriented, there is much to suggest that some of the "routing" and "rooting" practices have converged as digital media have become more accessible across the Jamaican media world. The increased participation crosscuts location, nation, and (particularly in mobile spaces) well-entrenched class and color lines, creating a space for new conversations about Jamaica and the role of the Jamaican diaspora. Through the medium(s) of digital media and technologies, these new voices will inevitably inspire debates and tensions over authenticity within the transnational worlds of Jamaican migrants.

NOTES

1. Jamaicans also argue that Jamaica's "special status" in God's eyes is attributed to the name of the island itself (pronounced JAH-MEK-YA), which can be interpreted as follows: "Jah" is the Rastafarian word for God, "mek" means "make," and "ya" is short for "you."

2. Research for this article is based on an extended ethnographic engagement with Jamaicans since 1994, which included research on two distinct research projects carried out in 2000–2001 and 2004–2005.

3. Plaza is considering how generation influences one's sense of identity and identification with Jamaica, Canada, West Indians, Caribbean, and other people of African descent (the African diaspora). In her study of transnational family networks, Olwig (2007) argues that an individual's relationship to a particular ethnic identity and desire to engage with others who share that ethnic identity vary significantly among individuals, particularly in relation to their geographical location as well as their "place" within the wider social hierarchy. In my research with Jamaican returnees, I also found that the extent to which individuals felt compelled to participate

in community organizations (in this case returning resident associations) was tied largely to their particular migration experience as shaped by the social, legal, and economic policies in each migration destination (see Horst 2007).

4. A vlog is a blog that uses video as the medium.

5. Bearing in mind that it is difficult to statistically decouple Jamaican migrants from the broader African American population, a recent Pew study (Horrigan 2009) suggests that broadband adoption has been on the increase among African Americans but still is lower than the national average, at 46 percent.

6. *Babyfather* is a term used in Jamaica to denote the father of a child when the mother and father are not formally married.

7. I do not want to suggest that the mobile phone is the great equalizer. However, given the difficulty of acquiring almost any form of a visa to travel or move outside of Jamaica, Jamaicans I have interviewed over the years genuinely appreciate the relative control over their participation in transnational relationships that has become possible through the mobile phone.

REFERENCES

Allen Consulting Group. 2002. The Jamaican ecommerce blueprint. Prepared for the Commonwealth Secretariat and Jamaican government, Sydney.

Basch, Linda, Nina Glick-Schiller, and Cristina Szanton-Blanc. 1994. *Nations unbound: Transnational projects, postcolonial predicaments, and deterritorialized nation-states.* Amsterdam: Overseas Publishers Association.

Cooper, Carolyn. 2004. *Soundclash: Jamaican dancehall culture at large.* London: Palgrave Macmillan.

Gilroy, Paul. 1993. *The black Atlantic.* London: Verso.

Hinrichs, Lars. 2006. *Codeswitching on the Web: English and Jamaican patois in e-mail communication.* Amsterdam and Philadelphia: Benjamins.

Horrigan, John B. 2009. Home Broadband Adoption Pew Internet and American Life Project. Available at http://pewinternet.org/Reports/2009/10-Home-Broadband-Adoption-2009.aspx.

Horst, Heather A. 2006. The blessings and burdens of communication: The cell phone in Jamaican transnational social fields. *Global Networks: A Journal of Transnational Affairs* 6, no. 2: 143–59.

———. 2007. "You can't be two places at once": Rethinking transnationalism through return migration in Jamaica. *Identities: Global Studies in Culture and Power* 14, no. 1: 63–83.

Horst, Heather A., and Daniel Miller. 2005. From kinship to link-up: The cell phone and social networking in Jamaica. *Current Anthropology* 46, no. 5: 755–78.

———. 2006. *The cell phone: An anthropology of communication.* Oxford and New York: Berg Press.

Internet World Stats Usage and Population. 2007. *Jamaica: Internet usage, broadband, and telecommunications reports.* Available at http://www.internetworldstats.com/car/jm.htm.

———. 2008. *Jamaica: Internet usage, broadband, and telecommunications reports.* Available at http://www.internetworldstats.com/car/jm.htm.

Jenkins, Henry. 2006. *Convergence culture.* Cambridge: MIT Press.

99calleri99. 2006. Dutty wine white. November 1. Available at http://www.youtube.com/watch?v=1SZH4IqyftU&NR=1.

Olwig, Karen Fog. 2007. *Caribbean journeys: An ethnography of migration and home in three family networks.* Durham: Duke University Press.

Plaza, Dwaine. 2006. The construction of a segmented hybrid identity among one and a half and second generation Indo- and African-Caribbean Canadians. *Identity: An International Journal of Theory and Research* 6, no. 3: 207–30.

Simmons, A., and Dwaine Plaza. 2006. The Caribbean community in Canada: Transnational connections and transformation. In *Negotiating borders and belonging: Transnational identities and practices in Canada,* ed. Lloyd Wong and Vic Satzewich. Vancouver: University of British Columbia Press.

Smith, Michael Peter, and Luis Guarnizo, eds. 1998. *Transnationalism from below.* New Brunswick, N.J.: Transaction Publishers.

Soccersocks23. 2006. Dutty wine Jamaican style. July 23. Available at http://youtube.com/watch?v=aqY8l3jYQYU.

Thomas-Hope, Elizabeth. 1992. *Explanation in Caribbean migration.* London: Macmillan.

2kharrier. 2007. Hurricane Dean Jamaica. August 19. Available at http://www.youtube.com/watch?v=XMl2gFZC598.

Vickerman, Milton. 1999. *Crosscurrents: West Indian immigrants and race.* Oxford: Oxford University Press.

9 Maintaining Transnational Identity
A Content Analysis of Web Pages Constructed by Second-Generation Caribbeans

DWAINE PLAZA

In 2000 there were an estimated 451 million Internet users worldwide, which represented 7.4 percent of the world's population.[1] By 2006 the number of users had jumped to one and a half billion, or approximately 25 percent of the world's population. The growth in Internet use between 2000 and 2006 has been especially dramatic in certain parts of the world. In Latin America and the Caribbean there has been an enormous 370 percent growth in the number of Internet users.[2] It is with this increase of the Internet as a communication and information medium worldwide that this study endeavors to explore how second-generation Caribbean-origin migrants living in the international diaspora construct and use Web pages as a symbolic bridge that connects familiar Creole cultural values and practices with the feelings of object loss and cultural mourning.

As social constructions, Web pages capture more than a moment in time. Unlike a still photograph, a Web page is a visual image that is growing and changing as the Webmaster adds or deletes images, text, sound, or hot links. Web pages like photographs can denote a certain apparent truth, provide documentary evidence, or tell the Web surfer a little about the individual or group who constructed and maintains it (Becker 1995). The connotative meanings of the Web page emerge from the social and historical contexts under which it was constructed, particularly in situations where the conventions are like road signs; just as we have learned to recognize the meaning of road sign symbols almost instantaneously, we have learned over the short period of time in which the Web has existed to decode the denotative and connotative content of Web pages.

In the past two decades there has been much scholarship on ethnic and cultural identities. Scholars like Anzaldúa (1995), Hall (1996), Rosaldo (1989), and Nagel (1994), working within a postmodern framework, theorize identity as hybrid, dynamic, fluid, and multilayered. They argue against essentialist notions of identity as fixed and bounded. Given that members of the Caribbean international diaspora live in a world of high modernity, they have created their own world that is reflected in music, fusion of language, food choices, styles of dress, and other markers of authentic transnational identity. Scholars like Maira (1999), Vertovec (2001), and Waters (1990) have pointed out that although the work of postmodernists has contributed significantly to dismantling essentialist notions of identity, it runs the risk of homogenizing the notion of hybridity and neglects to capture the complex view of the lived experiences of American, Canadian, or British Caribbeans.

Furthermore, cultural critics often neglect to take into account the diverse experiences of the second generation, particularly in terms of social class or ethnicity. Second-generation Indo-Caribbeans, for example, position themselves very differently from African Caribbeans in diaspora primarily because they are racialized in different ways. African-origin Caribbeans tend to elicit negative images in the consciousness of the dominant population, whereas Indian-origin Caribbeans are regarded as closer to the model minority (Plaza 2006). It is these sorts of ethnic differences and cultural fusions that we hope to explore in the Web pages that are constructed by second-generation Caribbeans living in the international diaspora.

Despite my desire to examine how all second-generation Caribbean men and women in the international diaspora are constructing Web pages, this paper will focus attention only on the activities of Caribbean student organizations based at university and college campuses in North America and Great Britain.[3] Caribbean-origin university students, who are mainly second generation,[4] are used for this research because they are the generation most likely to be comfortable with sending e-mails, downloading music, doing research on the Internet, surfing in cyberspace, participating in blogs, or maintaining their own Web sites. The Caribbean second generation is also likely to be living on the "hyphen"[5] and caught between two worlds while growing up in the international diaspora (Waters 1999; Kasinitz et al. 2002; Levitt and Waters 2002).

By constructing their own Web sites, second-generation Caribbean-origin men and women have in effect given themselves a new voice to disseminate information about their Creole culture, history of migration, and transnational lifestyles. Having a new agency over their circumstance in the international diaspora, the second generation who are attending college and university seem to derive a sense of relief from the cultural mourning and object loss they experienced while growing up in societies rife with radicalization and racism (Portes and Zhou 1993; Zhou 1999).

HISTORICAL AND THEORETICAL FRAMEWORK

Caribbean people have a long history of surviving economic adversity by moving to neighboring countries where jobs are more abundant. The contemporary Caribbean diaspora living in Britain, Canada, and the United States is a product of a "culture of migration" that developed as a survival strategy in the context of a long secular decline in sugar production and plantation agriculture starting in the early 1800s (Marshall 1982). Despite circulation within the Caribbean region that later expanded over time to include longer-distance movement, Caribbean people always brought with them aspects of their Creole cultural socialization (Lowenthal 1972).[6] This cultural baggage included a Caribbean diet, musical preferences, colorful language, superstitions, Caribbean myths and folklore, and unique living arrangements and family structures. Although each Caribbean territory had its own version of these cultural practices, there was a common Creole bond and a historical experience of being marginalized under the yoke of a white colonial master class, which unified Caribbean migrants, particularly as they faced racism and marginalization in the international diaspora (Henry 1994).

Since the 1960s a Caribbean international "diaspora" in some major cities in the eastern United States (New York, Boston, or Baltimore), in the United Kingdom (London, Manchester, or Birmingham), and in Canada (Toronto, Montreal, or Winnipeg) emerged. The formation of large Caribbean-origin migrant communities in these metropolitan cities and the resources that such immigrant communities provided to new migrants strengthened and transformed the Caribbean culture of migration. Caribbean migrants began to see themselves as both "here" and "there" in the Caribbean, although they were living abroad (Simmons and Plaza 2006). "Home" began to be viewed not just as the place where one was born or just where one lived, but more generally anywhere friends, relatives, and members of the cultural community were found. In effect, what began as a Caribbean culture of migration expanded over time to become a Caribbean transnational cultural community.[7] Thus, the culture of migration was retained as one key element in this geographically dispersed transnational community. Within the maturing transnational Caribbean community came the emergence of the second generation, who either were born in the new country of residence or arrived young enough that their major socialization years were spent in the new "home" country.

The second generation, like their parents, straddled two different locations—"here" and "there." For many, home became not necessarily the place where they were born or lived. Home could just as easily be the place where their parents were born or where they still had kin and fictive kin. Growing up in transnational households, it was not uncommon for the second generation to be exposed to the music, diet, superstitions, cultural values, and family structures that were distinctly Creole. At the same time, they would be deluged with the values, beliefs, and culture of the place where they were growing up. Many became caught in the position of the "marginal man"[8] while growing up on the hyphen. On the one hand, many were also pulled in the direction of the mainstream culture but drawn back by the cultures of their parents. By late adolescence, however, some continued to follow the linear assimilation model and identify themselves with the host culture, whereas others began to experience segmented assimilation. Those who felt marginalized often began to reestablish a strong coethnic identity, which drew on traditional Caribbean Creolized cultural ideologies and practices while at the same time including the cultural practices of the dominant society (Plaza 2006).

As a direct result of segmented assimilation, many second-generation Caribbeans over the years experienced a sense of cultural mourning. The idea of cultural mourning has its origins in the theories of object loss as conceptualized by Sigmund Freud (1939). In most cases of object loss individuals are able to mourn their loss in a way that prevents derangement. According to Volkan (1981) the mourner eventually finds "linking phenomena" that provide "a locus to externalize contact between aspects of the mourner's self representation and aspects of the representation of the deceased." Linking objects play a role in mourning in that they create "a symbolic bridge to allow the mourner to get over the situation" (Frankiel 1994). Linking objects in the case of second-generation Caribbeans might include eating authentic Caribbean food, listening to soca or reggae music, belief in the myths and folklore of Caribbean culture, or code switching in the use of language (Volkan 1981, 20).

Using the theory of cultural mourning, Ainslie elaborates on the theory to help explain the transition that oftentimes takes place with immigrant groups who find themselves living in a hostile foreign land: "When an immigrant leaves loved ones at home, he or she also leaves the cultural enclosures that have organized and sustained experience. The immigrant simultaneously must come to terms with the loss of family and friends on the one hand, and cultural forms (food, music, art, for example) that have given the immigrant's native world a distinct and highly personal character on the other hand. It is not only the people who are mourned but the culture itself, which is inseparable from the loved ones whom it holds" (1998, 287).

Ainslie draws on Winnicott's theory of the potential space to note that immigrants living abroad often find a space to engage in activities that "bridge the emotional gaps" created by their feelings of dislocation and loss (ibid., 289). This space allows first-generation immigrants and their children to restore the "object loss" they feel. This might include the engagement in activities that create the "illusion of restoration of what was lost" (ibid.). For Caribbeans in the diaspora this might include attending a Carnaval cultural show, reading a Caribbean newspaper online, joining a hometown or alumni association, attending a disaster-relief dance, or developing a Web site with Caribbean content. Ainslie further notes that immigrants tend to fill this potential "empty" space with activities, objects, or artifacts that keep alive the illusion of continuity with the homeland. In this regard, the potential space serves as a platform where immigrants can begin to negotiate their adaptation to the new environment.

It is with the theory of cultural mourning, segmented assimilation, and transnational lifestyles in mind that we set out to examine how second-generation Caribbean-origin university students are building Web sites as a way to alleviate their feelings of cultural mourning and to help bridge the emotional gaps of marginalization they experience in the locations in which they now live.

METHODS AND PROCEDURES

Sample

This study is based on a nonrandom quantitative content analysis. The sample is a convenience and judgmental one in that I looked at (n = 50) Web sites constructed and maintained by Caribbean student organizations at universities and colleges across the United States, Canada, and Great Britain. A Google search for Caribbean student organizations on November

TABLE 9.1 | SAMPLE OF SCHOOLS BY COUNTRY WITH CARIBBEAN STUDENT ORGANIZATIONS

UNITED STATES		GREAT BRITAIN	CANADA
Northeastern University	City College of New York	Bristol University	York University
University of Florida–Tampa	University of Buffalo	Lancaster University	McGill University
Stanford	Georgia Tech University	Reeding University	University of Waterloo
University MIT	Drexel University	University of Cambridge	Queens University
Georgia State University	Swarthmore College	Strathclyde University	University of Toronto
University of South Florida	Old Dominion	Durham University	University of Western Ontario
Virginia Tech	University of Florida	Cambridge University	Trent University
College Park	Notre Dame	Imperial College London	
University of Albany	Tufts University	Loughbourgh University	
Florida Atlantic University	Vanderbilt University	City University	
University of Central Florida	Syracuse University	East London University	
Columbia University	Big Hampton University	Oxford University	
SUNY–New Platz	Rochester Institute of Technology		
Renselear Poly-Tech Institute	University of Michigan		
University of Pennsylvania	Southern Illinois–Carbondale		
Embry Riddle Aeronautical University			

NOTE: n = 50.

13, 2005, yielded 199,000 hits. To be selected as part of the sample three main criteria were used. First, the Caribbean student organization had to be based at a college or university in the international diaspora. Second, the Caribbean student organization had to have a Web site that was active within the past year. Finally, there had to be evidence that the Web site was being managed by students. As a result of this sampling criteria, fifty schools were selected: thirty-one (62 percent) of the schools are in the United States, seven (14 percent) are in Canada, and twelve (24 percent) are in Great Britain. The fifty schools in the study are listed in table 9.1. The unit of analysis for this study is Caribbean student organizations' individual Web sites, the frames created within each site, text, sound files, active hot links, and any visual images found at the site.

Procedure

Since there was no existing coding scheme available to examine the Web pages constructed by Caribbean-origin university students, one was developed by adapting several content categories found in previous Internet and transnational research. This included the studies carried out by Donelan (2004), Allard and Vandenberghe (2003), Hine (2001), Horsfall (2000), Adams-Parnham (2004), and Horst and Miller (2006). Donelan (2004) examined Web sites maintained by hate groups and militia groups;

Allard and Vandenberghe (2003) examined the personal Web pages of ordinary users who were seeking to express their interests, passions, and hobbies; Hine (2001) explored how Web-page production has become a socially meaningful act for the individual Web-page developer and the institution concerned; Horsfall (2000) studied the ways in which five religious groups use Web pages to disseminate their doctrines; Adams-Parnham (2004) analyzed the Internet as a tool for Haitians in the diaspora to engage in civic deliberations and networking; and finally Horst and Miller (2006) studied cell phone records to examine the "link-up" transnational networking strategies employed by Jamaicans.

The categories were refined for the analysis by reviewing previous studies, which included content analysis of sport photographs, Internet pornography images, magazines, videos, and cartoons (Stanley and Plaza 2002; Mehta and Plaza 1997; Garcia and Milano 1990; Palys 1986; Winick 1985; Palmer 1979). I devised a three-part coding procedure. In the initial stage, I went through all of the Caribbean student Web sites in the sample in order to generate specific coding categories and operational definitions. From the initial eyeballing of the data I devised seventy-five categories that became the foundation for the analysis.

In order to develop a baseline that ensured intercoder reliability, I employed a senior undergraduate student to work independently to code the Web sites from within the sample. By initially working together and then apart we were able to iron out any questions about content categories and definitions. The content categories were dichotomously coded for either their presence or their absence of prescribed criteria. Concern about intercoder reliability led to a statistical testing for reliability between the two coders. Since the variables were dichotomously coded and nominally scaled, it was necessary to use kappa as a measure of agreement between the pair of coders. This particular measure was calculated using the crosstabulation function in SPSS PC for Windows (Version 11.0). The intercoder reliability test takes into account the amount of agreement expected by chance. According to Landis and Koch (1977), kappa values greater than 0.75 indicate excellent agreement beyond chance, values between 0.40 and 0.75 indicate fair to good agreement beyond chance, and values less than 0.40 indicate poor agreement beyond chance. I calculated mean kappas for each variable and for each of the coders. I dropped variables with mean kappas below 0.40 from subsequent analysis. Fifty of the seventy-five variables originally constructed for analysis were usable because they showed very

TABLE 9.2 | CONTENT CATEGORIES, VARIABLES CODED, DATA ANALYZED, AND THEORIZED INDICATORS

CONTENT CATEGORIES	VARIABLES CODED	DATA ANALYZED	THEORIZED INDICATORS
Membership	Gender, ethnicity, age, "race," size of organization, statements regarding the organization, rituals	Photographs, text	Segmented assimilation, Creole cultural socialization, transnational identity, linking objects
Social Activities	Religious parties, fashion shows, cultural show, banquets, semi-formals, interschool cultural exchanges, sports (cricket, soccer, basketball, dominos, cards)	Photographs, text, hot links	Segmented assimilation, transnational connections, cultural mourning, object loss, symbolic bridge, linking pbjects
Political activities	Political mobilization, Invited international speakers, sponsorship of Caribbean symposia and conferences, Caribbean-community outreach, mentorship program	Photographs, text	Transnational connection, object loss, symbolic bridge, cultural mourning, linking objects
Web Site Content	Language—Patois; Images—palm trees, Caribbean maps, beaches, coral reefs, waterfalls, rainforests, Caribbean flags, Rastafarian colors	Photographs, text, hot links, images	Segmented assimilation, transnational connections, cultural mourning, object loss, symbolic bridge, linking objects
Photographs	Parties, formal attire, attractive females, white students, sporting events, cultural events, liming (informal get-together)	Photographs, text, images	Segmented assimilation, cultural mourning, object loss, linking objects
Hot Links	Other Caribbean organizations, Caribbean-based newspapers, radio stations, tourist boards, parties	Hot links	Transnational connections, cultural mourning, object loss, symbolic bridge, linking objects
Ethnic Control	Themes of events, types of music listened to, political ideology, language used, social events	Photographs, text, images	Segmented assimilation, transnational connections, symbolic bridge, linking objects
Transnational activities	Projects done internationally, exchange programs, fund raising	Photographs, text, hot links	Transnational connections, symbolic bridge

strong agreement between the two coders (above 0.65). These fifty variables are the ones reported and theorized about in this chapter.

A brief explanation of the fifty variables used for carrying out the content analysis is outlined in table 9.2. The table shows the content categories, the variables coded for, the data analyzed in each Web page, and the theorized indicators for the variables coded for.

Findings and Discussion

Table 9.3 tells us a great deal about the Caribbean student organizations, and how they tend to function. An interesting trend is that women dominate most of them, particularly in the executive administration. We found that 36 percent of the sample had a woman president and a female dominance of virtually all executive positions. Another 20 percent of the sample had a female dominance with very little male presence in the organization. We found only 14 percent of the sample had a male president and male dominance of the executive positions. Also interesting is the fact that 10 percent of the organizations with a male president had all of the other executive duties being carried out by women. For the rest of the schools it was unclear as to the gender makeup of the executive administration. What we seem to be witnessing in the Caribbean international diaspora is the feminization of the student leadership roles at universities. This pattern may be a result of fewer Caribbean-origin men attending university in the diaspora. In Canada, Simmons and Plaza (1998) note the significant trend of fewer Caribbean-origin men undertaking university studies compared to Caribbean-origin women.

Ethnic control of these student organizations was another interesting trend that seemed to be fairly consistent in the international diaspora. Individuals of Jamaican and Trinidadian ethnic origin seemed to be in leadership positions. These two ethnic groups are among the two largest contributors

TABLE 9.3 | TRENDS IN CARIBBEAN UNIVERSITY STUDENT ORGANIZATION WEB PAGES

Organization executive administration	
Female president—Female dominance	36%
Female dominate—Few males	20%
Web site has no information about Administration	20%
Male president—Male dominance	14%
Male president—Females doing work in organization	10%
Ethnic leadership	
Jamaican	30%
Trinidadian	20%
Barbadian	5%
Guyanese	5%
Haitian	5%
Unknown Control	5%
Total	50

NOTE: The total (n = 50) represents the number of schools included in the study sample. The percentages will not total 100, because each school's Web site may have multiple or no indicators of a particular category.

of migrants from the English-speaking Caribbean to the international diaspora, but this should not explain their dominance of the student organizations. I speculate that these two countries are currently the major contributors to the musical and cultural tastes of the second generation, and as a result anyone who can trace their lineage to Trinidad or Jamaica might receive more social capital from the membership when election time comes around. See table 9.3 for a more detailed look at these trends from the Web sites sampled.

Table 9.4 informs us about the Caribbean student organizations' activities in the international diaspora. These activities provide evidence of the degree to which the membership experiences a feeling of object loss or cultural mourning. Cultural mourning is demonstrated by the kinds of activities in which the members participate. Fifty percent of the sample had an annual cultural show that involved the student members putting on a public display of their "authentic" Caribbean cultural origins before the university community. Eighty percent of the organizations had Caribbean-theme dances that featured soca and reggae music. In the larger schools, on the East Coast of the United States and in central Canada, internationally recognized Caribbean reggae and soca artists were invited to perform on campus. Coupled with this, 60 percent of the schools had an annual semiformal banquet, which was advertised as having an "authentic" Caribbean menu and ambiance. The reproduction of "authentic" Caribbean food, music, dancing, and cultural shows on campus all seemed to reduce the feeling of object loss and cultural mourning for the second-generation students. These events also suggest a continued transnational connection to "home" for the second generation.

Other activities that also suggest a continued desire for a transnational connection to the Caribbean region are evidenced by the fact that 54 percent of the sample engaged in social events, which included Caribbean students from other universities. In a few cases (like Montreal's McGill University or London's University of Western Ontario) this went further in the sense that university students from the United States were encouraged to attend the annual Caribbean cultural shows and sports events put on at these Canadian universities. This cross-border movement demonstrates that the second generation is participating and perpetuating a fluid transnational culture where the elements of being both "here" and "there" are somewhat blurry.

Being involved and abreast of politics in the Caribbean region was also very important to 48 percent of the student organizations in the sample. Related to this, 44 percent of the schools invited expert speakers to visit their campuses to talk about culture, underdevelopment, or political issues unfolding in the Caribbean region. Also interesting was the fact that 40 percent of the sample sponsored an annual symposium or conference that had a Caribbean theme. This was particularly the case for universities on the eastern seaboard of the United States, where large pockets of Caribbean migrants live.

Evidence of the membership sticking with their "own kind" comes from the photographs posted on various Web sites. Having undertaken a content analysis of the photographs it became evident that the membership in the schools sampled were mainly people of African Caribbean ethnicity (70 percent). Only 30 percent of the members appeared to be Indo-Caribbean, white European, or other mixed ethnicity. The posting of these photographs undoubtedly sends a latent message to outsiders visiting the Web site that although these organizations are open to anyone, in fact the membership is overwhelmingly for people who are of African ethnic origin. People of African ethnicity are often the only group that is seen as being authentically Caribbean (Plaza 2006).

Not surprisingly, a large number of the photographs posted on the Web sites (64 percent) featured attractive Caribbean-origin females in individual or group poses. This might be expected since 70 percent of the organizations had a significant imbalance in the number of female compared to male members. We speculate that the various Webmasters who manage the sites might be exercising an unconscious form of sexism whereby they post more photographs of women than men because these images are a tool for attracting Caribbean-origin men to attend future school events. See table 9.4 for an indication of these trends from the schools sampled.

Table 9.5 sheds light on the fluid transnational lifestyles of second-generation Caribbean-origin university students. Choice of music was without a doubt the most significant and consistent indicator of individuals seeking out a symbolic bridge to reconnect them with familiar Creole values. Soca and reggae musical references could be found on 80 percent of the Web sites. The primary language used at the Web sites was overwhelmingly English (95 percent), but within the English text we found (58 percent) code-switched patois as part of the posted messages to the membership.[9] The ability to code switch between two distinct cultures and languages seems to allow second-generation Caribbeans to feel like they are still very much connected to the multiple worlds they live in.

The scenic images posted on the Web sites provide more evidence that second-generation Caribbean-origin students experience a sense of cultural

TABLE 9.4 | CARIBBEAN STUDENT ORGANIZATION ACTIVITIES AND IMAGES

Activities and events	
Dances soca/reggae theme	80%
Semiformal banquets	60%
Interorganization events	54%
Cultural shows	50%
Political meetings	48%
International speakers	44%
Fund-raising	42%
Fashion shows	40%
Symposia or conferences	
Sporting events	
Community-outreach mentorship programs	
Poetry readings	
Religious-themed events	
Images on the Web site	
Attractive females	64%
Parties, Dancing, socializing	46%
Whites and non-Africans at events	30%
Total	50

NOTE: The total n = 50 represents the number of schools included in the study sample. The percentages will not total 100, because each school Web site may have multiple or no indicators of a particular category.

mourning and object loss. The symbols and objects that are selected to represent what it means for the organization to be "authentically" Caribbean are a clear indicator of this. Sixty percent of the Web sites had palm trees, 54 percent had the image of an idyllic sandy beach scene, 25 percent had a coral reef, 22 percent had a rain forest or waterfall scene, while 44 percent had the flags of various Caribbean countries displayed. The choice of scenic images, symbols, or objects used in the construction of the Web page suggests the desire by the second generation to create a symbolic bridge back to what they value as idyllic representations of what it means to be authentically Caribbean.

In terms of fluid transnational linkages, we find strong evidence from the Web sites that the second generation continues to maintain an active Caribbean connection. Forty-two percent of the Web sites had hot links to newspapers that are produced daily in the Caribbean. These electronic newspapers contain the most recent political, social, and economic news about the countries in which they are produced. Forty percent of the Web sites had hot links to Caribbean-based radio stations doing real-time program streaming. This means that an individual can be tuned in real time to a popular Jamaican- or Trinidadian-based radio station while living anywhere. The twenty-four-hour-a-day music, news, or popular culture they listen to can in an instant become part of their own consciousness. Hence, the ability to be simultaneously living in the Caribbean or the international diaspora tends to reinforce the notion for the second generation that they are living fluid transnational lifestyles and choosing when to cross back and forth over an invisible cultural bridge.

Thirty-two percent of the schools sampled had established hot links to tourist-board Web sites in the Caribbean. These sites had mainly historic, geographic, meteorological, topographic, and cultural information about each country. These links appear to be an attempt by the second generation to pass on practical information about the region to anyone who visits their Web site. Ironically, the various Caribbean tourist-board Web sites tend to give a Westerner's viewpoint of what awaits the foreign visitor to the "exotic" territories. The realities of a poor educational system, abject poverty, a dilapidated infrastructure, and a lack of good public health care are neglected by the tourism-board Web sites. What is shown are the ideals of an island beach paradise where life is carefree and not regulated by time. We speculate that this neglect of reality puts the second generation in collusion with the Western gaze about their home countries as being idyllic playgrounds of white sand and waterfalls.

The final indicator of the second generation participating in transnational activities is the fact that 28 percent of the schools sampled are involved in international aid projects that are designed to improve social conditions in

TABLE 9.5 | CARIBBEAN STUDENT ORGANIZATION CULTURAL MOURNING AND TRANSNATIONAL INDICATORS

Cultural mourning indicators	
Music choice reggae/soca	80%
Caribbean food reference	68%
Palm trees	60%
Patois at Web site	58%
Beach image	54%
Flags of Caribbean	44%
Map of Caribbean	28%
Coral reefs	25%
Rain forest	22%
Transnational links	
Caribbean newspapers	42%
Caribbean radio stations	40%
Caribbean tourist boards	32%
Projects done in home country	28%
Total	50

NOTE: The total n = 50 represents the number of schools included in the study sample. The percentages will not total 100, because each school Web site may have multiple or no indicators of a particular category.

the Caribbean. These projects involve fund-raising to help local schools, assistance for the infirm, or disaster-relief initiatives. In a few cases, second-generation Caribbean students traveled back to the region on development aid projects. We found this transnational charity work taking place most frequently in schools that had a strong second-generation Haitian or Jamaican presence. Both Jamaican and Haitian second-generation university students seem to feel a special sense of responsibility for helping out family and fictive kin left behind, because as many Web sites noted, "They are the unfortunate victims of circumstances beyond their own control." See table 9.5 for an indication of these trends.

CONCLUSION

Using Web pages as a space to alleviate the pain of personal and cultural loss seems to be the practice taking place for second-generation Caribbean-origin university and college students in this sample. The Web pages constructed seem to act as a symbolic bridge that connects familiar Creole cultural values and practices with the second generation's feelings of object loss and cultural mourning.

From the content of the Web sites examined, it is also apparent that despite the fact that many second-generation Caribbean-origin university students are living both "here" and "there" on a transnational "hyphen," most seemed fairly comfortable navigating between both worlds. Unlike their parents, who are often stuck feeling like marginalized and radicalized outsiders in the countries they migrated to, the second generation in this study seems to have the intellectual, social, and cultural capital to take advantage of both worlds. They can use their Caribbean cultural background and values as an anchor for maintaining a sense of pride and high self-esteem, while at the same time they can use their American, Canadian, or British cultural capital to be successful in terms of settlement and future employment.

For many second-generation Caribbean-origin university students, their ethnicity and cultural identity seem to be fluid, situational, and volitional. They are based on a dynamic process in which boundaries, identities, and cultures are negotiated, defined, and produced though social interactions inside and outside their community. The construction of Web sites can be seen as the newest tool that allows these young people to participate in an evolving Caribbean transnational culture. The Internet has come to fit directly into the transnational orientation for many second-generation Caribbean-origin people because it provides them with a voice to express who they are, where they have come from, and what their cultural values are. Web sites have also given this cohort a feeling of "agency" that helps them to avoid feelings of object loss. This acquisition of agency is particularly important for Caribbean immigrants who have been traditionally powerless to have a voice in the global public sphere.

Undoubtedly, the cyberspace world of the Internet and the construction of Web pages have given Caribbean-origin university students a secure and safe place where the in-group discourse can include issues that might not be uttered in other public spaces. Their constructed Web sites also seem to provide a therapeutic, social, and psychological means for these young people to be able to maintain their Creole culture and continue to have a transnational identity in the international diaspora.

NOTES

1. The author would like to acknowledge the valuable contribution of John Davidson at Oregon State University. As an undergraduate student in sociology, Mr. Davidson provided valuable assistance in the coding and the gathering of the data for this research.

2. See http://www.internetworldstats.com.

3. A small proportion of Caribbean student organizations in the international diaspora have members who are classified as foreign-visa students (officially resident in the Caribbean but temporarily attending school in the diaspora). It is impossible to determine how much influence these foreign-visa students have on the administration or direction of Caribbean student organization Web pages worldwide. We do know that with the decline in local Caribbean economies (due primarily to the implementation of structural adjustment policies), the number of foreign-visa students originating from the Caribbean has sharply declined in the past ten years. Most university students living in the Caribbean have little choice but to remain in the region when pursuing their degrees. Of those who are able to travel to North America and Europe for their postsecondary training, many are likely to come from wealthy families or be the recipients of government-sponsored academic scholarships (Simmons and Plaza 1992).

4. In this chapter immigrant generations within the Caribbean-origin diaspora are classified. The second generation (or later generations) includes those persons who were either born in the diaspora or born in the Caribbean but arrived in the diaspora at preschool age (zero to four years). The first generation includes those who were born in the Caribbean but arrived in the diaspora over the age of eighteen and went directly into the workforce in the period before 1980. The inclusion of Caribbean-origin children who arrived at very young ages in the second generation is based on the assumption that these children share many cultural and developmental experiences with American-, Canadian-, or British-born children. Although

scholars may vary in their ways of defining the new second generation, they have generally agreed that there are important differences between the first generation and children of the second generation, particularly in their psychological developmental stages, in their socialization processes in the family, in their schooling experience, in their treatment in the society at large, as well as in their orientation toward their homeland (Zhou 1999).

5. Homi Bhabha's (1990) reading of the hybrid, as a second-generation Caribbean who resists the Euro center and engages in a constant search for what he calls "the third space," a truly interhyphen national space, a new creative space in the border zone. Bhabha's research also suggests that Caribbean second-generation men and women are creating a hyphenated space of transhyphen nationalism in the international diaspora. These links exist between and across home countries and the international diaspora.

6. Creolization is best thought of as the product of cultural mixing that began within the tropical, colonial plantation milieu and continues to the present. Initially, it embraced the politico-economic realities of conquest, slavery, and indentureship and involved indigenous inhabitants of the Caribbean region, Africans, East Indians, Chinese, and all manner of Europeans. The cultural forms that resulted from these encounters are living, and hence continually evolving, realities that respond to the historical and material circumstances within each discrete country in the region (Allahar and Varadarajan 1994).

7. Transnationalism is defined as the process by which immigrants forge and sustain multistranded social relations that link together their societies of origin and settlement through the creation of cross-border and intercontinental networks (Glick-Schiller 1998; Portes 1999; Vertovec 1999; Basch, Schiller, and Blanc 1994). Although transnationalism is not a new phenomenon, it has been facilitated more recently by space- and time-compressing technologies that include telephone, e-mail, and relatively easy, low-cost long-distance travel across borders.

8. Initially conceptualized by Robert Park and later formalized by Stonequist (1961), the "marginal-man" situation is one in which the individual who, through migration, education, marriage, or some other influence, leaves one social group or culture makes a satisfactory adjustment to another and finds himself on the margin of each but a member of neither.

9. Segmented identities are commonly manifested in language code switching: being able to converse in proper English when talking to teachers and authority figures while at the same time being able to slip into patois or other Caribbean dialect when among family, kin, or close friends. Code switching is an integral part of segmented social capital.

REFERENCES

Adams-Parham, Angel. 2004. Diaspora, country, and connection: Internet use in transnational Haiti. *Global Networks* 4, no. 2: 199–217.

Ainslie, Ricardo. 1998. Cultural mourning, immigration, and engagement: Vignettes from the Mexican experience. In *Crossings: Mexican immigration in interdisciplinary perspectives,* ed. Marcelo Suarez-Orozco. Cambridge: Harvard University Press.

Allahar, Antón, and Tunku Varadarajan. 1994. Differential Creolization: East Indians in Trinidad and Guyana. *Indo Caribbean Review,* no. 2: 123–40.

Allard, Laurence, and Frederic Vandenberghe. 2003. Express yourself! Personal Web pages: Between techno-political legitimization of expressive individualism and peer-to-peer reflexive authenticity. *Reseaux* 21, no. 117: 191–219.

Anzaldúa, Gloria. 1995. Re-thinking margins and borders: An interview with Gloria Anzaldúa. *Discourse* 18, no. 2 (Winter): 7–15.

Basch, Linda, N. Schiller, and C. Blanc. 1994. *Nations unbound.* Langhorne, Pa.: Gordon and Breach.

Becker, Howard. 1995. Visual sociology, documentary photography, and photojournalism: It's (almost) all a matter of context. *Visual Sociology* 10, no. 1: 5–14.

Bhabha, Homi K. 1990. Narrating the nation. In *Nation and narration,* ed. Homi K. Bhabha. New York: Routledge and Keegan Paul.

Donelan, Brenda. 2004. Extremist groups of the Midwest: A content analysis of Internet Websites. *Great Plains Sociologist* 16, no. 1 (Summer): 1–27.

Frankiel, Rita. 1994. *Essential papers on object loss.* New York: New York University Press.

Freud, Sigmund. 1939. *Civilization, war, and death: Selections from three works by Sigmund Freud.* Ed. John Rickman. London: Hogarth Press, Institute of Psycho-analysis.

Garcia, L. T., and L. Milano. 1990. A content analysis of erotic videos. *Journal of Psychology and Human Sexuality* 3: 95–103.

Glick-Schiller, Nina. 1998. The situation of transnational studies. *Global Studies in Culture and Power* 4, no. 2: 155–66.

Glick-Schiller, Nina, Linda Basch, and Christina Szanton-Blanc. 1992. *Transnational perspective on migration: Race, class, ethnicity, and nationalism reconsidered.* New York: Annals of the New York Academy of Sciences 645.

Hall, Stuart. 1996. Who needs identity? In *Questions of cultural identity,* ed. Stuart Hall and P. Du Gay. London: Sage.

Henry, Francis. 1994. *The Caribbean diaspora in Toronto: Learning to live with racism.* Toronto: University of Toronto Press.

Hine, Christine. 2001. Web Pages, authors, and audiences: The meaning of a mouse click. *Information, Communication, and Society* 4, no. 2: 182–98.

Horsfall, Sara. 2000. How religious organizations use the Internet: A preliminary inquiry. *Religion and the Social Order* 8: 153–82.

Horst, Heather, and David Miller. 2006. From cell phone to link-up: Cell phones and social networking in Jamaica. *Current Anthropology* 46, no. 5 (December): 755–78.

Kasinitz, P., M. Waters, J. Mollenkopf, and M. Anil. 2002. Transnationalism and the children of immigrants in contemporary New York. In *The changing face of* home: The transnational lives of the second generation, ed. P. Levitt and M. Waters. New York: Russell Sage Foundation.

Landis, John, and G. G. Koch. 1977. The measurement of observer agreement for categorical data. *Biometrics,* no. 33: 159–74.

Levitt, Peggy, and Mary C. Waters, eds. 2002. *The changing face of home: The transnational lives of the second generation.* New York: Russell Sage Foundation.

Lowenthal, David. 1972. *West Indian societies.* London: Oxford University Press.

Maira, Sunaina. 1999. The paradoxes of an Indian-American youth subculture (New York mix). *Cultural Anthropology* 14, no. 1: 29–60.

Marshall, Dawn. 1982. The history of Caribbean migrations. *Caribbean Review* 11, no. 1: 6–9.

Mehta, Michael, and Dwaine Plaza. 1997. Pornography in cyberspace: An exploration of what's in Usenet. In *Culture of the Internet,* ed. Sarah Kiesler. Mahwah, N.J.: Lawrence Erlbaum Associates.

Nagel, J. 1994. Constructing ethnicity: Creating and recreating ethnic identity and culture. *Social Problems* 41, no. 1: 152–76.

Palmer, C. E. 1979. Pornographic comics: A content analysis. *Journal of Sex Research* 15: 285–98.

Palys, T. S. 1986. The social content of video pornography. *Canadian Psychology* 27: 22–34.

Plaza, Dwaine. 2006. The construction of a segmented hybrid identity among one and a half and second generation Indo- and African-Caribbean Canadians. *Identity: An International Journal of Theory and Research* 6, no. 3: 207–30.

Portes, A. 1999. Towards a New World—the origins and effects of transnational activities. *Ethnic and Racial Studies* 22, no. 2: 463–77.

Portes, A., and Min Zhou. 1993. The new second generation: Segmented assimilation and its variants among post-1965 immigrant youth. *Annals of the American Academy of Political and Social Science* 530: 74–98.

Rosaldo, Renato. 1989. *Culture and truth: The remaking of social analyses.* Boston: Beacon Press.

Simmons, Alan, and Dwaine Plaza. 1992. International migration and schooling in the eastern Caribbean. *La Educación* 34, no. 107: 187–213.

———. 1998. Breaking through the glass ceiling: The pursuit of university training among African-Caribbean migrants and their children in Toronto. *Canadian Ethnic Studies* 30, no. 3: 99–120.

———. 2006. The Caribbean community in Canada: Transnational connections and transformation. In *Negotiating borders and belonging: Transnational identities and practices in Canada,* ed. Lloyd Wong and Vic Satzewich. Vancouver: University of British Columbia Press.

Stanley, Kathleen, and Dwaine Plaza. 2002. Camaraderie and hierarchy in college football: A content analysis of team photographs. Special issue, *Sociology of Sport on Line* 5, no. 2 (November–December). Available at http://physed.otago.ac.nz/sosol/v5i2/v5i2.html.

Stonequist, Everett. 1961. *The marginal man: A study in personality and culture conflict.* New York: Russell and Russell.

Vertovec, Steven. 1999. Conceiving and researching transnationalism. *Racial and Ethnic Studies* 22, no. 2: 447–62.

———. 2001. Transnationalism and identity. *Journal of Ethnic and Migration Studies* 27, no. 4: 573–82.

Volkan, Vamik. 1981. *Linking objects and linking phenomena: A study of the forms, symptoms, metapsychology, and therapy of complicated mourning.* New York: International Universities Press.

Waters, Mary. 1990. *Ethnic options: Choosing identities in America.* Berkeley and Los Angeles: University of California Press.

———. 1999. *Black identities: West Indian immigrant dreams and American realities.* New York: Russell Sage Foundations.

Winick, C. 1985. A content analysis of sexually explicit magazines sold in an adult bookstore. *Journal of Sex Research* 21: 206–10.

Zhou, Min. 1999. Coming of age: The current situation of Asian American children. *Amerasia Journal* 25, no. 1: 1–27.

10   Tidelike Diasporas in Brazil
From Slavery to Orkut

JAVIER BUSTAMANTE

The diaspora phenomenon has become a component of globalization. It is a complex, systemic, and nontemporary phenomenon that is not localized geographically speaking but rather worldwide. Given its complexity, and the different forms it may take, we speak not of *one* diaspora but rather of *several* diasporas, and digital diaspora is perhaps one of the most interesting of them. Brazilian diasporas are not unidirectional movements; rather, they are dynamic and multidirectional flows. According to de Lucas (2003)—following the thought of Mauss—diaspora is a complex social phenomenon that involves multiple relationships (work, legal, technological, cultural, political, economic, and so on). Consequently, it cannot be reduced only to its cultural, political, or economic aspects. For that reason diasporas and migrations should be looked at in a new way without falling into the trap of stereotypes that fit our own interests depending on the role we play in those processes.

Due to its continental size and the racial diversity that has characterized the country since it came into being, Brazil is an extraordinary social laboratory that allows us to study how diasporic flows, over time, shape a country. Practically half of Brazil's population is of full or partial African descent, and some of the most distinguished Brazilian anthropologists, like Ribeiro (1995), stress that Brazil was born from a mixture of races. In the 1960s French sociologist Bastide (1979) already described Brazil as *the world's largest racial democracy* because it was the destination of multiple diasporic movements that regarded it as the promised land. As we will see, three hundred years of slavery—a forced diaspora—deeply marked the country's demographic features. After the slave trade, the different waves of migration added new features to the country's identity.

The incorporation of cultural features of different ethnic groups did not come about in an atmosphere without tensions. Until not long ago, some cultural expressions from an African or native background like *candomblé* or *umbanda* were discriminated against. *Capoeira*, a blend of African martial arts and dance, was also persecuted during many generations. Until the mid-twentieth century, samba, which is now one of the most characteristic elements of Brazilian culture abroad, was prohibited by the authorities as it was regarded as the music of the *poor people* (de los Santos 2001).

This ethnic and cultural heterogeneity is highlighted by one of the fathers of its independence, José Bonifácio (1973), as one of the main problems that existed even before the birth of Brazil as a country. However, in the end, it has become one of the country's main intangible assets. This ethnic and cultural blend has become a strategic resource in an increasingly globalized world. Brazil has now balanced its diasporic flows. Instead of being a destination for diasporic flows, there is now a large Brazilian diaspora overseas that invests heavily in the country and exports its cultural industry, especially its music. Internal diasporas are also reshaping the country. Brazilians overseas adopt forms of solidarity and communication with their home communities through the paradigmatic means of digital diaspora. The extensive use of social networks, especially Orkut, shows us how the concept of diaspora is evolving, and how the standards for communication and types of social relationships are changing within communities abroad. This tidelike movement of Brazilian diasporas is this chapter's leitmotif.

THE HISTORY OF BRAZIL AS A HISTORY OF DIASPORA

First Identity: Destination for the Diaspora of Slavery

The words of García Fajardo (2002) aptly emphasize the relationship among slavery, diaspora, and migrations when he ironically states that immigrants are very polite people who return the visits that Europeans made for five hundred years. They already know the way; all they need to do is retrace the steps of the conquerors, missionaries, and colonizers who occupied and cultivated their land, uprooted their traditions, and subjugated or enslaved them using as an excuse the three mythical c's invoked by King Leopold II of Belgium and adopted by the 1885 Berlin Conference: "civilization, Christianity, and commerce."

Between the sixteenth and nineteenth centuries, slave traders took between eight and eleven million blacks from Africa to the New World, mainly to Brazil and to the Caribbean. However, the slave trade also took place in other regions of Latin America and British North America (Salvador 1981). The African slave trade began in Brazil in the mid-sixteenth century out of the need for cheap labor in a country too large to be developed through the work of European laborers alone. This forced African diaspora responded to the demands of three long cycles in the country's economic history: sugar-cane plantations and sugar mills in the sixteenth and seventeenth centuries, the mining industry in the eighteenth century, and coffee in the nineteenth century (N. Soares 2001).

Williams (1944) states that the slave trade was not the only option considered to replace the native workforce, seen as not docile enough or adequate for tedious production work. Indians were occasionally subjugated to slavery but not in a systematic way. Sometimes it was as a punishment. White emigrants were seen as an alternative to native labor. Before leaving their homeland they were forced to sign a work contract in return for their boat fare. They were known as *indentured servants* in the British colonies. A second type of laborers was the *redemptioners* who traded a deficiency payment for boat fare, but if they failed to fulfill their commitment they could be auctioned off by the ship's captain. Finally, convicts were sent to serve their sentence in the New World for a specified period of time. The mercantilist theories of the day favored voluntary and nonvoluntary migration. Their aim was to reorganize the demographics of the metropolis in the Old Continent by expelling criminals and convicts and reducing the number of the poor and unemployed, finding them more profitable tasks in the colonies. According to Haar (1940), indentured servitude, which was incredibly popular in the British colonies in the Americas, was born as a result of the combination of two complementary but opposed forces: the appeal of the New World but also the repulsion of the Old World (see also Williams 1944).

Brazil maintained one of the longest-lasting systems of black slavery in the world. In fact, it became the last country of the Americas to abolish it with the Golden Law in 1888. The slave trade, a type of forced diaspora abolished in 1850 when the Eusébio de Queiroz Law was passed, was a significant moment in the history of Europe, the Americas, and Africa. For three centuries, colonial slavery was the leading type of immigration in several important areas of the New World. The political, social, and economic history of several countries in the Americas was based on the trade of people that characterized the old institution of slavery. Consequently, diasporas and migratory flows in many historical events in Europe, the Americas, and Africa can be understood only if they are analyzed in the context of the practice of slavery (N. Soares 2001).

Second Identity: Destination for Diasporas

The shift from slavery to a European immigration to Brazil had little to do with an economically based diaspora. It basically found its roots in an ethnic and racist philosophy, such as that of Handelmann (1982), according to which Brazil's future *depended* on European migration given the innate limitations of the country's native population. In fact, the Handelmann ideology was put into practice immediately after the abolition of slavery. The massive arrival of European immigrants was as a result not only of the growing demand for laborers to work on coffee plantations but also of an ill-concealed desire to change the ethnic composition of the country's population.

The Industrial Revolution also resulted in the disappearance of slave labor as it was gradually replaced by wage-earning jobs. Industrial society had little interest in slave labor. On the one hand, assembly lines needed workers in good physical shape and with a degree of professional training that slave labor could not offer. On the other hand, mass production demanded workers with purchasing power, a middle class that could handle the increase in productivity. There would be no one to buy the goods produced without wages. Both of those factors, qualified workers and workers with purchasing power, plus a scientific division of labor were necessary for the industrial cycle to be effective, elements clearly far out of step with the logic of slavery.

The abolition of slavery and the Industrial Revolution at the end of the nineteenth century brought the first waves of immigration. In twenty years—from 1884 to 1903—1.7 million immigrants entered the country, mainly from Germany, Italy, Portugal, and Spain. The Italians—more than half of them—settled in São Paulo and in the southern states. The Portuguese settled mainly in Rio de Janeiro. The Spaniards went to Rio, São Paulo, Bahia, and Pará. The first to arrive when the country opened its ports to *friendly nations* (1808) and after the country's independence (1822) were the Germans. They settled in the southern part of the country, particularly in Santa Catarina and Rio Grande do Sul, where they established their first colony, San Leopoldo. The first Japanese arrived in 1908 as emigrants subsidized by their own government, and they settled mainly in the region from São Paulo to the southern part of the country as farm laborers replacing the recently abolished slave labor. The first immigrants were treated by their masters similar to the way in which the old slaves had been treated. Obviously, this gave rise to dissatisfaction, and many of them returned to their homeland or tried to find other ways of making a living, as this was not what they had expected from the New World (N. Soares 2001).

Between 1882 and 1934, 4.5 million immigrants arrived. Starting in the 1930s, the big migrations gave way to internal diasporas. The demand for laborers was filled by laborers from the northeastern states and from Minas Gerais who were forced to move to São Paulo and the southern region of the country in search of work. The only immigrants who arrived in significant numbers were the Japanese. In spite of the initial opposition to this ethnic group—as it was considered desirable to *whiten* the population with the European immigration—they were eventually accepted as an alternative to the restrictions imposed on subsidized immigration by Italy. Between 1932 and 1935, almost a third of all immigrants were Japanese. In more recent years, other nationalities, namely, Argentineans, Uruguayans, Chileans, Bolivians, Syrian-Lebanese, and Koreans, joined the aforementioned nationalities to form part of the Brazilian national identity (Fausto 2008).

The tidelike movement of Brazilian diasporas is also portrayed within the country. Brito and Carvalho (2006) point out that socioeconomic and regional disparities in a country the size of a continent have made migration a characteristic feature of Brazilian culture. The most significant internal migratory flows involve an exodus from the most impoverished areas to the big cities. The rural exodus of the 1970s stands as good proof. In only ten years the city's population increased by 30 million people. The migratory flows take place between states (e.g., from the Northeast to São Paulo) and from the less populated areas in a state—the so-called interior—to the big cities. In fact, the character of the *retirante* is vividly portrayed in the country's literature, as a northeasterner migrant who leaves a homeland characterized by the poor and sunburned soil of the *catinga* forest, fleeing hunger and drought with his family in search of a better future. Graciliano Ramos's novel *Barren Lives* (1938) sharply portrays that situation from an existentialist point of view. However, this trend has reversed itself in the past three decades. Permanent internal migration has given way to *returned migrants*, who go back to their homeland years later instead of staying in the usual places. Thus, an amazing change has taken place in Brazil's internal migration patterns (Brito and Carvalho 2006).

Third Identity: Brazilian Diaspora Overseas

From the 1930s until 1960, immigration gradually decreased while internal migratory flows increased. In the 1980s the target of migratory flows changed mainly in the direction of the United States, Japan, and Paraguay. In this decade a curious phenomenon appeared, called the "me-20" by anthropologist Margolis (1998). It involves Brazilians traveling with a tourist visa who stay to work for one or two years and then go back to their homeland for a similar period of time. After that time, they return to the United States again as tourists, and they repeat the process when they need money (Amaral, Friedrich, and Fusco 2005). In the 1990s there were already more than 1 million Brazilians abroad. At the beginning of the twenty-first century, Brazil found itself firmly placed on the international migratory scene for both internal and external migration (Baeninger 2003). According to the 2004 official records of Itamaraty (the Foreign Affairs Office), there are around 1.8 million Brazilians living legally and officially overseas. However, according to an estimate based on the consular services demanded by emigrants with no valid documentation—considered illegal immigrants in their host country and unrecognized emigrants for the authorities at home—the total number increases up to 2.5 million. The Inter-American Development Bank estimates that Brazil takes in about $5.8 trillion from emigrants abroad, half of the total foreign investment in the country. In the end, this community is the biggest overseas investor in Brazil.

These are records of a two-way diaspora. It is estimated that around 400,000 Japanese have arrived in Brazil since 1908. In the last decades many of those immigrant descendants have taken the opposite route. The word *dekasegi* refers to Brazilians of full or partial Japanese descent who immigrate to Japan to work mainly in the industry sector. They start out on their journey with the dream of making money since salaries are higher. However, they find only long working hours and increasing social isolation. For many of them, the vision of the new El Dorado vanishes once they realize that their Brazilian cultural references are stronger than their Japanese ones and that Japanese society does not always regard the descendants of Japanese emigrants as full citizens.

SOCIAL NETWORKS, ORKUT, AND THE BRAZILIAN DIGITAL DIASPORA

A social network consists of a set of actors and nodes connected by a particular type of relationship. Each type of relationship corresponds to a different network, even though the actors are the same. The regular patterns established between a network's actors and the dynamic flow of relationships that determines the position each actor has structurally within the network are the main points of study in the analysis of social networks. International migratory networks are social networks, and, as such, they have not a static nature but a dynamic and procedural one. These migratory networks are not limited to one dimension or to any one scale. They are local, regional, interregional, and international. This multiple-scale nature is not the result of the opposition between different scales but rather the combination of said scales (W. Soares and Rodrigues 2004; Vainer 2002). Virtual or digital social networks are the latest expression of the aforementioned emerging complexity because they incorporate groupware elements that make new ways to experience diasporas and migrations, new types of communication and solidarity, new ways to understand social roles in the country of emigration and the country of immigration all possible. They are subsets of social networks, and they benefit from the network-analysis models developed thus far. Thus, the study of virtual social networks allows us to observe new types of expression of the Brazilian digital diaspora.

Orkut is a social network created by Turkish Orkut Buyukokkten and promoted by Google since 2004 that is particularly linked to the Brazilian digital diaspora. It was designed to help people maintain existing social relationships, regain contact with those they had been linked with in the past, and start new relationships based on common interests according to the different categories on Orkut. In the beginning, it was necessary to be invited by one of its registered users to join, but since October 2006, it has been open to any Internet user with a Google account. As far as social networks go, only MySpace has close to the same number of users.

The way Orkut works is that users classify others using five categories: "Best Friends," "Good Friends," "Friends," "Acquaintances," and "Haven't Met." Unlike other networks such as MySpace, in which profiles of those who do not belong to a user's network of personal contacts are hidden, all users can see all profiles as long as the visitor is not included in the "Ignore List" of the particular user being visited. Apart from MySpace, Facebook and Ning are Orkut's most important competitors. Ning may be a competitor in a more special way since it allows for the creation of personal networks similar to communities on Orkut, according to Wikipedia. The profiles show direct and indirect connections with other users and with communities that they are connected to. Communication in Orkut consists of messages to friends, messages to friend's friends, and messages to forums in the communities (Recuero 2004, 7).

I would like to present some demographic data about Orkut provided by its own Web site as of 2008. In spite of being an American initiative, this Google social network has become an institution in Brazil. In fact, 54.27 percent of its users are Brazilian. A long way from Brazil is India, the second country with the most registered users (16.77 percent). The United States is only in third place with 15.18 percent, and Pakistan is in fourth place with a much lower percentage (1.17 percent). The United Kingdom, Japan, Afghanistan, Portugal, Germany, and Canada are on the top-ten list of countries in which Orkut has been a success. Their percentages range from 0.55 to 0.37. These first ten countries sum up 89.9 percent of all Orkut registered users. According to Alexa.com, 69.0 percent of all Orkut users are Brazilian, and 14.7 percent are from India. In addition, 67.5 percent of the information flow also belongs to Brazil (data from March 2008). In all, there are more than 120 million users, evidence of the enormous importance that Orkut has in the Brazilian diaspora.

As far as the users' age range is concerned, 61.57 percent users are between eighteen and twenty-five years old. The total for available data (users eighteen years old or older) is 88.18 percent, which leads one to deduce that the rest have not declared their age (Orkut does not allow registration for those under eighteen years of age). In any case, it is significant that Orkut does not explain this classification further.

The users' interests in this social network are divided into four categories: friends (64.22 percent), dating (20.66 percent), activity partners (19.86 percent), and business networking (19.42 percent). Regarding marital status, 37.08 percent of the users claimed to be single, 11.02 percent married, and 7.91 percent dating. Also, 0.28 percent declare to be in an open marriage, and 2.63 percent not to be in one. It is interesting to see that 41.07 percent of its users did not complete this survey, carefully keeping all information related to their marital status or the nature of their sentimental relationships closely guarded.

Jobst (2005) and Recuero (2004, 2005, 2006) point out some of the problems about Orkut when it comes to creating a real social network. In the first place, a user's popularity or social success is measured by the number of contacts. This implies that there are users who set out to get the highest number of contacts, whether they are friends or not. These false connections become a real obstacle for the development of strong relationships in Orkut communities. "Friend collectors" appear, and they act as hubs within the network. This results in the *Matthew Effect* in which already popular users (because of their large number of contacts) become even more popular, with the same effect in communities. This cumulative effect is also important as far as the software level is concerned, but it is not useful when it comes to creating a real social network. In the Barabási model (2003) there is a nonrandom pattern in the dynamics of network structuring. This is the so-called Rich Get Richer law.

According to Recuero (2004) the existence of a social relationship implies costs in terms of time, dedication, commitment, and so on. Consequently, it is not possible for those hubs to maintain real interaction with such a large number of people. We could say that Recuero's argument is relevant regarding hubs with a very large number of contacts. However, this trivializing effect on relationships does not seem to apply in the case of intermediate-level hubs. As a general rule, and until a critical number of contacts is achieved, the larger a hub's number of contacts is, the greater his ability to make new contacts, resulting in an increase in the usefulness of his personal network. This critical number refers to the level at which interaction is reduced because of lack of time and resources, weakening social ties.

This phenomenon particularly affects the so-called friend collectors, or "everybody's friend." These are users who try to get a higher rating and more popular within the network by gathering the largest possible number of contacts. Since it is simply a matter of requesting and accepting friends, no social link is necessary for someone to appear as a "friend" within another user's network. It is for this reason that Recuero doubts that we are talking about a real social network. However, these friend collectors play an important role in spreading very useful information about communities that newcomers would be hard-pressed to find on their own. Although a strong relationship may not develop between veteran users and newcomers, they could still serve as communicative elements in a triad, bringing nodes closer together that, without their intervention, would have no significant probability of meeting.

For Jobst (2005), the previous statements show that Orkut is a social network with the potential for several levels of interaction but whose interaction potential has not been satisfactorily exploited. This is mainly due to the enormous amount of information on the network and the multiplication of contacts resulting from the system's topology. The larger the number of communities and friends within the communities in which one participates, the greater the difficulty in maintaining an adequate level of interaction with all of them. The result is that the user is limited to a reduced circle of friends and to a low level of participation in most of the communities in which he is registered. The larger the number of relationships, the lower the level of maintenance of these ties.

Building a community of citizens in cyberspace is a parallel problem. Sassen (2001, 2003) suggests a "presence politics" and a de facto citizenship to overcome a concept of citizenship that, on one hand, is linked to space (a feature that has no place in digital diasporas) and, on the other, is linked to nationality and formal work. The objectives are to empower groups that are marginalized and to explain the paradox that the increasing importance of social groups (including digital diasporas) represents with an increasing strength for political action in spite of a lack of full citizenship. De Lucas (2003) suggests the concept of multiple or multilateral citizenship as an application of a participatory, inclusive, and pluralistic democracy, along the same political lines as Sassen's presence politics and Baubӧck's transnational citizenship. De Lucas advocates a citizenship that goes beyond its formal technical dimension, and guarantees the rights of people stably residing in a territory. What needs to be replaced is a concept of citizenship based on nationality by another one based on residence, avoiding in that way the difficulty that liberal proposals face in trying to overcome the cultural and ethnic roots of the republican model of citizenship. Taking a step further—and considering cyberspace as a territory with singular characteristics—the previous analysis could be applied to those digital diasporas that live and express themselves in that space; this is a step toward a digital citizenship.

SOCIAL NETWORKS AND DIGITAL DIASPORAS: TERRITORIALIZATION AND DETERRITORIALIZATION

Digital diasporas are also identity-building processes. Chambers (1994) uses the Walkman (a more modern version would use an MP3 player) as a metaphor to explain the process for identifying an individual in terms of a compilation of life stories. The physical description of a man (or woman) walking (the walk-man), in motion, is in his portable sound device (the Walkman). The choice of songs he listens to on his Walkman is an intensively private experience with respect to his physical environment. The music chosen is, therefore, a collage of sounds that redefines the environment (soundscape and landscape) that, in turn, defines the physical territory of his body in the experience of motion. However, this also deterritorializes him because his identity remains in his Walkman, in that collage of sound through which he composes his own soundtrack that both determines his identity and by which he chooses to identify himself. It is a changeable, portable identity in motion and, thus, diasporic (Contador 2001). Something similar happens with Orkut and Brazilian emigrants. The set of contacts or communities in which one is registered makes up more than one soundscape or mediascape, a set of components of identity that reflect desires, wishes, values, membership of, and association with social groups. These reflect one's degree of isolation, integration, and level of commitment to collective initiatives. Thus, this mediascape is not only an intensely private experience but also a deeply collective one.

Gallissot (1987) uses this identification process to explain the impossibility of second-generation emigrants developing an integrated identity in their host country. In this case, beyond the prevailing cultural expressions, other cultural references—or mock cultural references—will serve as resources to be employed in building an identity. However, this will not be an identity in the classical sense but rather a modulation matrix of cultural references inherited from their parents or acquired in their host country. Digital diaspora experiences—and Orkut among them—belong to this cybercultural mediascape within the Walkman/walk-man metaphor with new, changeable, and mobile references through which they regard themselves as citizens of several worlds at the same time (see also Contador 2001).

We should not confuse this virtualization process with a deterritorialization process. Information and communication technologies create new territorializations in a context of constant deterritorialization that are characteristic of a globalized society. In fact, digital diasporas are also processes of deterritorialization (the loss of one's home) through which one later territorializes once again a new *locus* of that which was displaced through virtual networks. Ortega y Gasset in his classic work *Meditación de la Técnica* (1939) states that technology is an amazing prosthetic machine that allows us to live in the world; it is our mechanism for evolution. According to Ortega y Gasset, animals evolve by adapting to the environment in which they live. However, man transforms his environment so that it adapts to him. According to Lemos (2003, 2007), when a human being is deterritorialized, he employs technology to become territorialized again, thus building his own habitat. In the end, we are dealing with a basic diasporic strategy: transform the new space into a more comfortable place for the migrant to live in.

Both Weissberg (2004, 117) and Lemos (2003, 2007) reflect on the problem of territorialization and deterritorialization in networks. For Weissberg, at first glance, networks completely separate themselves from territory, as allowing for a virtual multiplication of presence makes the concept of territory obsolete. However, we can also appreciate the fact that local communities establish themselves online to find new ways to survive and strategies to reinforce their ties, thereby increasing the number and intensity of their meetings in the real world. Consequently, territory is not completely separated from networks, and location is not stripped of its importance (see also Pithan and Timm 2007b). We have already seen how Orkut's *orkontros* (face-to-face meetings of Orkut's users) are ways to strengthen virtual links in the real world. We have also seen how many Orkut communities focus on regaining contact with old friends. This also reminds me of some Brazilian native communities who use Google Earth in order to control those territories that are threatened by lumber companies or by *garimpeiros* (gold and precious-stone prospectors). They use satellite photographs taken every so often and compare them on a computer screen to determine which areas are being deforested. It is in the virtual world where they can find the best weapons to control their territory. There is nothing in deterritorialization that would require acting in cyberspace with the zeal for protecting living spaces that is found in the real world.

Lemos (2007) shows us how society's deterritorializing dynamics are not solely the result of physical diasporas. New deterritorializing phenomena brought about by digital media, fluctuations in cultural and subjective borders, and compression movements affect space and time. Politics gets deterritorialized with the emergence of new power agents like transnational corporations. The economy is deterritorialized through globalization, job dislocation, and transborder data flow. Cyberculture is a culture of deterritorialization. However, Lemos does not overlook the fact that cyberculture not only delocalizes and destroys hierarchies but also reorganizes and creates new types of power and control. This juxtaposition of physical and virtual space is what makes possible new territorializations: mapping, control, vigilance, and so forth.

Digital diasporas are examples of the phenomenon described by Lemos in that they promote new types of territorialization that destabilize power structures, modify national identities, and—in the interaction between cyberspace and urban space—transform the meaning of today's cities. The concept of the marketplace as a public space and the concept of home as a private space are destroyed. An Internet user connected to different virtual communities from home can experience nomadic and diasporic processes. He or she can be a stay-at-home cosmopolitan while sites become a space for absolute anonymity and isolation. Lemos considers the idea of using cyberspace to disconnect from physical space an exaggeration. For instance, in using Orkut as a tool for digital diaspora, the ties established in this seemingly nonexistent place allow emigrants to reterritorialize their living space.

Maffesoli (1995, 1997, 1999) has developed a comprehensive analysis of social relationships in virtual networks. The author considers these relationships to be characterized by banality, superficiality, and fragmentation. Their ephemeral and fragmentary categories lead to overall behavioral patterns like fashion, hedonism, erotic transgression, celebrity worship, and tribalism. All of this boils down to a concrete and changeable identification with social groups. Large human communities have developed historically around common problems, transcendental problems, political ideals, nationalities, professions, religions, and so on. In postmodernism the metanarratives that define modernity are revitalized, and a step forward is taken toward the creation of a new model in more flexible relationships. Human beings are more prone to mimicry, transforming according to the situation, context, and relationship with the group they belong to or are associated with. Friendships change throughout one's life, as it is not common to die in one's homeland, and the location of home changes frequently. The feeling that online communication is instant communication provides the feeling of presentness. The idea of a settling-in process in the here and now in which we extrapolate from the physical world forms of being typical of the virtual world is destroyed (Pithan and Timm 2007a). Also Rüdiger (2004) shares this analysis when he points out that the Internet confirms the disappearance of a unified social personality when human beings integrate themselves into a grid of infinite open relationships in constant flux. According to the author, this grid provides obvious benefits but also risks in different planes of its existence: either one's life becomes entirely Internet based, a focal point that gives meaning to one's experiences is given up, or technology is provided with the power to activate one's relationships in ephemeral or fractional circles of interaction.

According to Pithan and Timm (2007a), Orkut is a paradigmatic expression of these trends pointed out by Lemos, Rüdiger, and Maffesoli, as it promotes the recreational, erotic, hedonistic, and spiritual aspects of postmodern culture. However, they take a step even further in their analysis. It is true that virtual environments allow users to keep several divided identities at the same time. However, at the same time Orkut responds to a search for interaction and social integration, thereby adding value to shared interests. These interests can be seen in virtual coexistence, the sharing of advice and ideas, and the display of personal data that are spread and shared all over Orkut. In the end, Orkut is based on the effective search for social connection. Orkut is technology that allows for the creation of totems around which people can assemble. In diasporic communities these factors are extremely important in order to keep a cohesive identity, far from set borders, as this collective identification is based on the search for and supply of support. A desire to see one's existence reflected in the face of others lies behind the ambition for popularity within the network. Even though the themes of Orkut communities are usually superficial and taken from everyday life, they provide information that can be useful to an individual at a particular moment—advice that may add value to one's lived experience. Those topics can also help to revitalize and increase a social group's awareness of a particular subject or create weak links between users but allow for the creation of triads through which improbable relationships become increasingly possible.

ORKUT, BRAZILIANS, AND THEORIES OF NETWORKS

The study of social networks can be traced back to the eighteenth-century Swiss mathematician Leonhard Euler who solved the famous Königsberg bridge problem (i.e., how to cross the city's seven bridges and get back to the starting point without crossing any one bridge more than once). Euler's solution is considered to be the first theorem in graph theory. He regarded space as a set of nodes connected by lines or, in other words, a network. In fact, the study of networks first began with the study of graph properties, but it is also important to remember that the analysis of social networks is based on the structural analysis of the 1960s and 1970s. Recuero (2004) does a great job presenting new approaches to social networks particularly applicable to the study of digital diasporas: the *Random Network Model,* the *Small-World Model,* and the *Scale-Free Model.* These three models regard structure as a dynamic element in constant change. Consequently, if we want to understand how a network works, we need to understand its evolutionary dynamics.

The *Random Network Model,* created by Erdös and Rényi (1959, 1960), assumes that complex networks have a tendency to grow randomly. The authors use a cocktail party as an example. Only one link between each of the guests will be necessary for all of them to be connected in the end. The more connections that are then added, the greater the probability for the creation of more interconnected clusters within the network. This addition process would be random because each cluster in a network would have the same probability of receiving new connections. Orkut has some features that follow this model because up until October 2006 it was necessary to be invited by one of its registered users in order to gain access to it. Even though it is no longer necessary to be invited, the multiplication of contacts through mechanisms shared by all nodes shows an evolutionary dynamics that at least appears to be random.

The *Small-World Model* is meant to answer the question of how many degrees of separation exist between nodes in a network, or, for instance, how many intermediaries are necessary in a given community to connect us to those who are the furthest from us. In the 1970s, Stanley Milgram proved that for a letter to reach a final addressee unknown to the sender, it inevitably passed through a reduced number of hands. In a "small world" all individuals have few degrees of separation in this way. In virtual social networks like Orkut, this "small world" feeling is constantly experienced because it takes only a few jumps to find, link after link, a community whose members are brought together by a criterion relevant to us, a criterion that we also share and that allows us to build and expand our own network using ourselves as the center.

This model also shows the importance of networks in which there are different levels of friendship or closeness, with some links being stronger than others, particularly thanks to the contribution of Granovetter (1973). The author discovered that weak links in a network are more important than the strong ones, even though the latter are the ones that tend to capture the interest of sociologists. Strong links are established with people who are closest to our world and with whom we share a high degree of affinity. They form concentric circles with a high degree of communication and stability. Weak links put us in contact with people (nodes) who have a strong attachment with other clusters. Consequently, if people were not connected by weak links, the probability of connecting with other clusters would decrease. This is one of the explanations for why an emigrant ultimately ends up joining mutual-help networks in a setting that is not very familiar to him. Furthermore, the role of triads in the revitalization of social networks is emphasized. If two people who do not know each other have a weak relationship with a person whom both of them know, then the probabilities of meeting are much greater than if that relationship did not exist. This is why the building, maintenance, and spreading dynamics in a diasporic network do not only follow random patterns (Recuero 2004).

In the *Scale-Free Model,* the aforementioned Barabási law—the Richer Get Richer law—is fully represented in opposition to the random dynamics of law development. The larger the number of contacts a node has, the greater the probability that new ones are accumulated. Therefore, new nodes will prefer to connect to nodes that are already very connected. This feature is called preferential attachment, and it generates networks in which few nodes have many connections while many others have very few. These networks are called no-scale networks (Recuero 2004). This phenomenon can be clearly seen in Orkut, where some communities have an extremely high number of members and a large number of communities have a rather reduced number of members.

Furthermore, two other factors go against the random dynamics of network development. First, a large percentage of members have to look for their friends or acquaintances in the real world. In other words, the idea is to make this online community mirror the community circles off-line that one belongs to. The second factor becomes especially clear in the case of digital diaspora: new contacts and joining new communities are focused around the person's social context in the country of emigration as well as in the country of immigration but more so around the latter. Emigrants search for information about the country in which they plan to establish themselves or support groups in the towns or cities in which they have arrived. It is not important that the online interactions within these emigrant communities are not abundant or profound. They open the door, making off-line relationships possible later in the real world.

It is for this reason that I do not believe it is possible to analyze the impact of virtual networks without taking into the account the off-line links they generate and the activity that they lead to outside of the network. For instance, many Orkut users schedule face-to-face meetings with other local users from their communities, or *orkontros*. These meetings serve not only to strengthen the links between these users but also to build new relationships with people outside Orkut who are somehow connected, by some other type of link, with a member of the network. This reminds us once again of Erdös and Rényi's cocktail-party metaphor.

Orkut faithfully reflects certain characteristics of a digitally integrated Brazilian society and especially those of the Brazilian diaspora who actively use cyberspace. Pithan and Timm argue, "The power of technology over society and running high individualism, absolute values of modern age, the world's rationalist approach mean that features and values that are characteristic from modern age are substituted by new features and values: community, a subjective vision of the world, fragmentation, the construction of automatic myths, tribalism, the exhibition of oneself . . . features of a new social condition that is still being defined. Orkut is a tool of social integration that fits with the profile of homo aestheticus, the cyberculture man" (2007a).

A more detailed analysis shows how well Orkut works to help users regain contact with people with whom they shared some period of their lives (childhood, school, or college) or to strengthen friendships with people outside of their physical environment. In this sense, the appearance of Brazilian forums or communities abroad—interest groups within Orkut—is significant. Groups are largely classified according to geographical distribution (according to continents or countries) but also according to professions or interests (Brazilian doctors in Spain, Brazilian surfers in Japan, and the like). The multiplication of these communities reflects the complexity of the Brazilian digital diaspora as opposed to traditional emigration. As previously stated, the past two centuries reflect Brazil's history as a country that took in immigrants, whereas its present social structure has been shaped by different migratory waves. From Italians and Japanese to even Germans, each immigrant community has contributed different cultural features to what makes Brazil unique, integrating them into the national identity. Brazil is also an emigrant country. The military dictatorship brought about the exodus of a large number of political opponents. Among them, an intellectual and artistic elite influenced by the culture of their host countries returned years later and shaped the present Brazilian identity.

Brazilians who are involved in the diasporic movement use Orkut quite a bit to exchange all kinds of information related to emigration: discussions about integration, searching for a job, bureaucratic procedures, the promotion of Brazilian culture abroad, leisure activities, and so on (National 2005). On Orkut these Brazilians demonstrate the openness and fun-loving and sociable nature that they are known for. Hundreds of communities on Orkut have been created by Brazilians abroad, covering practically the entire world. Perhaps this is what makes it so easy for Brazilians to humanize a virtual network. A uniting element for these users is living out similar experiences. Many experiences in everyday life abroad become a real adventure, which creates strong bonds. Beyond the cold statistical analysis of the numbers and content of communicative exchanges within the network, we should pay attention to the importance attributed to such exchanges by emigrants far from home. An emigrant may experience them more intensely than those individuals who find themselves in a more familiar environment. The latter are more likely to be naturally integrated and their social lives more complete; online relationships are just complementary. Orkut is a relevant element of the Brazilian experience of living abroad.

REFERENCES

Amaral, Ernesto Friedrich, and Wilson Fusco. 2005. *Shaping Brazil: The role of international migration.* Available at http://www.migrationinformation.org/Profiles/print.cfm?ID=311.

Baeninger, Rosana. 2003. O Brasil na rota das migrações internacionais recentes. *Jornal da Unicamp.* Available at http://www.unicamp.br/unicamp/unicamp_hoje/ju/agosto2003/ju226pg2b.htm.

Barabási, Albert-László. 2003. *Linked: How everything is connected to everything else and what it means for business, science, and everyday life.* Cambridge: Plume.

Bastide, Roger. 1979. *Brasil, terra de contrastes.* Rio de Janeiro: DIFEL.

Bonifácio, José. 1973. Representação à asembléia constituinte e legislativa do império. In *Obra política de José Bonifácio.* Brasília: Centro Gráfico do Senado Federal.

Brito, F., and J. A. Carvalho. 2006. As migrações internas no Brasil: As novidades sugeridas pelos censos demográficos e pelas PNADs recentes. In XV Encontro Nacional de Estudos Populacionais, Caxambu-MG. In *Anais do XV Encontro Nacional de Estudos Populacionais,* 1–16.

Chambers, Iain. 1994. *Migrancy, culture, identity.* London: Routledge.

Contador, António Concorda. 2001. A música e o processo de identificação dos jovens negros portugueses. *Sociologia,* no. 36: 109–20.

de los Santos, José Fernando. 2001. Punto de vista. *Estado de Minas,* no. 42.

de Lucas, Javier. 2003. Inmigración y globalización: Acerca de los presupuestos de una política de inmigración. REDUR, no. 1: 43–70.

Erdös, P., and A. Rényi. 1959. On random graphs, I. *Publicationes Mathematicae* 6: 290–97.

———. 1960. The evolution of random graphs. *Magyar Tud. Akad. Mat. Kutató Int. Közl* 5: 17–61.

Fausto, Boris. 2008. Histórico da imigração no Brasil. Available at http://www.diasmarques.adv.br/pt/historico_imigracao_brasil.htm.

Gallissot, René. 1987. Sous l´identité, le procès d´identification. *L´Homme et la Société* 83: 12–27.

García Fajardo, José Carlos. 2002. Los inmigrantes quieren ser globalizados. *Solidarios* (September 19). Available at http://www.morfonet.cl/secciones/informe/0111.htm.

Granovetter, Mark. 1973. The strength of weak ties. *American Journal of Sociology* 78: 1360–80.

Haar, C. M. 1940. White industured servants in colonial New York. *Americana:* 371.

Handelmann, Heinrich. 1982. *História do Brasil.* Itatiaia: Belo Horizonte.

Jobst de Aquino, Maria Clara. 2005. Interação mútua e interação reativa no Orkut: Uma Abordagem do sistema como rede social e campo interativo. Presented at the Intercom 2005, XXVIII Congresso Brasileiro de Ciências da Comunicação.

Lemos, A. L. M. 2003. *Cibercultura: Tecnologia e visa social na cultura contemporânea.* Porto Alegre: Sulina.

———. 2007. Ciberespaço e tecnologias móveis: Processos de territorialização e desterritorialização na cibercultura. In *Imagem, visibilidade e cultura midiática,* ed. Ana Silvia Médola, Denise Araújo, and Fernanda Bruno. Porto Alegre: Sulina.

Maffesoli, M. 1995. *A contemplação do nundo.* Porto Alegre: Artes e Ofícios.
———. 1997. *Du nomadisme: Vagabondages initiatiques.* Paris: Livres de Poche.
———. 1999. *No fundo das aparências.* Petrópolis: Vozes.
Margolis, Maxine. 1998. *Invisible minority: Brazilians in New York City.* Boston: Allyn and Bacon.
National. 2005. Brasileiros que vivem nos EUA criam rede de relacionamento. *Jornal National* (February 11). Available at http://www.braziliansuperlist.com/noticia/brasileiros_que_vivem_nos_eua_criam_rede_de_relacionamento_no_orkut.
Orkut. n.d. http://www.orkut.com/MembersAll.aspx.
Ortega y Gasset, José. 1939. *Meditación de la técnica.* Madrid: Espasa-Calpe.
Pithan, Flávia Ataide, and Maria Isabel Timm. 2007a. *Características das relações interpessoais na contemporaneidade: Um estudo sobre o Orkut.* Available at http://www.versoereverso.unisinos.br/index.php?e=12&s=9&a=99.
———. 2007b. *A auto-regulação interna do Orkut pela ação dos usuários.* Available at http://www.utp.br/interin/EdicoesAnteriores/03/artigos/art_livre_02_pithan_timm.pdf.
Ramos, Graciliano. 1938. *Vidas secas.* Reprint, Rio de Janeiro: Record, 1998. English version, *Barren lives.* Austin: University of Texas Press, 1971.
Recuero, Raquel. 2004. Teoria das redes e redes sociais na Internet: Considerações sobre o Orkut, os Weblogs e os Fotologs. In *Intercom-XXVIII Congresso Brasileiro de Ciências da Comunicação.* Porto Alegre: Sulina.
———. 2005. Um estudo do capital social gerado a partir de redes sociais no Orkut e. Available at http://www.pontomidia.com.br/raquel/arquivos/composraquel recuero.pdf.
———. 2006. A dinâmica das Redes Sociais na Internet: Estudo do Orkut, Weblogs e Fotologs. Projeto de Pesquisa do Núcleo de Pesquisas em Comunicação Social da Universidade Católica de Pelotas, March 1, 2005–December 31, 2006.
Ribeiro, Darcy. 1995. *O Povo Brasileiro—A formação e o sentido do Brasil.* São Paulo: Companhia de Letras.
Rüdiger, F. 2004. *Introdução às teorias da cibercultura.* Porto Alegre: Sulina.
Salvador, J. G. 1981. *Os magnatas do tráfico negreiro.* São Paulo: Pioneira-EDUSP.
Sassen, Saskia. 2001. *¿Perdiendo el control? La soberanía en la era de la globalización.* Barcelona: Bellaterra.
———. 2003. *Contrageografías de la globalización: Género y ciudadanía en los circuitos transfronterizos.* Madrid: Traficantes de Sueños.
Soares, Ney. 2001. El tráfico de esclavos africanos hacia el nuevo mundo, sus implicaciones éticas, los efectos de la tecnología en su extinción y su influencia en la formación de la nacionalidad brasileña. Master's thesis, Universidad Complutense de Madrid.
Soares, Weber, and Roberto Nascimento Rodrigues. 2004. *Uma leitura dos vínculos entre as trocas migratórias internas e a emigração Internacional de Valadares e de Ipatinga segundo a perspectiva egocentrada da Análise de redes.* Available at http://www.abep.nepo.unicamp.br/site_eventos_abep/PDF/ABEP2004_75.pdf.

Vainer, Carlos Bernardo. 2002. As escalas do poder e o poder das escalas: O que pode o poder local? *Cadernos IPPUR/UFRJ* 15, no. 2 (January–July).

Weissberg, J. L. 2004. Paradoxos da teleinformática. In *Tramas da Rede,* ed. A. Parente. Porto Alegre: Sulina.

Williams, Eric. 1944. The origin of negro slavery. In *Capitalism and slavery.* Available at http://www.shunpiking.com/bhs2007/200-BHS-RD-EW-capnsla.htm.

11 Salvadoran Diaspora
Communication and Digital Divide
JOSÉ LUIS BENÍTEZ

The phenomenon of transnational communities in the context of international migration flows and globalization processes has introduced new perspectives for understanding the dynamics of diasporas, communication practices, and the intricacies of the global digital divide. Indeed, recent communication research has focused on the intersections between migration and communication, particularly in the ways in which migrants use new information and communication technologies (ICTs) and interact with identity-formation processes. In this way, the field of communication contributes with new themes and innovative methodological approaches to the emergent and crucial area of transnational studies (Karim 2003, 1–18).

Nowadays immigrant groups around the world create and maintain diverse economic, social, political, religious, cultural, and communicative processes and practices that enable a new sense of deterritorialized communities or diasporas. This transnational social space entails "social practices, artifacts, and symbol systems that span different geographical spaces in at least two nation-states" (Pries 2001, 18). In this manner, one significant area of communication research is to understand how immigrant or diasporic communities make use of the mass media and ICTs as a way to sustain family and social networks, reproduce local and collective identities, and participate in a new transnational public sphere. Likewise, it is imperative that transnational communication studies promote consciousness and accountability about the global digital divide and its consequences, particularly for immigrant communities. These ongoing dissimilarities of Internet access and usage between developing and industrialized societies, and within different groups in the same society, have critical consequences in terms of cultural, social, and economic development.

Thus, this chapter ponders some articulations between communication and migration, some examples of the Salvadoran diaspora's usage of Internet and ICTs in relation to the production of collective and national identities, and new implications for tackling the transnational digital divide. Although these considerations are based on empirical data from Salvadoran immigrant communities, I hope these issues can be helpful to other Latino diasporas and immigrant networks, particularly in the promotion of transnational public policies that undertake the challenges of the global digital divide (Benítez 2005, 279–318).

THE SALVADORAN DIASPORA

Some studies suggest that Salvadorans migrated to the United States as early as 1941 (Montes and García 1988, 36). During the 1950s and 1960s, thousands of Salvadoran peasants migrated primarily to Honduras, and thus by 1969 when the war between those two countries took place, there were approximately 300,000 Salvadorans living in Honduras (White 1973, 31–100). This war, known as the "soccer war," lasted only one hundred hours, but the Salvadoran president at that time argued that the principal reason for the war had been the defense of Salvadoran migrants in Honduras (Anderson 1981, 128).

During the civil war (1980–1992) between the Salvadoran government and the FMLN (Farabundo Martí National Liberation Front in its Spanish acronym), there were at least three distinct migratory processes: refugees within the country, people who migrated to and asked for political asylum in other countries, and a mass migration flow, predominantly to the United States. Individuals and communities were granted asylum in countries such as Honduras, Nicaragua, Guatemala, and Mexico, while others migrated to Canada, Italy, Australia, Sweden, and Spain. Since the end of the war and the 1992 Peace Accords, the flow of international migration, largely to the United States, has not stopped. Several factors have maintained this migration process, especially the conditions of poverty and economic difficulty, the consequences of socionatural disasters, the environment of delinquency and insecurity in the country, as well as the search for family reunification and the support of transnational social networks to migrate.

According to the 2007 Salvadoran Ministry of Foreign Affairs' report, there was a total of 2.2 million Salvadorans living abroad, which constitutes approximately 35 percent of the total population of 5.8 million estimated by the 2007 census. These statistics suggest that 1,842,100 Salvadorans live in the United States, mostly in California and the Washington, D.C., metropolitan area. Other countries with large numbers of Salvadoran immigrants are Canada (135,500), Mexico (36,049), various Central American and Caribbean countries (137,449), Italy (32,130), Australia (18,755), Spain (3,200), and Sweden (2,320) (see Ministerio de Relaciones Exteriores de El Salvador n.d.). These figures include not only an estimate of undocumented migrants but also second-generation Salvadorans registered at the consulates.

A critical question is whether these large numbers of Salvadoran immigrants around the world can be called "diaspora." Even though the concept of diaspora is controversial in the field of social sciences, it is useful as a metaphor for explaining the transnational communication and social practices between immigrants and their home communities and countries (Karim 2003, 1–18). Indeed, some studies about Salvadoran immigrants in the United States propose the figurative notion of a Salvadoran diaspora, which entails important considerations about the transformation of national and collective identities, the crisis of ethnic solidarity among immigrants, and the production of new hybrid personal and collective identities (Mahler and Pessar 2001). As Clifford argues, the term *diaspora* is a "signifier, not simply of transnationality and movement, but of political struggles to define the local and distinctive community, in historical contexts of displacement" (1994, 308).

Undoubtedly, the experience of migration implies a physical and cultural dislodgment, and the organization of a diasporic community engenders new sociocultural conditions in which social class, race, religion, gender, ethnicity, generation, and other structural properties overlap in the configuration of hybrid and diasporic identities (Hall 1996, 134). In the same vein, it is crucial to analyze how the mass media and ICTs such as television and radio programs, popular music, newspapers, home videos, cell phones, Internet communications, and teleconferencing, among others, create and maintain symbolic resources for reproducing and negotiating personal and collective identities in the transnational social space. Thus, it is critical to explore how these processes of production, circulation, and appropriation of mediated messages and new technological resources engage migrants and diasporic communities toward new forms of social interaction and integration (Giddens 1984, 139–44). As a consequence, it is essential to evaluate to what extent the local and transnational mass media and ICTs promote a new democratic platform for diasporic communication and a participatory public sphere. Or, on the contrary, do these technologies reinforce inequalities and new features of a transnational digital divide among immigrant communities and their home countries?

DIASPORIC MEDIA AND COMMUNICATION STUDIES

Giddens's structuration theory provides important theoretical frameworks for analyzing not only human action and social structures but also the significance of time-space contexts of interaction. Giddens proposes the notions of "disembedding" and "reembedding" mechanisms in social systems, which describe the lifting of "social relations and the exchange of information out of specific time-space contexts, but at the same time provide new opportunities for their reinsertion" (1991, 141).

Likewise, other scholars suggest that these concepts resemble the notions of "deterritorialization" and "reterritorialization" of sociocultural practices (García Canclini 2001, 96). In this way, the mass media and ICTs can be considered as reembedding mechanisms of modernity that reinsert symbolic systems and mediated discourses across local, transnational, and global communities (Hjarvard 2002, 71). There are three levels of interactional situations produced by the use of communication media, according to Thompson: face-to-face interaction (copresence), mediated interaction, and quasi-mediated interaction (1995, 84). Following this perspective, it is possible to identify these levels in the diasporic media and communication processes.

First, the contexts of copresence might include the ethnic or local mass media among immigrant communities (newspapers, magazines, and local radio and television programs). Second, the mediated interaction dimension across the transnational social space can take into account transnational media (diasporic radio and television programs, diasporic Internet Web sites, and the use of ICTs such as cell phones, home-produced videos, or teleconferencing). Third, the level of quasi-mediated interactions can comprise the global media (satellite communications and interactive television, among others) that provide to global audiences, including specific diasporic communities, symbolic representations, cultural messages, and discourses.

DIASPORAS AND THE DIGITAL DIVIDE

Drawing from diasporic media studies it is crucial to address the impact of the global digital divide on immigrant communities at the levels of face-to-face, mediated, and quasi-mediated interactions. Moreover, the phenomenon of the digital divide entails complex issues of symbolic systems, power relations, and normative orders that preclude or allow the participation of transmigrants (those who are engaged in transnational practices) and immigrant communities in the network society. In other words, as Norris (2001, 32) emphasizes, the understanding of the digital divide requires three levels of analysis—the macro level (technological and economic environment), meso level (context of political institutions), and micro level (individual resources and motivations)—that affect online participation.

Originally, the term *digital divide* indicated the different inequalities of access to the Internet in terms of geography, income level, gender, ethnicity, education, language, generation, and technological connectivity. However, Castells (2001, 265) argues that what is at stake in the global digital divide is also the unequal distribution of knowledge, power, and network capacity in diverse aspects of social activities. Thus, the consideration of the digital divide among immigrant communities should also include the different levels of social interaction and communication at local, transnational, and global levels. Furthermore, there are some particular features in the experience of international migration and the production of transnational practices that elucidate new dimensions of the digital divide. Then, the implications of the digital divide are much more than just the access and usage of the Internet and ICTs, but also the possibilities of reembedding mechanisms of social practices across the transnational space. In this sense, as Castells points out, the Internet can be a medium that "brings people into contact in a public agora, to voice their concerns and share their hopes" (2001, 164).

The complexities of the digital divide also demand the consideration of the concept of cyberculture. As Escobar notes, this term originates "in a well-known social and cultural matrix, that of modernity, even if it orients itself toward the constitution of a new order—which we cannot fully yet conceptualize but must try to understand—through the transformation of the space of possibilities for communicating, working and being" (2000, 57). Therefore, a cyberculture research agenda has to include the examination of new discourses, methodological approaches and practices, and political economy trends developed in cyberculture. Similarly, in the context of the diasporic media and communication process, it is fundamental to understand how the Internet and ICTs constitute reembedding mechanisms of symbolic orders and technological resources and how they set up normative regulations that intermingle with immigrants' discourses and practices in their everyday lives.

SALVADORAN DIASPORA AND INTERNET USAGE

Internet studies about immigrant communities have focused on self-representations of immigrant communities on the Internet and how Internet communications support the production and maintenance of collective identities in cyberspace (Mitra 2001). Even though the notion of a "virtual community" is ambiguous, it is helpful for examining how diasporas are engaged in transnational communication processes and practices through the Internet and other ICTs.

Tyner and Kuhlke propose four spatial categorizations of Internet communications in the experience of diaspora: intradiasporic, interdiasporic, diaspora and homeland, and diaspora and host society (2000, 241). The intradiasporic level includes Web sites and Internet usage by immigrants primarily in their local contexts of interaction. For instance, a Salvadoran immigrant in the United States comments about the Internet, "I use it a lot in my job and also personally: to communicate with my friends through the Internet, and also I have all my things there."[1] For this young woman the Internet is an important tool not only for her work but also to keep in touch with information and social networks in her immediate context.

One example of a Salvadoran diasporic Web site in the intradiasporic level is Centro Deportivo (http://www.centrodeportivo.com). This site was developed in 2001 by a Salvadoran immigrant who resides in the Washington, D.C., metropolitan area. The original idea was to create a Web site focused on soccer information; however, the acceptance of the Web site motivated the expansion of the scope to other social, cultural, musical, and political activities organized by Salvadoran and other immigrant groups in the Washington, D.C., area. It seems that many people visit this Web site because they can find various photos of community and social activities. In this way, the pictures on the Web site constitute a new space for collective recognition, and a manifestation of how a new visual cyberculture intersects with traditional cultures prevailing among some Salvadoran immigrants. This Web site also promotes the configuration of a sense of collective identity among some Salvadoran and Latino immigrants. At the same time, it serves as a means of communication at the interdiasporic level because Salvadoran immigrants in other geographical areas of the United States also visit this Web site.

Similarly, there are some Web sites that intend to be a virtual space of encounter for the Salvadoran diaspora around the world. One example is Guanacos.com. This Web site was launched in 2004 and has been an important site allowing Salvadoran immigrants in different countries to keep in touch. The Web site has information about migration issues, health, tourism, culture, sports, music, and real state. Guanacos.com constitutes a useful electronic portal with a link to a comedy program, *La Charamusca* (http://www.lacharamusca.net), a weekly show produced in the United States. This show focuses on cultural events and Salvadoran traditions such as local celebrations, Christmas memories, and personal experiences of Salvadorans around the world. Lacharamusca.net offers an audio and video podcast and a blog, receives e-mails and voice mails, and has a virtual map of listeners around the world. Although this map is not sophisticated, it intends to locate the different listeners dispersed around the world.

Another multimedia Web site is http://www.salvadorenosenlinea.com. This Web site includes general information about El Salvador, traditional foods, and folk characters and legends. Moreover, it has the capacity to stream audio and video from a local radio and television channel in Usulután, El Salvador. Another diasporic Web site, http://cipotes.net/, is a platform for chatting and participating in online forums about different events in El Salvador. Some of these forums address social events and political topics, and they have a range of 100 to 300 comments. By January 2008, this Web site had a total of 734 members, primarily composed by Salvadorans around the world. These Web sites not only represent virtual spaces of individual interaction and mediated contact but at the same time reinforce particular elements of the Salvadoran national identity.

Another Web site that intends to interconnect the Salvadoran diaspora is http://www.salvadorenosenelmundo.org, which was launched by the asociación Salvadoreños en el Mundo (Salvadoreans Around the World). This Web site includes varied information about El Salvador, the Salvadoran diaspora, the economic contribution of immigrants through remittances, and current news from El Salvador. One important objective of this portal is the promotion of online forums for the Salvadoran diaspora. By January 2008, there were forums about the following topics: Salvadoran politics, Salvadoran organizations and associations around the world, the expatriate vote, art and culture, women's issues, and immigration; however, the participation in these online discussions is very low. On the other hand, this Salvadoran organization has been very active in organizing Salvadoran diaspora conventions and promoting the possibility that the Salvadoran diaspora around the world can participate in the Salvadoran electoral process. In this respect, it seems that the major political parties in El Salvador acknowledge the crucial economic contributions of the Salvadoran diaspora but are not willing to recognize the fundamental right of immigrants' political participation.

In the realm of the mass media, there are some radio programs produced by Spanish-language radio stations especially in Los Angeles and the Washington, D.C., metropolitan area. These programs oriented toward Salvadoran immigrant communities are also broadcast over the Internet, and thus Salvadorans in other countries can listen to them. In the Washington, D.C., area, for instance, there are two leading radio stations: Radio America (http://www.radioamerica.net) and Radio La Campeona (http://www.lacampeona1420.com). These radio stations, since they stream their programming over the Internet, have received phone calls and e-mails from Salvadoran immigrants in cities not only in the United States but also in other countries, such as Australia.[2] A Salvadoran disc jockey at Radio La Campeona comments about these interdiasporic interactions: "We have received phone calls from people out of the state, from North Carolina, California and Houston. Also our listeners recommend us to others, and people like the fact that we have the camera system so you are streaming audio and video at the same time."[3]

Another example of interdiasporic communication was launched in 2006 by the Casa de la Cultura de El Salvador en Los Angeles (Cultural House of El Salvador in Los Angeles). This organization supports an online radio station called Radio Pipiles (http://www.radiopipiles.org), which intends to be a cultural linkage for the Salvadoran diaspora around the world. This online radio station is located in Los Angeles but seeks to reach the Salvadoran community dispersed in Southern California and other countries. This project has been endorsed by the Salvadoran Ministry of Foreign Affairs, social organizations, and community volunteers.

On the other hand, there are several ways in which individuals, immigrant groups, and associations use Web sites and personal blogs in order to interconnect the diaspora with the homeland. For instance, one Salvadoran immigrant living in the United States recognizes that she uses the Internet "for writing letters. I communicate with my sister in El Salvador and two nieces in Mexico. In the beginning I read a lot *El Diario de Hoy* and *La Prensa Gráfica* [two Salvadoran newspapers], but I do not read them much any more, my job consumes a lot of time."[4] For this immigrant, the Internet is primarily a way to maintain family communication and, secondarily, be updated about the news in El Salvador, specifically by reading two of the Salvadoran newspapers available on the Internet. In fact, these two newspapers have incorporated since 2007 a new feature: the e-paper, which allows an electronic view of the print edition of the newspaper. This decision came as a result of the great influence of the Salvadoran diaspora. Some assessments consider that about 70 to 80 percent of the daily visits to these newspapers'

Web sites come from places outside of El Salvador. Furthermore, the Salvadoran diaspora demands from the national newspapers more local news from specific provinces and cities in El Salvador.

This perspective emphasizes the relevance of local identities for many Salvadoran immigrants and the search to maintain close ties to their hometowns or specific geographical regions. One example of this perspective is the Web site http://www.intipucacity.com, which interlinks news, photos, social events, opinions, and chat rooms primarily between the city of Intipucá, located in the eastern department of La Unión in El Salvador, and the large number of immigrants from that city who live in the Washington, D.C., area. The Webmaster of Intipucacity.com describes how he updates the Web site: "I have some sponsors and one person who send me information from Intipucá. People uses the chat, but more from there to here. The chat is used mostly by youths."[5] Thus, this Web site, which can be described as a "cyber village," is not only a platform for personal interactions but also a way to promote and reinforce transnational collective processes such as local development projects, translocal political activities, and the promotion of popular music and cultural expressions like the local patron-saint celebration.

In the same level of diaspora-homeland communication, some associations of Salvadoran immigrants in different countries have created Web sites that intend to interconnect this time-space context. One example of these Web sites is Salvadoran Diaspora in Canada (http://www.elsalvadorencanada.com). This Web site was launched in May 2006 and aims to be a useful portal for Salvadorans living in Canada and other countries. By January 2008, it had a total number of 231 registered members. Moreover, this Web site provides the capacity to upload videos, download music, and participate in chat rooms and forums. Salvadorans living in Sweden have also launched a Web site, http://www.conase.se. This Web site provides news from El Salvador and Latin America, local events, photo galleries, and general information about El Salvador and Sweden. Another Salvadoran community that has a presence on the Internet is the Salvadorans living in Italy, mostly in the city of Milan: http://web.tiscali.it/es_it/comunidad/gruvisal.htm. The Web page is very basic, with general information about the community's religious, cultural, and sports activities.

In the United States, there are some Salvadoran associations that also have their own Web sites, including one in Los Angeles known as ASOSAL (http://www.asosal.org). This Web site includes information about immigration issues, culture, community events, education, and fund-raising for solidarity actions in El Salvador. Another example is the Web site of the International Salvadoran Women Association, also based in California (http://www.aimsal.org). This association was founded in 1996, has membership in various countries, and aims to help Salvadoran communities and hometown communities in El Salvador. For instance, the association supports educational scholarships in the United States and in El Salvador.

Other Salvadoran groups are also using blogs as a way to communicate their political views and concerns about El Salvador and Latin America, including http://elsalvadorcomite.blogspot.com. This blog includes news, comments, and articles that criticize the current Salvadoran government and express new political and symbolic discourses. Another blog that reaches the Salvadoran diaspora is http://salvadorenosenelmundo.blogspot.com. This blog proposes to be a space for discussion and debate about different issues relevant to the Salvadoran diaspora.

In the level of interaction between the diaspora and the host society, there are a few examples in the case of the Salvadoran diaspora. Most of the Web sites have been developed only in Spanish and aim primarily toward intradiasporic, interdiasporic, and homeland relations. Nonetheless, there are some Salvadoran immigrants who use Internet communication resources such as blogs, Facebook, and MySpace as mechanisms for interaction with other people. For instance, there is a group of young Salvadorans living in the United States who have created a site on MySpace: http://www.myspace.com/salvadorenosenelmundo. This site is in English and provides information about community events, photo galleries, and links to videos about the Salvadoran events available on YouTube. Moreover, this site has a link to Homies Unidos (United Homies), based in Los Angeles, which according to its own description is a community-based organization committed to developing creative alternatives to youth violence and drugs. Likewise, there is another site, http://groups.myspace.com/salvadorpride, which by January 2008 reported 6,094 registered members. This space is mainly a platform for exchanging photos and establishing friends and social networks, which is also a key element for defining individual identities in the new cyberculture. The incorporation of virtual ethnographies could be an appropriate methodology for grasping the details and dynamics of these cyber groups.

Definitely, these Web sites, blogs, and other Internet communication tools illustrate how Salvadoran migrants and diasporic groups use these online spaces as a way to articulate their voices and sociocultural representations and to sustain transnational networks among relatives and friends. Nevertheless, the number of Web sites and their usage is still very limited. This fact is supported by a 2007 Pew study, Latinos Online, which reports that only 56 percent of Latinos in the United States go online, compared to 71 percent of non-Hispanic whites and 60 percent of non-Hispanic blacks. The same study highlights the relevance of age, education, and language as key factors determining the digital divide among Latino first, second, and third generations. In this sense, "several socio-economic characteristics that are often intertwined, such as low levels of education and limited English ability, largely explain the gap in internet use between Hispanics and non-Hispanics" (Pew Internet and American Life Project 2007).

Additionally, for some immigrant workers in the United States, the problem of the digital divide is not only a matter of Internet connectivity but also the lack of basic technological skills. This is a reality among many Salvadoran immigrants in the United States, particularly from the first generation. Likewise, the 2007 Pew study emphasizes that "Latinos who trace their origins to Central America (which is comprised of El Salvador, Nicaragua, Guatemala, Belize, Honduras, Panama, and Costa Rica) are among the least likely to use the Internet. Fifty percent of Latinos who trace their origins to Central America use the Internet" (ibid.). This study considers that the causes of this disparity are linked to the lower levels of education and English proficiency among immigrants from Central America.

On the other hand, it is necessary to understand the conditions of the digital divide in the diaspora's home country, because these elements can help to explain some particular characteristics of the transnational digital divide. The historical and technological context of El Salvador experienced a decisive moment in 1996, when the national telecommunications sector was privatized and several reforms were introduced by the Salvadoran government through its neoliberal economic program. These reforms were supposed to generate open competition among transnational corporations and bring better services to consumers. Although there are critical considerations to this privatization process, by 2006 there were eleven companies offering Internet services, ten companies offering fixed phone services, and five companies offering cellular phone services (SIGET 2006).

According to the International Telecommunication Union (ITU), by 2006 El Salvador had 127,410 Internet subscribers (1.85 per 100 inhabitants), 637,000 Internet users (9.26 per 100 inhabitants), and 42,300 broadband subscribers (0.61 per 100 inhabitants). On the other hand, the 2005 government statistics estimate the existence of 1,115 cybercafes, primarily in the capital of San Salvador (VII Censos Económicos 2005). Despite some government initiatives, the country continues with very low rates of Internet connectivity and usage. In this context, the Salvadoran government produced in 1999 a document called "Strategy for Building a Learning Society in El Salvador" (http://www.conectando.org.sv/English/index.htm). This document incorporates important dimensions of the digital divide in the country and recognizes the key role of the Salvadoran diaspora in building a learning information society. This strategy points out how Internet usage and other ICTs are becoming widespread, and in the short run might intensify and cheapen connectivity among Salvadorans in El Salvador and abroad. Some of these proposals have materialized through the national network Infocentros (http://www.infocentros.org.sv), which include communication forums, chat rooms, and e-commerce services specially designed for the Salvadoran diaspora.

Although this strategy identifies key participants for building transnational networks, it seems that Salvadoran immigrant communities have not had a relevant role in this project. In 2004, the government launched a new e-government strategy for El Salvador, and in 2005 the program *e-pais* (e-country). However, these proposals have not progressed due to the current government's lack of political will to promote public policies in the realm of technologies, public information access, and communication. Furthermore, as Castells argues, to address the digital divide in each country not only is a matter of knowledge and political will but also "depends on managerial capacity of the economy, on the quality of the labor force, on the existence of social consensus, based on social redistribution, and on the emergence of legitimate political institutions rooted in the local and able to manage the global" (2001, 270–71).

SALVADORAN DIASPORA, CELLULAR PHONES, AND VIDEO TELECONFERENCING

Different communication studies analyze the relevance of cellular phone usage among immigrants, and how this technology has become a crucial linkage of transnational families and communities. Thus, Deuze proposes that "the cell phone can be seen as a case in point for the experience of life, work, and play in the mediapolis: a wireless device, instantaneously connected to a regional or even global network" (2007, 13). In this sense, the study of cell phone usage among immigrants needs to include the constitution of symbolic boundaries between private and public spaces of communication, the intensity and maintenance of transnational relations, the new possibilities of time-space mobility, and the mechanisms of social integration in everyday life.

Cell phone technology allows immigrants to renegotiate family, employment, and sociocultural networks in the copresence contexts of interaction as well as transnational-mediated social relations. Transnational families dispersed across national borders make use of distance communications, especially through cell phones, chat sessions using Webcams, text messaging, and teleconferencing, in order to nurture and strengthen their ties. On the other hand, cell phones are also used for sustaining close social networks, and perform entertainment activities in everyday life, particularly among youth. The technological trends suggest that the cellular phone is the new convergence platform of different media: it can integrate text messaging, a digital photo camera, an Internet connection, a video recorder, a music player, games, and a radio receiver, among other possibilities. As a consequence, cell phones and the Internet promote a new convergence of work and leisure time but might also reshape some structural features of the transnational digital divide among diasporic groups.

The 2007 Pew study Latinos Online reports that Hispanics in the United States are more likely "to consider the cell phone a necessity, rather than a luxury. Fully 59 percent of Hispanics consider them a necessity, compared with fewer than half of non-Hispanic whites (46 percent) and non-Hispanic blacks (46 percent)" (Pew Internet and American Life Project 2007). According to the same Pew study, 59 percent of Latino adults in the United States have a cell phone, and 49 percent of Latino cell phone users send and receive text messages. Moreover, the Pew study considers that similar factors impact Latino Internet and cell phone usage. Additionally, Salvadoran immigrants in the United States have influenced the adoption of cell phones among their relatives in El Salvador.

The Salvadoran government statistics indicate that by 2006 there were 1,035,177 fixed lines and 3,851,611 mobile lines in the country (SIGET 2006). This is a substantial growth of 59.7 percent from the 2.4 million cellular phones reported in 2005 (Panagakos and Horst 2006, 112). Consequently, 55.1 percent of the Salvadoran population has a cell phone, very similar to the figure of Latino cell phone users in the United States. By the year 2006, there were two companies in El Salvador offering satellite telephone services, and the maximum rates were twenty-two cents per minute between domestic cell phones and ten cents per minute for calls to the United States.

Besides the maintenance of social and family networks, some studies underline how immigrants use cell phones as a way to exercise social control and surveillance of their relatives. In this sense, regular interpersonal communication through cell phones might develop a new "virtual-copresence" context of interaction among transnational families and social networks. In this process of transnational phone communication in El Salvador, there has been a disproportionate amount of traffic between incoming and outgoing international phone calls. In 2006 there were 2,507,146 minutes of incoming phone calls and only 526,451 minutes of outgoing phone calls from El Salvador (SIGET 2006). Thus, some people in El Salvador might have a cell phone but mainly receive calls from their relatives abroad, especially between spouses or close relatives. On the other hand, it is cheaper to make phone calls with prepaid telephone cards available in other countries such as the United States, and normally Salvadoran immigrants can spend more money in phone calls than their families and friends in El Salvador.

In short, the Internet and ICTs such as cell phones are shaping new symbolic flows of sociocultural networks in the transnational space, but as Tufte notes among diasporic groups, "While the mobile phone has more to do with negotiating social networks and striving for social integration, the use of the Internet provides more opportunities for experimenting with social and cultural, collective and individual identities" (2002, 258). At the same time, the magnitude of international phone calls in several countries with large number of immigrants is affecting the political economy of the global information infrastructure, and the access and possibilities offered by cellular phones can enhance the economic opportunities for marginalized groups such as the example of the Grameen Phone Project (http://www.grameenphone.com), linked to the microcredit pioneer the Grameen Bank in Bangladesh.

Video teleconferencing is another component of ICTs that Salvadoran immigrants use as a mechanism for establishing mediated reunification of transnational families. There are some small businesses that offer these communication services, particularly between El Salvador and various cities in the United States. Even though the cost of this service is more expensive than phone calls, this mediated interaction allows a more complete experience of encounter, especially for Salvadoran immigrants who cannot return home due to various circumstances.

Other Salvadoran immigrants now use chat rooms on the Internet that incorporate Webcams and allow for real-time audiovisual communication. Likewise, programs such as Skype are growing in popularity among some transnational families. Not only do immigrants and their family networks disrupt geographical borders through their use of these ICTs, but these communicative practices also engender new transnational experiences. Through online chats between the grandmother and her nephew or homemade videos that travel back and forth, transnational families participate in the rites of passage of their relatives, have a glimpse of cultural and religious celebrations, and exchange images, narratives, stories, and symbols that interplay with the configuration of diasporic personal and collective identities.

CONCLUSIONS AND CHALLENGES FOR THE DIGITAL DIVIDE

The Internet and ICTs, considered key communication resources, constitute a new global medium that interlinks the local and the global, represents reembedding mechanisms, and allocates the reorganization of time-space contexts of interaction. These elements are even more crucial in the experiences of diasporic groups, particularly in the case of the Salvadoran diaspora. Definitely, Salvadoran immigrants, particularly in the United States, are one of the ethnic groups with less access to and usage of the Internet. The factors of lower levels of education, limited English proficiency, and age play a critical role in determining who is connected to the Internet. It would be relevant to compare these elements with Salvadoran diasporic communities in other industrialized countries such as Canada and Australia.

It is important to include in the analysis of the digital divide the issue of immigrants' migratory status, particularly in the United States, where there has been an increase in anti-immigrant policies in recent years. Thus, undocumented immigrants might have a different usage of the Internet and ICTs, especially as mechanisms for building social networks and interacting with their relatives back home. Consequently, for some undocumented immigrants, mediated interactions such as video teleconferencing or homemade videos might constitute the only ways they can see their relatives and friends. Similarly, undocumented immigrants might appropriate differently the diasporic media programs—for instance, radio programs—that interconnect El Salvador with specific cities in the United States and Canada.

To understand and deal with the implications of the digital divide among immigrant communities, it is indispensable to also attend to the conditions in the home country. In this sense, El Salvador, in terms of Internet connectivity and usage, is currently behind other Central American countries. Despite Salvadoran government proposals and the ongoing influx of technological remittances by Salvadoran immigrants, there is very limited progress in terms of overcoming the digital divide. On the other hand, the rapid growth of cellular phone usage in El Salvador, the digital convergence of this technology, and the maintenance of communication flows with the Salvadoran diaspora can offer new possibilities for undertaking transnational digital disparities.

Although there are some Web sites, electronic portals, blogs, and local media created by Salvadoran immigrants and specific groups around the world, there is very limited communication at the level of the interdiaspora.

Certainly, the Internet and ICTs make possible this kind of interdiasporic communication, but the Salvadoran diaspora does not have a recognized Web site or portal that integrates different initiatives. It is, probably, not necessary that Salvadoran immigrants have one specific Web site, but it would be valuable if there were an electronic gateway that interconnected the diasporic online sites. Otherwise, Salvadoran diasporic Web sites may encounter the problem of dispersion in cyberspace.

Furthermore, the interdiasporic initiatives should emerge from the immigrant communities without the participation of the government. This can ensure that certain ideological perspectives among immigrant communities do not disturb their common needs and interests. Undoubtedly, the possibility of more interdiasporic communications in an environment of dialogue and pluralism can reinforce not only the sense of a Salvadoran diaspora but also more participation in the home country. If Salvadoran immigrants were more united, there is no doubt that they would have been granted the right to participate in the upcoming Salvadoran elections. Besides the acknowledgment of this fundamental right, the Salvadoran government would have to promote and sustain transnational cultural and media policies for Salvadoran immigrant communities. Conversely, the experiences and competencies accumulated by Salvadoran immigrants in different countries can have an important impact on the fragile democratic process in El Salvador.

In conclusion, it is evident that the Internet and other ICTs intertwine at different levels of personal and collective identity formation, particularly among immigrant groups. In the transnational social processes, these technologies and communicative practices uphold new forms of social integration of diasporic communities. Thus, the discussion about new technologies and digital disparities is related not only to accessibility but also to how the usage of these technologies alters social and cultural practices, family life, and distant affective relationships, among other dimensions in the transnational space. Indeed, a vital consideration about the digital divide is how these technological disparities exclude a large number of people from a new public sphere and, as Martín Barbero indicates, a "new sensorium, that is, new ways of being together" (2002, 638).

Finally, the following points intend to be a contribution for building an agenda to tackle the digital divide in Salvadoran society. First, it is imperative to incorporate the Internet and information and communication technologies as key components of development. This implies that the government needs to create and implement public policies in the realm of technologies and communication. It is also necessary to negotiate more cooperation for development and international aid in the area of new technologies. Likewise, these public policies should be combined with a social consensus about the relevance of communication technologies for local and national development; private corporations, nongovernmental organizations, professional associations, and universities should participate in this effort.

Second, an agenda for the digital divide has to recognize that El Salvador and its diaspora constitute a new transnational society. Nowadays, it is impossible to face the crucial problems of El Salvador without taking into account the Salvadoran diaspora and the process of economic and cultural globalization. For this reason, transnational public policies can be successful only if they are linked to Salvadoran immigrant organizations, hometown associations, and professionals in the diaspora. Moreover, Salvadoran entrepreneurs in the diaspora can support small businesses in the technology sector.

Third, to overcome the digital disparities in national and global levels, it is essential to implement educational strategies for a new network society. Sometimes, the problem is not the lack of technologies but limited skills and knowledge about the possibilities of the Internet and ICTs. Therefore, it is also necessary to implement media educational programs, with the perspective that people not only consume but also produce their own digital content on the Internet and with other ICTs. In this sense, transnational social networks such as organizations, unions, mass media, students, churches, professional associations, soccer teams, and musical groups, among other sectors of society, can encourage the use of the Internet and ICTs. The grassroots productions of digital and multimedia content will enhance debate, freedom of expression, and pluralism in the transnational public sphere.

Fourth, public policies addressing the challenges of the digital divide have to promote more access to the Internet, broadband and wireless connectivity, and updated national telecommunication regulations. Because communication technologies are crucial for development, this issue should receive priority as another basic public demand. On the other hand, these policies should encourage open access to the Internet and ICTs. In this manner, communication projects such as Infocentros, community cybercafes, and public spaces with wireless connection to the Internet must be considered. Certainly, the digital divide will not be narrowed only by private corporations and the free-market economy. The government has to take leadership and social responsibility.

Finally, communication policies should include strategies for implementing e-government processes at local and national levels, legal mechanisms to ensure open access to public information, and technological resources for citizen participation in a digital democracy. Ultimately, the solutions to the digital divide are intertwined with addressing citizens' demands for socioeconomic development and political participation in the network society.

NOTES

1. Alexia Martínez, personal communication, 2004.
2. Herbet Baires, personal communication, 2004.
3. Vanessa Parada, personal communication, 2004.
4. Aracely Martínez, personal communication, 2004.
5. Carlos Velásquez, personal communication, 2004.

REFERENCES

Anderson, Thomas. 1981. *The war of the dispossessed: Honduras and El Salvador, 1969.* Lincoln: University of Nebraska Press.
Benítez, José Luis. 2005. Communication and collective identities in the transnational social space: A media ethnography of the Salvadoran immigrant community in the Washington, D.C., metropolitan area. Ph.D. diss., Ohio University.
Castells, Manuel. 2001. *The Internet galaxy: Reflections on the Internet, business, and society.* Oxford: Oxford University Press.
Clifford, James. 1994. Diaspora. *Cultural Anthropology* 9, no. 3: 302–38.
Deuze, Mark. 2007. *Mediawork: Digital media and society series.* Cambridge: Polity Press.
Escobar, Arturo. 2000. Welcome to Cyberia: Notes on the anthropology of cyberculture. In *The cybercultures reader,* ed. David Bell and Barbara Kennedy. London: Routledge.
García Canclini, Ernesto. 2001. *Consumers and citizens: Globalization and multicultural conflicts.* Trans. George Yudice. Minneapolis: University of Minnesota Press.
Giddens, Anthony. 1984. *The constitution of society.* Berkeley and Los Angeles: University of California Press.
———. 1991. *Modernity and self-identity.* Stanford: Stanford University Press.
Hall, Stuart. 1996. Politics of identity. In *Culture, identity, and politics: Ethnic minorities in Britain,* ed. Yunas Samad and Ossie Stuart Terence Ranger. Aldershot: Avebury.
Hjarvard, Stig. 2002. Mediated encounters: An essay on the role of communication media in the creation of trust in the "global metropolis." In *Global encounters: Media and cultural transformation,* ed. Gitte Stald and Thomas Tufte. Luton: University of Luton Press.
Karim, Karim H. 2003. Mapping diasporic mediascapes. In *The media of diaspora,* ed. Karim H. Karim. London: Routledge.

Mahler, Sarah, and Patricia Pessar. 2001. Gendered geographies of power: Analyzing gender across transnational spaces. *Identities* 7, no. 4: 441–59.

———. 2002. Identities: Traditions and new communities. *Media, Culture, and Society,* no. 24: 638.

Ministerio de Relaciones Exteriores de El Salvador. n.d. http://www.rree.gob.sv/sitio/sitiowebrree.nsf/pages/ssalvext_asuntoscomunitarios.

Mitra, Ananda. 2001. Marginal voices in cyberspace. *New Media & Society* 3, no. 1: 29–48.

Montes, Segundo, and Juan J. García. 1988. *Salvadoran migration to the United States: An exploratory study.* Washington, D.C.: Hemispheric Migration Project, Georgetown University.

Norris, Pippa. 2001. *Digital divide: Civic engagement, information poverty, and the Internet.* Cambridge: Cambridge University Press.

Panagakos, Anastasia, and Heather Horst. 2006. Return to Cyberia: Technology and the social worlds of transnational migrants. *Global Networks: A Journal of Transnational Affairs* 6, no. 2: 109–24.

Pew Internet and American Life Project. 2007. http://www.pewinternet.org/pdfs/Latinos_Online_March_14_2007.pdf.

Pries, Ludger. 2001. The approach of transnational social spaces: Responding to new configurations of the social and the spatial. In *New transnational spaces: International migration and transnational companies in the early twenty-first century,* ed. Ludger Pries. London: Routledge.

SIGET. 2006. *Boletín Estadístico.* Available at http://www.siget.gob.sv/documentos/telecomunicaciones/estadisticas/boletin_estadistico_20060.pdf.

Thompson, John B. 1995. *The media and modernity: A social theory of the media.* Cambridge: Polity Press.

Tufte, Thomas. 2002. Ethnic minority Danes between diaspora and locality–social uses of mobile phones and Internet. In *Global encounters: Media and cultural transformation,* ed. Gitte Stald and Thomas Tufte. Luton: University of Luton Press.

Tyner, James A., and Olaf Kuhlke. 2000. Pan-national identities: Representations of the Philippine diaspora on the World Wide Web. *Asian Pacific Viewpoint,* no. 41: 231–52.

VII Censos Economicos. 2005. http://www.censos.gob.sv/sitepoblacion/index-2.html.

White, Alastair. 1973. *El Salvador.* New York: Praeger Publishers.

12 3D Indian (Digital) Diasporas

RADHIKA GAJJALA

There are several entry points into online South Asian digital formations. Some privilege the cultural and social practices; some privilege the economic routes. The several "routes" crisscross in layers. Neither the cultural nor the economic routes are mutually exclusive. It is just that the cultural entry point tends to precede the economic layer sometimes, and the economic quest for jobs tends to precede cultural transformations. The economic, cultural, and social always intersect with the personal in some way, and as a critical feminist scholar, my stated point of entry into any project is through a revealed personal location.[1] In this essay, therefore, I interweave narrations that alternate between privileging personal routes through narration based in my experience as a "nomad," a cultural route through Second Life (a 3D—tridemensional—gamelike environment; see http://www.secondlife.com) (Karamcheti 1992) that follows Indian-looking clothing stores, dance clubs linked with Bollywood music, and related symbolism. In doing so, I explore and describe ways to conceptualize mediated spaces of diaspora—as communicative spaces of imagined community, as virtual community, as digital diasporas, as identity production, and as an economic quest.

COMMUNICATIVE SPACES OF DIASPORA

My journey into the communicative spaces of diaspora has always been initiated through a search for community. My physical journeys away from my country of origin began at age three, and although I have always had a home address in India to return to, and although most of my schooling happened in that geographic location, "community" for me has always been built—never taken for granted—that is, put together from those around me and from memories, things, media, and communication tools. As far back as I can remember I have always had to work at being a community member or at gaining community support for myself. So I have never thought of community as existing without my participation in it in some form. Much of my early knowledge of a "home" in India came to me through letters from my siblings—read out loud (before I could read) by my father. It was the active reading of these letters and the enacting of emotion through his narration that brought my siblings' worlds into mine. They existed because of the letters and because the content of the letters were communicated to me by parents who were invested in my connecting with my siblings and their community. Likewise, through stories about my parents' experience with the Gandhian way of life, their investment in a nation free from British Rule was conveyed to me. But their investment in remaining connected to that nation was visible to me in the everyday through observing their media-consumption patterns and through consuming the same media as they did (since there were no ipods in those days, I had no choice but to share in these acts of consumption unless I was outdoors climbing trees).

Thus, my envisioning (which to this day remains mostly an envisioning) of Bombay came from black-and-white photos, stories narrated by my parents, and the letters written by my siblings. My affective understanding of "home," at that time, came from my mother's singing, my father's reading and chanting of Sanskrit and Telugu texts, the nostalgic stories they narrated, records of varieties of music from India, and radio broadcasts on a shortwave radio from "All-India Radio." In the 1960s and 1970s, the television, radio, newspapers, letters, telegrams, and the telephone were some of the ways in which we would connect to some sort of "community" from "home." Yet for a child who had spent as much time outside of this home-nation as within it during her first sixteen years of life, home and community were always clearly mobile and always in the affective space of imagination constructed through other people's memories.

In embodied form, this "community" came together in Indian–South Asian potluck parties with their aromas, incense, and multiple tongues spoken while some form of music from that continent played in the background. Very early in life, then, I learned community must be built through choice and circumstance. Circumstance determined where I was physically placed. Choice allowed me to choose or be chosen to engage in specific social, educational, and economic activities that made me part of a community.

So when the Internet and the notion of e-mailing messages became a possibility in my adult life, and I happened to be living in the United States at this time, it was but logical that I looked actively for "community" through that screen as well. Since the teleology of events had always been disjointed and nonlinear for me in a sense, community was always to be put together in the day-to-day through active participation, production, and redefinition of self through consumption and communication. Therefore, constructing community in Internet space, as scattered bytes came together to form messages, seemed a logical way of seeing the world. Thus, in 1992, the first thing I sought to do through this communication medium was to build community through encounters and collaborative meaning-making with the people I came in touch with.

My past and continuing research quests come from this urge to build and connect to "community" through multiple locations. Therefore, my work has involved building, joining, and participating in the online community while also engaging in extended periods of off-line ethnographies in different "locales" (Northwest Ohio and South India). I examine online formations that are referred to as virtual communities and digital diasporas through the lenses enabled by what I learned in these online and off-line journeys.

SOUTH ASIAN–INDIAN DIGITAL DIASPORAS

When home is no longer a concrete geographical place and exists within the "two-dimensionality of memory and nostalgia" (Karamcheti 1992), as is the case for the "overseas Indian" or diasporic South Asian, cyberspace allows for contact zones allowing for "cyborg diasporas" or "digital diasporas." "Digital diasporas" occur at the intersection of local-global, national-international, private-public, off-line–online, and embodied-disembodied. In digital diasporas, a multiplicity of representations, mass-media broadcasts, textual and visual performances, and interpersonal interactions occurs. The material and discursive shaping of community through such digital encounters indicates nuanced and layered continuities, discontinuities, conjunctures, and disjunctures between colonial pasts and a supposed postcolonial present. Thus, "Indian" digital diasporas occur within racially, geographically, culturally, ecologically, and socioeconomically marked configurations of the local, which in turn exists within a power structure that conflates a certain specific sociocultural, urbanized way of living as "global." As various transnational subjects travel through cyberspace—that is, through mouse clicks and keyboard taps, multitasking between various online and off-line activities, conversations, and "windows"—they negotiate an online existence within such technological environments in different ways.

The digital encounters of interactive meaning-making in these digital diasporic spaces produce not only social and digital spaces of cultural representation but also contact zones of cultural contestation. Such a notion stems from Mary Louise Pratt's codification. Pratt defines her contact zone as "social spaces where disparate cultures meet, clash, and grapple with each other, often in highly asymmetrical relations of domination and subordination" (1992, 4). These contact zones are "the space[s] of colonial encounters, the space[s] in which peoples geographically and historically separated come into contact with each other and establish ongoing relations, usually involving conditions of coercion, radical inequality, and intractable conflict" (ibid., 6). This notion of contact zone is predicated on the unequal power structures of colonial encounter that usually involve (white) Westerners and non-Western cultures in the era of colonization. The asymmetrical relations between the West and the East, the United States and India, nonresident Indian and resident Indian, and so on are lived via digital diasporas in the age of digitization, circulation, and globalization of specific kinds of cultural and material capital embedded in sociocultural processes of meaning-making shaped through the social and economic logics driving the proliferation of digital technologies and labor.

Diasporic communities the world over, of course, seized upon the opportunity to connect globally to form various virtual communities. The question is, then, what is the implication of being able to form such connections and of being able to sustain them via online networks? What sorts of cultural practices are reproduced and sustained? What sorts of histories and memories are captured and stored? How are digital diasporas reenvisioning cultural, social, and religious pasts and futures? What might be the implications for future generations of the nations that such digital diasporic communities claim as "home"? In addition, how do the technological interfaces available via the Internet-based digital media shape the reproduction of such identities and community practices? Anderson (1991) wrote of "imagined communities" shaped through print capitalism. What we see now in digital diasporas are an extension and transformation of such a logic—multiple and nuanced.

My research, since 1993, has focused on what is referred to as the "South Asian diaspora" in relation to gender and nation. I came to choose terms such as *South Asian* and *diaspora* to refer to my work examining immigrant Indian women living outside of India, specifically the United States, because these were the labels under which much of the existing literature is located. Even as these terms are problematized and contested by postcolonial theorists, they continue to be used as a point of reference. Since the mid-1990s, when India's economy "opened up" to the free market, the term *diaspora* became attractive to business worlds invested in the "free" flow to and from India of transnational capital and labor as well as consumer goods produced by multinational corporations. Indian businesses began to celebrate the "Pravasi Bharatiya," and the High-Level Committee on the Indian Diaspora (HLC) set up by the government of India recommended that the ninth day of January be celebrated as the "Pravasi Bharatiya Divas" (Overseas Indian Day). "While receiving the Report of the HLC at a public function at Vigyan Bhavan in New Delhi on 8th January 2002 Prime Minister accepted the recommendation and announced that 'Pravasi Bharatiya Divas' (PBD) will be celebrated in India and abroad on the ninth day of January every year. The choice of the date is significant as it was on this day that Mahatma Gandhi himself a Pravasi Bharatiya in South Africa for almost two decades finally returned to India in 1915 to lead India's freedom struggle" (http://indiandiaspora.nic.in/pbdivas.htm). The Internet and digital finance, of course, in addition to the "brain recirculation" of immigrant Asian populations, played a large role in the establishment of this date (see V. Gajjala 2006). Thus, notions of a "digital diaspora" gained prominence in the business sectors.

Just as in the 1980s and 1990s, the notion of a "South Asian diaspora" was simultaneously used as a strategic term of coalition by corporate interests as well as postcolonial and other critical theorists, the notion of an "Indian digital diaspora" began to emerge as simultaneously significant for both groups of researchers after the dot-com era, when Indian programmers were called to work for transnational corporations via the Internet. The proliferation of this notion coincides with the emergence of outsourcing and crowdsourcing as ways to cut costs and recruit Internet technology (IT) workers around the clock in the service of transnational business.

In what follows, I revisit the notions of "virtual community" and "digital diasporas," while continuing to interrogate the celebratory notion of a "digital diaspora," which is increasingly being touted as a favored global way of doing business and doing community. I use this framework to look at my ethnographic investigations of Second Life in order to see what sorts of South Asian–India digital diasporas are currently manifested in such 3D communicative mediated environments.

VIRTUAL COMMUNITY AND DIGITAL DIASPORA: IN REAL LIFE VERSUS VIRTUAL LIFE?

In his article "Why We Argue About Virtual Community," published in the mid–1990s, during the heyday of the notion of "virtual community," Nessim Watson carefully unpacks the use of this phrase. He argues that using the label "virtual" community for groups of people interacting with each other online makes it seem as if the community "is not actually community." Watson writes, "The term 'virtual' means something akin to 'unreal' and so the entailments of calling online communities 'virtual' include spreading and reinforcing a belief that what happens online is like a community but isn't really a community. My experience has been that people in the offline world tend to see online communities as virtual, but that participants in the online communities see them as quite real" (1997, 129). Although communities online are different from communities off-line, he argues, they are communities nonetheless.

Watson also examines the uses of applying or denying the metaphor of community for groups of people interacting online. He discusses Neil Postman's critique of the notion of virtual community, which is centered on the belief that online groups enable the separation of the "real" from the "virtual." Postman objects to the use of the term *community* within online contexts because online groups "do not contain the stake that exists in 'real' communities (Castranova 2005). . . . [T]hey lack the essential feature of a common obligation. More accurately, online communities lack the consequences of not meeting or participating in the common obligation of most communities" (Watson 1997, 122).

What both of these viewpoints fail to ask is the "So what?" question. What's at stake in using the term *virtual community*? Why is virtuality always referred to as separate from "reality"? We are still not asking this question often enough because we have great difficulty in framing the issue. When the point of entry into research about what we do with and around a computer starts at "the Internet," at the "New Technology" question, at the "digital diasporas" fascination, this scholarship, a generation after the Internet generation, seems to arrive at the virtual-real binary. Thus, when we talk of our practices in relation to Internet socioeconomic connectivity, we often stop there. For instance, even recent works such as Castranova's (2005) sophisticated analysis of economic practices in computer gaming falls into the trap of reproducing these binaries.

Nancy Baym, another virtual community scholar, points out, "Although in many ways research has become more sophisticated, the continuing debates over the nature and worth of the virtual community belie an ongoing presupposition that there are two types of communities, one authentic and the other virtual" (2000). It is necessary that researchers examine digitally mediated contexts relationally rather than as "exclusive," stand-alone phenomena. For instance, Miller and Slater suggest that "what is really required, therefore, is a move from asking about "the nature of online relationships and identities" to asking an entirely different question: "What do people do online?" (2000, 539).

Although differences do exist between an online interaction and the off-line spatial conditions under which one interacts online and the sharing of physical and temporal space talking to someone face-to-face or in a group, the question of what people *do* online shifts our examination to online community habitus (Bourdieu 1977), which in turn is shaped by sets of online and off-line practices from various geographic, cultural, social, religious, and political locations.

Communities are made up of group practices, discourses, and structures. Likewise, virtual communities are made up of group practices and discourses fully embedded in off-line community formations, and the hierarchies in virtual communities (online networks) are embedded within these very power structures and ideologies. Ethnographers such as Don Slater have repeatedly emphasized the need to study online networks as a part of everyday off-line community life. We need to problematize this online and offline binary and the assumptions behind it and shift our examination back to the practices of everyday life. This will lead back to the centrality of the computer and the Internet in the everyday lives of specific located groups of people and allow us to ask questions that take us beyond the celebration of virtuality and "innovation" as freedom and democracy.

Thus, while it is true that the mainstream ideology behind the whole global information highway encourages the formation of niche virtual communities, it is also a fact that social relations and interpersonal exchanges within virtual communities cannot escape their connection with "RL" (real-life) political, economic, social, and cultural material practices. Further, what is implicit in the utopic celebration of "virtual community" is the assumption that community formation starts with the individual and is rooted in the individualist rhetoric that pervades the technological imaginary. The dystopic view of virtual community, on the other hand, ignores the pervasive everyday nature of our engagement with online sociocultural environments. Both these extremes treat Internet technologies as "new" and somewhat outside of the everyday. What the Utopian as well as Dystopic vision of cyberspace and online technologies overlook is the fact that the individual is embedded within the practices, structures of power, and discourses that make up the community (Williams 1985), and these practices work through interpersonal negotiations and economic and cultural practices through which inclusions and exclusions within such communities are determined.

In any community, the relationship between that community and the individuals that constitute it is established through continual negotiation and renegotiation of common goals and ideologies. The relationship of the "I" to the "we," which in turn is situated in relation to a "they" and also to a context, ideology, culture, and politics that the "they" position the community with or against. The community and individuals within it therefore are bound by imagined and real accountability links. Structures of domination and hegemony within communities rely on reconfigured and reshaped networks of distributed power and redistributed accountability linkages, which both reinstate and reconfigure the status quo.

Contrary to utopian visions of community, communities are not peaceful havens but are often collectivities of individuals struggling for, with, and against power. The discursive records of community struggle and negotiation of common terms and group identity are available in the archives of most e-mail lists. In the case of Internet communities, it is possible to retrieve the actual archives and study the process, whereas in face-to-face communities or in the case of other types of imagined communities, there is no long-lasting exact record of the actual interactions. Most often community histories are written from hegemonic viewpoints. Dissenting voices are either not represented at all or labeled as troublemakers who work against communal harmony. Voices are suppressed. Raymond Williams points out, "Community can be the warmly persuasive word to describe an existing set of relationships, or to describe an alternative set of relationships. What is most important, perhaps, is that unlike all other terms of social organization (state, nation, society, etc.) it seems never to be used unfavourably, and never to be given any positive opposing or distinguishing term" (1985, 76).

For the members of the community who are continually silenced or unable to express their experience or viewpoint within the given discursive framework or power structure, community is not a positive experience. If a member of a community is dissatisfied with the community, why not leave and form another community? This is definitely possible in certain social networks online, but when these social networks are heavily linked to the off-line economic-based realities and job-based teams (such as the many Linux communities that cluster together online in groups while working on a variety of open-source products), it is not altogether possible to "leave" and not always strategic to be silent. Therefore, "virtual communities" in this sense are not virtual at all. The accountability links and material consequences of community formations are as real in communities formed through the use of online technologies. In fact, the use of online communication to enable, foster, continue, and extend community formation leads to time-space continuums, relational formations, and material links that are simultaneously global and local, online and off-line. These, in turn, are firmly situated in processes of economic globalization, for economic practices and necessities are never outside of the social, just as the social is never fully separate from the economic.

Although the binary of virtual and real has been unpacked by several researchers, the discourse based in that binary continues in everyday life, perhaps because it works to reinstate the "newness" of computer-based technologies over and over again with each new generation of software.

LOOKING FOR "INDIA" ON SECOND LIFE, 2003–2008

Prior to Second Life, my experience with 3D communicative environments came from having tried out what was known as "Alphaworld" in 1997, and from watching my son play computer games in the late 1990s and early 2000s. My experience with text-based MOOs (multiuser domains object oriented) and MUDs (multiuser domains) allowed me to understand the logic of building selves and worlds within online multiuser environments. In 2003, I got an account on Second Life, mostly because I needed something that functioned like a MOO but also had a graphical user interface. In my classes on performing digitally mediated identities I had been using MOOs (specifically Linguamoo and PMC MOO) to make my students understand the notion of constructing identity and community in online settings. The use of MOOs pedagogically allowed me to make them see how identity and context functioned together and how existing hierarchies shape such identities and contexts. In addition, MOOs provide a way to map the social codes, practices, and cues inherited by Web 2.0 social networking systems and instant-messaging practices.

Once I entered Second Life the best way for me to explore this environment was to look for anything that seemed Asia-like, while trying to make my avatar look more ethnic. Apart from a place called Sone Ki Lanka with a Buddhist stupalike structure and pillars with textures made of images that looked very much like the sculptures from the temples at Belur and Halebid in Karnataka, I found very little non-European-looking environments there. Nor was I able to modify my avatar to look properly raced or ethnic. In those initial months, however, Second Life was comparatively slow and had very little social life for me to explore. I reinvented myself and got another avatar and gained some more insight into the communicative environment. I would log on periodically to check what was going on and slowly lost interest.

But in early 2006 when I logged on after a gap of a few months, there seemed to be all sorts of interesting activity. So I reinvented myself yet again and "rad Zabibha" was born on Second Life. She went looking for Bollywood and actually found "India" (albeit a very laggy blank space at that time) on Second Life. Indian environments on Second Life took various forms. "Indianness" appeared in various symbols, clothes, avatars, and places in Second Life. This time I decided I was going to live in Second Life for a longer time and even set up my own business, selling "handloom." I was going to explore the possibilities for raced and ethnicized presences and practices on Second Life. In what follows I draw on rad Zabibha's explorations, experiences, and observations on Second Life.

Although I have not actually quoted from any Second Life avatar's conversations with me, I will be discreet in referring to places and avatars. In order to maintain confidentiality and adhere to the Second Life social codes, I will not use any real names, neither the first-life real names nor the Second Life names. I will also try to avoid using the names of actual Second Life locations when the mention of them might reveal either the Second Life identities or first-life identities (or both) of any of the Second Life avatars I refer to here. But some locations and Second Life characters are Second Life celebrities—public figures whom you will find mentioned on several Second Lifers' blogs and other relevant Web sites. I will mention their actual Second Life names.

SECOND LIFE BOLLYWOOD: A NEW WORLD BECKONS

In 2007, Indusgeeks produced a machinima (which Marino [2004] defines as "animated filmmaking within a real-time virtual 3-D environment") called "India" in Second Life—shot on location on Bollywood Island (see http://www.youtube.com/watch?v=GIg4XMhh14Y&feature=related). "A new world beckons" is the very first line that appears on the screen, while young men and women from a variety of vocations are shown rising up in a trance and running toward what turns out to be Bollywood Island in Second Life. While the exoticization of the Indian is a very obvious point to note and may not carry much significance in itself in a place like Second Life that is fashioned around the production of exoticized selves, what was striking to me was that there was an attempt to bring rural Indian-based avatar presences onto Second Life as well. For instance, there was a potter dressed in village garb (a dhoti) who was one of the entrance group of people running to Second Life.

This notion of New World resonates with a discourse that is emerging regarding "Bollystan" as "India's diasporic democracy" (Khanna 2005) with the likes of Aishwarya Rai (whose profile on MySpace emerged around the time that Rupert Murdoch bought the social networking site) and Sharukh Khan being proclaimed as ambassadors of Bollystan. In this mediated environment Bollywood becomes representative of "India" for much of the second and later generations of diasporic Indians. Although it is true that mobile generations of South Asian youth hang out in social networking systems and on blogs such as LiveJournal, Facebook, Orkut, and Hi5 (masked in semianonymity) and that even remix videos posted on YouTube blur notions of transnational sexuality and notions of "Indianness," as they hide behind and digitally manipulate Bollywood and other pop icons and music (V. Gajjala 2006), this particular video with its characterization of a "New World" beckoning in Second Life in fact recodes these transgressions through the heteronormative perfection of three-dimensional imagery.

In other Web 2.0 venues there is a continuing play on gender and identity as the Bollywood icons produced in such communities are subjected to a gaze that blurs the boundary between heteronormative idolization of Bollywood stars and queer pleasure, while also producing uncertainty about geographic location, as they appear to multitask between work, fun, and off-line and online formations of friends. In addition, YouTube videos of queer and transgender performers from India and other regions where Indian diasporas reside are shared and commented on. While the sharing of YouTube videos in the mainstream of Indian diasporas tends to be highly heteronormative and "Bollywoodized," there are videos uploaded from locales in India where transgender performers perform in semitraditional art forms. This once again provides a very intriguing intersection for the study of "digital diasporas."

However, the characterization of a Bollywoodized three-dimensional New World, evident in the machinima described above, reinstates hierarchies and binaries while also reproducing a very specific euphoric vision of development. It is not within the scope of the current chapter to discuss and elaborate on how YouTube and other Web 2.0 social networking tools are used in South Asian digital diasporas. There is work being done on how fan communities and remix Bollywood music perform South Asian digital diasporas (see, for instance, Zuberi 2008), but my focus in this chapter is on Second Life and Indian digital diasporas. Even as, in actuality, there is much gender bending and "cross-dressing" of avatars with great ease, the actual avatars produced—whether or not their owners sitting at the computer are of the same sex or gender—are made to look, dress, and behave in a highly heteronormative way following a post-1995 Bollywood ideal of female and male embodiment. So who is this vision of Bollystan–Second Life being presented to? And who is presenting it? Where are these future visions emerging from in relation to Second Life? As these kinds of representations attempt to showcase status quo heteronormativity and Westernized development in this "New World," how, if at all, are these codings and reinstating of norms being negotiated? Further ethnographic investigation may reveal more answers.

DANCE CLUBS

In this section I draw my analysis from visits to dance environments, and the observations are based on my experience of them. I must make clear that my understanding of these comes from a "deep hanging out" in various dance clubs on Second Life in order to understand some apparently sociotechnically scripted codes for behavior in such environments. Thus, I have visited dance clubs that self-describe as Hispanic clubs, as Middle Eastern, as reggae, as jazz, as "desi," as Bollywood focused, and so on.

From this deep hanging out, I have drawn out instances and made connections with performativity of gender and identity based in frameworks developed in earlier work, which I term "epistemologies of doing" (see R. Gajjala, Rybas, and Altman 2007). These environments are described based on specific visits that I made to a few Indian-theme dance clubs. In Second Life no "place" stays static for long. Groups and individuals are continually rebuilding and relocating; therefore, the dance club experiences I discuss here can be located only in my affective experience and memory of the events. Thus, my reading of performative cues is what I rely on as I describe these clusters of activity.

All these dance clubs have basic scripted objects—a dancing ball or a floor with dance scripts—to animate the avatars.[2] They all have streaming media set up, where the songs are streamed from a server (such as Shoutcast or something else) onto the Second Life location. Most of them have tip balls or some form of money-collecting scripted object, and some also have exploding objects that are scripted to allow visitors to enter into a competition to win the jackpot by making a money contribution. Some clubs have theme dances and competitions for dancers (this is also done through scripted object), where fellow dancers get to vote for the "best-dressed female in pink" or some such theme decided by the club owners.

In dance clubs focused on Indian interests, there is often a stock set of Bollywood remixed music streaming in. The Second Life avatars in these clubs are dressed in a variety of clothes, but more and more of them (since 2007) are dressing in ethnic-seeming garb mostly modeled after Bollywood characters' dress. Since 2007 Second Life residents and business owners have noticed more and more sari designers, for instance. Since the practice of designing a sari using Photoshop requires a certain amount of dedicated patience and continued effort in trying to get the detail and shape just right so as to make it look like a sari that an Indian from India would recognize as a sari, this increased production of saris to me indicates at least an increased interest in the consumption of such attire by Second Life avatars. Although the Goreans, a role-playing group that inhabits Second Life (see http://en.wikipedia.org/wiki/Gorean for a quick description of the Gor), are interested in saris, they have not been as particular about how Indian sari-like it should be.

The Second Life avatar of Indian origin tends to be more knowledgeable of a particular set of practices around sari wearing, and this seems to influence what they consider to be apparel worthy of the name "sari." This is reflected in their consumption and production of saris and comments about saris when their avatars are wearing them. Interestingly, then, this female attire becomes a certain symbol of "authenticity" of Indianness at the same time that it is still exotic and sexualized through its use by the Goreans on Second Life. There are, of course, instances where both the Gor market and the Indian market are targeted by a sari producer. Second Life avatar LP, for instance, has at least one store in a Gorean-focused mall and a store in at least one India-focused dance club and shopping center. She, the sari producer, also sells her creations at a now famous store that sells international apparel and is owned by the clothing designer OT. In 2006, OT, LP, and one or two others were the only sari and Indian-apparel designers easily found on Second Life. Now, there are several—both well known and not so well known. However, the perception of authenticity of saris on Second Life has more to do with the off-line practice of making the sari through digital imaging and textures from the original textile-based sari material than it does with the authentic Indianness of the Second Life avatar wearing the sari or the person behind the avatar. But I digress—the investigation of the authenticity of saris and Indianness on Second Life in relation to sari wearing is a topic to be explored more in depth in other writing.

As far as the dance clubs with Bollywood music are concerned, Indianness is established mainly through familiarity with the music being played, which is demonstrated in conversation among avatars in the club as they dance. Once again, Bollywood is invoked as representative of India in such mediated environments.

IS SHE "MORE" REAL BECAUSE I "KNOW" HER ON ORKUT?

In 2007, I met a young lady (or so the avatar said she was) who told me she was on Second Life because she had heard of the jobs you could get on Second Life and had seen an advertisement in a regional vernacular newspaper in India. She started to type to me in a roman script version of my mother tongue, saying she felt more "at home" on Second Life now that she had found someone who understood the same vernacular Indian language as the one she spoke in her everyday off-line life. When she told me where she was logging on from, I was more than mildly surprised. Not that the region she was logging on from was remote or rural, but it was not one of the hi-tech cities like Bangalore and Hyderabad or the more elite cosmopolitan cities like Mumbai, Delhi, Kolkatta, and Chennai. She had found a job on Second Life that would pay her about the equivalent of a dollar a week. She was annoyed at all the male-type avatars who kept asking her for sex, she said. She thought all that was silly and ridiculous. She wanted to learn all there was to learn about scripting and building in Second Life. Along the way she was certainly making some interesting friends. She has visited many India-centric places and has confessed to not feeling too comfortable in the dance clubs. She did not say clearly why.

Certainly, if we are to believe the avatar's story, here is a person behind the avatar who is clearly in diasporic space through Second Life and not because she has physically traveled outside of her home region. She was encountering versions of "America," "China," "the Netherlands," and even "Australia" as she interviewed for jobs. Some of these interviews were done through the voice feature on Second Life, and the accents of the people behind the avatars came through to her, and that was her way of identifying where they were from, based on her knowledge of the geographic location of such an accent (gleaned from exposure to other media such as television and film).

Do I believe the "truth" of the story about this young lady I met on Second Life? What are the truths I believe about her and why? Does the fact that she linked to my profile on Orkut, where she has several friends from the region she claims to hail from and they all seem to think she is a young woman, mean that she is "real"?

What is "real" in this instance? My experience and her experience are certainly real. That I chatted with someone who understands my mother tongue is real. That there was an advertisement in the regional paper the avatar mentioned is real (I found a copy of the newspaper on the Internet, so it must be real). The fact that she is working for Linden dollars (Second Life has its own economy and a currency, and Linden dollars (L$) are the currency with a real-life exchange rate of about L$280 to US$1) on Second Life and building a shop and designing saris and jewelry is real. So why should her stories about her off-line life not be real? But that in itself does not matter to the present article and our understanding of digital diasporas in this framework. That the avatar has certain specialized knowledge of a specific geographical context and has language skills specific to a region attest to a certain kind of authenticity. She or he is a real Indian. Does it matter if she may be a he or that she may be someone who has recently traveled physically away from the region she claims to be from? Not for the understanding of digital diasporas in the framework I write from. But certainly it is of great importance that she is authentic in terms of Indian origin, and that she is looking to make Linden dollars on Second Life. And certainly it is important information that she has given me when she tells me that Second Life is being advertised in various media in India. Further, it is important and relevant since this information can be verified.

What all these truths about this Indian woman's presence on Second Life point to is the economic pull of Second Life for young IT-interested people living in India. This makes sense in relation to all the talk about crowdsourcing and outsourcing via Second Life. Businesses such as Wipro and IBM India have moved into Second Life to recruit and train. A visit to these areas reveals that they are fairly deserted at the moment, but the very fact that these big companies have announced their presence on Second Life draws more young job-seeking Indians and other Asians into Second Life, thus changing the cultural, visual, and interactive climate within this three-dimensional reality. As digital diasporas from these regions increase in size, the demographics and practices in Second Life will shift. In future work, I will be exploring these issues along with an examination of globalization and multiculturalism in Second Life.

This chapter has laid out a framework for extending and nuancing earlier generations of research around conceptualizations of virtual community and digital diasporas in at attempt to show how these ideas work within a "Web 3D" setting. This article was exploratory in that sense and is meant to start further conversations in relation to digital diasporas in 3D worlds.

NOTES

1. In reality there is no other way to start viewing any phenomena except through one's own reality and perception and then moving toward making observations and even generalizations, so I am not suggesting that only feminists speak as persons. What I am saying is that as a critical feminist scholar, I privilege speaking through revealing this personal location as it intersects with the phenomena I am researching.

2. The "Kelly's World" blog (http://www.kgadams.net/2006/06/11/my-second-life-deflowering) describes scripted objects as follows: "Objects a user creates can have scripted behaviors—a table could have a fold out extension, or those ears I mentioned could wiggle. Even more intriguing, an object's behavior could be based on something outside the game: virtual weather in an area could be based on real-world weather reports, for example—or a soccer ball could move based on telemetry from a real-world soccer ball."

REFERENCES

Anderson, Benedict. 1991. *Imagined communities.* London: Verso.

Baym, Nancy. 2000. *Tune in, log on: Soaps, fandom, and online community.* London. Sage.

Bourdieu, Pierre. 1977. *Outline of a theory of practice.* Cambridge: Cambridge University.

Castranova, Edward. 2005. *Synthetic worlds: The business and culture of online games.* Chicago: University of Chicago Press.

Gajjala, Radhika, Natalia Rybas, and Melissa Altman. 2007. Epistemologies of doing: E-merging selves online. *Feminist Media Studies* 7, no. 2: 209–13.

Gajjala, Venkataramana. 2006. The role of information and communication technologies in enhancing processes of entrepreneurship and globalization in Indian software companies. *Journal of Information Systems in Developing Countries* 26.

Karamcheti, Indira. 1992. The shrinking Himalayas. *Diaspora* 2, no. 2: 269.

Khanna, Parag. 2005. *Bollystan: India's diasporic diplomacy.* London: Foreign Policy Centre.

Marino, Paul. 2004. *3D game-based filmmaking: The art of machinima.* Scottsdale, Ariz.: Paraglyph Press.

Miller, Dan, and Don Slater. 2000. *The Internet: An ethnographic approach.* New York: Berg.

Pratt, Mary Louise. 1992. *Imperial eyes: Travel writing and transculturation.* London and New York: Routledge.

Watson, Nessim. 1997. Why we argue about virtual community: A case study of Phish.Net Fan community. In *Virtual culture: Identity and communication in cybersociety,* ed. Steve Jones. London: Sage.

Williams, Raymond. 1985. *Key words: A vocabulary of culture and society.* Oxford: Oxford University Press.

Zuberi, Nabeel. 2008. Sampling South Asian music. In *South Asian Technoscapes,* ed. Radhika Gajjala and Venkataramana Gajjala. New York: Peter Lang.

13 The Internet and New Chinese Migrants

BRENDA CHAN

THE CHINESE DIASPORA—OLD MIGRANTS, NEW MIGRANTS

The Chinese diaspora is one of the major global diasporas, and Chinese communities exist in many corners of the earth, from Oceania and Africa to Europe and America (Cohen 1997, 85–94; Pan 1998). Emigration from China was most significant from the 1850s to 1920s, when massive numbers of Chinese, usually men of peasant origin, left China for Southeast Asia and other parts of the world as indentured labor, working in tin mines and on plantations (Wang 1991, 6).

Between the 1950s to the late 1970s there was little movement in and out of the People's Republic of China (PRC) because of restrictive policies under the Communist government. Migration resumed only after 1979 with the implementation of economic reforms and the opening up of China (Skeldon and Hugo 1999, 335–36). Departures began to increase after 1984, when new laws permitted Chinese nationals to study abroad if they paid their own fees (Guerassimoff 1998, 145), but only a minority of these self-financed students returned to China (ibid., 150). A significant number of Chinese students also remained in foreign countries after the Tiananmen incident in 1989 (Nyíri 1999, 29). The revival in Chinese emigration from the PRC accelerated in the 1990s, as population pressure intensified competition for jobs, educational opportunities, marital partners, and social status in the emerging consumer society among urban communities (ibid., 28). Students in China are also caught up in the fervor to pursue foreign education, with some twenty-five thousand students going overseas for education annually (H. Liu 2005, 294–96).

Scholars have used the term *xin yimin,* or "new migrants," to designate PRC citizens who emigrated after 1979, when the economic reforms began in China (Nyíri 2001, 145; H. Liu 2005, 293). Hong Liu (2005, 293) has identified four main categories of new migrants from mainland China: students-turned-migrants (those who study abroad and may subsequently take up residence in their host countries), emigrating professionals, chain migrants (who join family members or relatives who are living abroad), and illegal immigrants.

However, these new migrants or *xin yimin* are different from Chinese migrants from Hong Kong and Taiwan and from the overseas Chinese who settled over several generations in Southeast Asia as well as other parts of the world. Unlike the earlier waves of Chinese migration that consisted mainly of unskilled labor hailing from South China, there is a sizable segment of highly educated professionals among the *xin yimin,* alongside a proletarian diaspora (ibid., 304). The native-place origins of *xin yimin* are more diverse and can be traced to many different cities and provinces all over China (ibid., 299). Moreover, most of the *xin yimin* were raised in Communist China and carry with them "a collective memory of mainland China prior to their migration" (Sun 2002, 143–44)—with the nationwide famine in the early 1960s and the Cultural Revolution as two significant events etched permanently in their collective memories. Sun (ibid., 9) prefers to designate these *xin yimin* as "paradiasporic" in their transnational condition, because of the continual engagements and attachments that they retain with the "motherland." Their identities, which are so deeply conditioned by the experiences of Communist China, will be further complicated as they confront the foreign cultures that they have entered for work and study.

FROM IMMIGRANT ORGANIZATIONS TO CYBER-COMMUNITIES

Ethnic Chinese communities outside of China have relied on three important institutions in maintaining a collective sense of Chinese identity: voluntary organizations (such as chambers of commerce and clan associations), Chinese-language education, and Chinese-language media (Sun 2005, 68).

Increasingly, the Internet—in the form of Web sites, online magazines, bulletin board systems (BBS), newsgroups, and so on—complements ethnic newspapers, radio stations, and satellite television as a component of diasporic Chinese media, providing new Chinese migrants with the latest news of social and political affairs in China and keeping them entertained with popular cultural products in the Chinese language: "Media images, in the form of DVD, VCD, MTV, such as films, television dramas, music videos, not to mention the Internet and satellite TV, have also proliferated and multiplied, reinforcing, destabilizing, and challenging prior understanding of what it means to be Chinese" (ibid., 66).

Moreover, the Internet, as a technology, is able to combine information storage, interactivity, and broadcasting, thus overcoming the limitations of traditional media. Mass communication is typically impersonal—one-way—and rarely involves its receivers as communicators. The Internet, on the other hand, offers its users the opportunity to generate and disseminate their own content. With its interactive environment, the Internet facilitates reciprocity between individual communicators (such as in e-mail) as well as participation in various discussion groups (Holmes 1997, 36), giving rise to new social formations called virtual communities, which are based on common interests rather than physical proximity (Rheingold 1993, 24).

Together with ethnic media, clan associations and other community organizations have traditionally been "important players in defining Chinese identity in diasporic Chinese communities, in mediating between native-place ties and the environment of the host society" (Chan 2006, 23). New Chinese migrants are more educated, mobile, and technology savvy than the earlier waves of unskilled Chinese immigrants at the turn of the twentieth century. They tend to form voluntary organizations based on alumni networks and professional ties rather than primordial kinship or native-place affiliation (H. Liu 2005, 306), and they also form virtual communities on the Internet (such as newsgroups, bulletin boards, and discussion forums), which are often able to perform functions similar to the off-line voluntary organizations (Chan 2006, 23–24).

Wenli Chen's (2006, 79–91) study on mainland Chinese immigrants in Singapore affirms that new Chinese migrants actively seek and provide social support on the Internet in their process of adapting to the host society—for instance, seeking advice on how to use the public transportation system, exchanging computer software, and finding friends to play online games with. Migrants turn to virtual communities to obtain information, emotional support, and companionship, because the Internet is easily accessible in Singapore and affords anonymity to users. This takes away any social stigma or threat to the migrant's "face" when seeking advice, counsel, and assistance (ibid., 145). Virtual communities also play an important role in rallying Chinese immigrants to fight for social or political justice when immigrants encounter discrimination and unfair treatment in the host society. For instance, computer networks formed by Chinese students collected online signatures and mobilized Chinese students to protest against a CBS News story in 1994 that called the Chinese students in the United States "the biggest spy network in America" (Wu 1999, 85).

Most of the past studies on virtual communities formed by new Chinese migrants are centered on the North American case, with Chinese News Digest as the premier example of the Chinese transnational cybercommunity (D. Liu 1999, 195–206; G. Yang 2003, 469–90; Sun 2002). CND was originally a news-distribution network founded by a group of mainland Chinese students and scholars in the United States and Canada in 1991, growing out of a concern about political events in China after the Tiananmen Square incident. It circulated China-related news to Chinese readers all over the world via weekly English publications and disseminated the first Chinese-language Internet magazine, *Huaxia Wenzhai* (Sun 2002, 118). Today, the CND Web site continues to host *Huaxia Wenzhai* and various discussion forums, and it also maintains a database containing a virtual library on modern Chinese history and a number of online archives (Wong 2003; G. Yang 2003, 474). The permanent virtual archives created by the CND feature online articles and photographs for three traumatic national events in the history of modern China: the Nanjing Massacre, the Cultural Revolution, and the June 4, 1989, Tiananmen Square incident (Sun 2002, 120).

In her analysis of CND's virtual museum on the Nanjing Massacre, Wanning Sun concludes that the Internet amplifies the dispersal and displacement of the new Chinese migrants, as shown in their pursuit of an essentialized notion of Chinese identity in cyberspace, in order to reconcile their fragmented identities in real life:

> In the case of Chinese scholars now living in North America, Japan, and other Western countries, in order to remain Chinese, one has to keep telling stories of being Chinese. Collective memories of China are kept alive by these self-exiled Chinese . . . by the constant retelling of familiar national stories. . . . What seems uncanny is that in spite of, or perhaps because of, the tension between new technology, which is memoryless and deterritorialized, and memory, which is bound by a

specific notion of time and place, the Internet and its attendant cyberspace prove to be hugely enabling in articulating a strategically "pure" collective subjectivity. (ibid., 132–33)

This essentialist Chinese identity that dominates the discourses in newsgroups and Web sites by new Chinese migrants tends to assume a kind of distinctive Chineseness built on the imaginary of a unitary territorial nation of China and a shared history of being oppressed by an imperialist "Other," such as Japan (Sun 2002, 122–31).

While Chinese students and new Chinese migrants retain a strong attachment to China and have a strong sense of Chinese identity, Wenjing Xie (2005, 399–403) argues that virtual communities formed by these migrants help them to understand the lifestyles and practices of the host society, thereby facilitating adaptation to the new environment. However, Srivinas R. Melkote and D. J. Liu (2000, 499) suggest that although use of the Chinese Ethnic Internet (consisting of Internet magazines, bulletin boards systems, and Web sites) can help Chinese students and scholars in the United States learn about and adapt to the American lifestyle (such as dress and food), the greater the dependency on Chinese Internet programs, the lower their acculturation of American values and the higher and more sustained their level of Chinese values. Thus, we can see that the Internet as a "global technology can contribute to a strengthening of cultural distinctiveness, and despite the placelessness of the Internet, it can serve to reinforce place" (Mackay and Powell 1998, 215).

THE INTERNET AND THE DIASPORIC PUBLIC SPHERE

Arjun Appadurai (1996, 21–23) has postulated that electronic media and communication technologies have facilitated the emergence of *diasporic public spheres,* as migrants engage in the work of postnational imagination, through consumption of popular cultural products produced in their homelands and through exchanges and interactions with fellow migrants as well as those who remain at home. Public spheres in the present day, he argues, are no longer confined to the boundaries of the nation but are constituted by migrants across various sites and nodal points on the globe.

Guobin Yang (2002) posits that Web sites, bulletin board systems, Internet magazines, newsgroups, chat rooms, and other online spaces in Chinese language form a virtual or online Chinese cultural sphere, with users drawn from ethnic Chinese communities all over the world. He argues that the online Chinese cultural sphere approximates a transnational diasporic public sphere, for it allows dispersed individuals and groups to gather and interact with one another to articulate personal, local, and global problems, and facilitates various forms of political activism and collective action.[1]

However, Guobin Yang (2003, 470–71) concedes that his definition of the public sphere is one that is broadly conceived as an open communicative space, and is not meant to conform strictly to the Habermasian model of rational-critical debate by a bourgeois public within a nation-state. This is because online spaces in the virtual Chinese cultural sphere are diverse in form and content (e.g., interactive BBS versus noninteractive Internet newsletters), engendering multiple and partial publics that transcend national boundaries (G. Yang 2002). Because of the transborder nature of the Internet, members of the Chinese diaspora can access online spaces that are hosted in the homeland, while people in China can enter the online spaces hosted in foreign countries that are created by the diaspora population (ibid.). The discourse communities that emerge from these heterogeneous online spaces are nonetheless connected, as illustrated by the cross-posting of messages between China-based and North America–based BBS and the rise of popular well-known Web sites frequented by users from various parts of the world (ibid.). Although membership in these discourse communities is often transient and discourses in the online Chinese cultural sphere are often fraught with contradictions and irrational outbursts (ibid.), Guobin Yang is optimistic that the online Chinese cultural sphere promotes democratic participation in its function as a communication network for public expression, civic association, and transnational protests (2003, 481–82).

In the next section of the chapter, I will critically examine Guobin Yang's (2003) framework of the transnational online Chinese cultural sphere, by drawing on my past research in online communities formed by new Chinese migrants in Singapore, as well as where I live and work as a locally born Chinese woman. The original purpose of the research was to examine how new Chinese migrants in Singapore use the Internet in the construction and maintenance of their cultural identities and build solidarity as a diasporic community. Singapore presents a unique and interesting case study because the new Chinese migrants in Singapore are living in a host society where ethnic Chinese already constitute the majority of the population. This is unlike the situation of Chinese students and immigrants in almost all other countries in the world, where they are an ethnic minority in the host society and face a hardened boundary between the Self and the Other in terms of race, ethnicity, and culture.

Singapore is a tiny island republic in Southeast Asia with a land area of only 690 square kilometers. Originally a British trading colony founded by Sir Stamford Raffles in 1819, Singapore attained independence in 1965 after a brief merger with Malaya. During the colonial period, the establishment of a free entry port in Singapore by the British attracted large numbers of immigrants from China, India, and the neighboring regions (Vasil 2000, 1–4). The early Chinese immigrants in Singapore were coolies and traders who saw themselves as sojourners who would eventually return to China. By 1849 the number of Chinese had exceeded that of the indigenous Malays, and the former became the majority of the population (Chiew 1995, 42). Today, with a resident population of about 3.3 million, Singapore remains a multiethnic society made up of ethnic Chinese (76.8 percent), Malays (13.9 percent), Indians (7.9 percent), and other ethnic minorities (1.4 percent) (Singapore Department of Statistics n.d.).

Since the 1990s, the Singapore government has adopted a selective migration policy with attractive initiatives to draw skilled professionals and entrepreneurs, to alleviate the problems of an aging population and the need to maintain economic competitiveness in a knowledge economy (Low 2002, 409–25; Yap 1999). The Singapore government also offers scholarships to mainland Chinese students with good academic backgrounds, which allows them to pursue tertiary education in Singapore, after which the scholarship recipients are obliged to work in the host country for a number of years (Woguo shi zhongguo 2006). These favorable policies attract mainland Chinese to move to Singapore for study and employment.

There are altogether some thirty-three thousand students from mainland China who are enrolled in public schools, private schools, polytechnics, and universities in Singapore (Chan 2006, 9). Singapore also imports about a hundred thousand migrant workers from China, most of whom are employed in the construction industry (R. Yang 2002). These migrant laborers are typically not granted permanent residence or citizenship in Singapore. There are also an increasing number of PRC nationals hired in the service sectors as wait staff, sales assistants, and so on (Toh and Sudderudin 2007). Besides students and migrant workers, there is another group of transients known as *peidu mama,* or "study mothers." These study mothers accompany their children to Singapore on a special visa, for the latter's primary school or secondary school education. They are allowed to stay in Singapore and may apply for work passes, subject to approval by the authorities (Huang and Yeoh 2005, 386–88). The Singapore government does not release official statistics on the exact number of new migrants from mainland China, but it is estimated that the country hosts between two and three hundred thousand new Chinese migrants (H. Liu 2005, 295).

Despite their growing presence, the new migrants from mainland China are not entirely welcome by the local Chinese community with open arms. The local-born Chinese Singaporeans have acquired a local way of life and a strong sense of Singaporean national identity (Chiew 2002, 34–35), and therefore perceive themselves as being very different from the Chinese from the PRC. Perhaps more documented in mass media and popular fiction (rather than academic research) is the animosity between the local-born Chinese Singaporeans and the new Chinese migrants from the PRC. The former think that new Chinese migrants are competing with local-born Chinese Singaporeans for jobs. For instance, during his National Day Rally speech in 2001, then prime minister of Singapore Goh Chok Tong criticized the attitude of a young Chinese Singaporean, who wrote to the largest English newspaper in Singapore, *The Straits Times,* labeling mainland Chinese migrants as "cheena" (a derogatory term somewhat similar to "chink" in the United States) and describing them as "a crude lot" (2001). Furthermore, female migrants from mainland China (often dubbed as *xiaolongnu,* or "dragon girls," by local-born Chinese Singaporeans) are stereotyped as gold diggers who seduce Singaporean men and break up Singaporean families.

On the other hand, new Chinese migrants in Singapore complain that they have been looked down upon or discriminated by the Chinese Singaporeans, as Chinese Singaporeans hold misconceptions that China is a poor and backward country. Although both Chinese Singaporeans and mainland Chinese perceive each other as belonging to the same ethnicity, the mainland Chinese in Singapore see themselves as the "authentic bearers of Chinese culture" (in terms of fluency in Mandarin and knowledge of China's history and geography); they consider Chinese Singaporeans a "substandard 'synthetic' kind of Chinese" (Loo 1997, 4, 42). Therefore, integration into the local Chinese community in Singapore is not an automatic process for the new migrants from the PRC, despite moving into a society with a Chinese majority.

I selected two virtual communities for case study: Springdale and AutumnLeaves.[2] Both Springdale and AutumnLeaves are Web sites set up by mainland Chinese students in Singapore, and they provide information about studying and working in Singapore. Both Web sites host a bulletin board system with several discussion forums covering a wide range of topics, such as current affairs, travel, creative writing, photography, gaming, life in Singapore, and so on (Chan 2006, 10–17). The discussion forums hosted on these Web sites are where virtual community emerges over time. Springdale was established by PRC students at the National University of Singapore. The bulk of its members are PRC students studying in two government-funded universities in Singapore, thus making it a more "campus-based" virtual community. Users of AutumnLeaves, on the other hand, present a more diverse profile. Its Singapore-based users include PRC students studying in various private schools and polytechnics in Singapore, permanent residents, employment-pass holders, and even *peidu mama*. It also has a larger proportion of China-based users than Springdale (ibid., 11).

Arjun Appadurai has talked about mass-mediated communities formed by migrants, in which the enmeshing of local experiences in taste, pleasure, and politics can possibly give rise to translocal social action (1996, 8). He notes that diasporic public spheres are "frequently tied up with students and other intellectuals engaging in long-distance nationalism (as with the activists from the People's Republic of China)" (ibid., 22).

The Internet has enabled the new Chinese migrants to establish virtual communities where these migrants are able to meet friends who share the same hobbies and interests. At the same time, these online communities have been producing multiple discourses of the Chinese nation and have been engaged in the imagination of the homeland, through BBS or discussion forums that talk about politics and current affairs in China (Chan 2005). This shared imagination of the homeland could become a "staging ground" for collective action by the migrants (Appadurai 1996, 7).

I analyzed the messages posted on the discussion forums on the Springdale and AutumnLeaves Web sites, based on selected discussion threads or topics. In addition, I also conducted in-depth face-to-face interviews with various members of the virtual communities and carried out participant observation during some off-line activities organized by the two communities. My quasi-ethnographic research on these two virtual communities was carried out during the period of February to October 2003, which coincided with the outbreak of the severe acute respiratory syndrome (SARS) pandemic in China and in various other Asian countries, including Singapore. This offered an opportunity for me to observe how virtual communities formed by mainland Chinese migrants in Singapore respond to a situation of the "homeland in crisis" and to learn whether the Internet might promote long-distance nationalism.

THE SARS EPIDEMIC AND LONG-DISTANCE NATIONALISM

SARS is a type of atypical pneumonia. It was believed that SARS first appeared in Guangdong, China, as early as November 2002, although the disease was first officially reported in February 2003. The disease later spread to Beijing and other areas in China. SARS also spread via international air travel to Vietnam, Singapore, Hong Kong, Taiwan, and Canada (World Health Organization 2003, 1–2). By May 2, 2003, afflicted areas in China included Beijing, Tianjin, and the provinces of Hebei, Inner Mongolia, and Shanxi (World Health Organization Representative Office in China 2003).

During this period, mainland Chinese students in Singapore rallied around the crisis of the SARS outbreak, seeking to raise funds to aid in the anti-SARS efforts in China. Participation from virtual communities in these efforts was uneven, and limited in terms of influence. The Singapore Chinese Student Committee, an off-line group made up of PRC student representatives from various educational institutions in Singapore, co-organized a charity concert with the Singapore Chinese Orchestra. The charity concert, called Youzi Qing, was held on June 7, 2003, to raise funds for the Red Cross Society in China, in aid of anti-SARS efforts in the mainland. Springdale and AutumnLeaves were invited as official Web sites for the charity concert, and each Web site sent a representative to sit on the organizing committee of the concert.

However, the Internet played only a marginal role in the concert, merely as one of the means of publicity. AutumnLeaves initially created a special Web page for the concert with a forum, but the Web page was underutilized by the organizing committee of the concert. It was never updated with information about the concert, and there were only two posts in the forum. Ticket sales were pushed almost entirely over the telephone, via interpersonal networks of mainland Chinese students in the polytechnics, universities, and private schools. As a virtual community AutumnLeaves had intended to go beyond the use of its Web site as a publicity vehicle for the concert. A group of AutumnLeaves members, who had formed a rock band, wanted to perform at the concert in the name of the virtual community. Unfortunately, the rock band failed to get through the auditions for the concert program. This created a lot of unhappiness among the AutumnLeaves members. Those who had initially intended to render support to the rock band even threatened to boycott the concert.

While the Internet can play a role in supporting long-distance nationalism, we must not assume that the Internet will always be a powerful and effective medium in mobilizing migrants into collective action. We have to consider the attitudes that people carry toward Internet technology that could prevent full exploitation of the Internet. We also have to take into account divisive factors within and outside of cyber-communities formed by migrants, which could limit the influence and power of these communities in organizing large-scale activities with off-line impact.

During the SARS crisis, a particular incident attracted more attention from the new Chinese migrants in Singapore and sparked a fervent debate in the forums of Springdale and AutumnLeaves. In April 2003, Goh Chok Tong, then prime minister of Singapore, canceled plans to visit China because of the SARS outbreak. He made the decision upon advice by his doctor. From my analysis of the discussion threads in Springdale and AutumnLeaves surrounding this event, it was evident that the new Chinese migrants found it difficult to understand why Goh, an ethnic Chinese, would cancel the official visit (Chan 2005). The decision by Goh stood in contrast to the actions of the then French prime minister, Jean-Pierre Raffarin, who visited China during the same period in spite of the SARS epidemic (BBC 2003).

The mainland Chinese migrants had assumed that local-born Chinese Singaporeans would have a stronger sense of affinity and empathy with China due to common ancestry with the Chinese people. Hence, many were puzzled and offended by Goh Chok Tong's cancellation of his trip to China. In the new Chinese migrants' online discussion of this incident, there were "outbursts of crude nationalism" in the messages posted to the forums in Springdale and AutumnLeaves (Chan 2005, 356). Some forum participants saw China as a regional power and called for China to impose sanctions or take military action against Singapore. For instance:

> China should punish all those who oppose her, like the U.S. does. (Chumeiren, Springdale)
>
> Bomb Singapore till she sinks! (pinwheel, AutumnLeaves) (Chan 2005, 355)

In online debates such as these, Internet discussion forums become spaces in which new Chinese migrants can indulge in a fantasy of the homeland as a superpower, as a means to challenge the perceived hegemony of the United States in the international political economy, and to vent their frustrations of being marginalized in the host society (Chan 2005, 362–63).

Of course, the Internet places a heavier burden on the agency of the user, to seek out the information he or she wants, compared to the audience in traditional media. Moreover, the Internet tends to foster communities based on shared interests rather than physical proximity, drawing together like-minded people in fragmented groups (Wellman and Gulia 1999, 172). Therefore, jingoistic messages posted by migrants online may be exposed to only a portion of the population that has Internet access; there may not be a critical mass around such discussion that will directly affect off-line interaction between migrants and the "natives."

On the other hand, if these messages were brought to the attention of the general public in the host society, it could lead to resentment and enmity between the migrants and the natives. For AutumnLeaves, its worst nightmare came true when an English tabloid in Singapore ran a lengthy news story in April 2004, quoting a posting on its forum by an eighteen-year-old PRC student. In his posting the PRC student called Singaporean students stupid, saying that "Singaporeans' brains are fed with pigswill" (Chia 2004). The news article also recorded other negative comments about Singaporeans that were made by the forum participants. The story, which was subsequently picked up by the Chinese press in Singapore, resulted in a jump in visitor traffic on the AutumnLeaves Web site, as many Singaporeans and PRC migrants engaged in a flaming war on the forum surrounding the issue. In an online statement to its users, AutumnLeaves administrators blamed the media for irresponsible reporting, arguing that the journalist had completely ignored other posts in the forum that were friendly toward Singaporeans.

This incident showed that once abusive and jingoistic messages are leaked out to the mainstream mass media beyond the confines of cyberspace, the friction between mainland Chinese migrants and Singaporeans can be exaggerated or magnified to unrealistic proportions. This will make it harder for the two groups to reconcile with each other and could add to the migrants' reluctance to integrate into mainstream society.

THE LIMITS OF VIRTUALITY

In the preceding section I have shown how online BBs formed by new Chinese migrants fall short of the normative conditions of rational-critical debate that should prevail in the public sphere, as idealized by Jürgen Habermas (1989, 27–31). Whether the Internet facilitates the formation of a public sphere remains a debatable issue today, as some scholars argue that online debate and interaction are fragmented into virtual groups of people with similar ideological positions, gathering together to reinforce each other's opinions (Dahlberg 2007, 828). As Kevin A. Hill and John E. Hughes (1997, 17–20) have concluded from their study, discussion groups on the Internet practice a form of ideological policing, in that people tend to post messages in newsgroups that are inclined toward a similar political stance as their own, and ideologically dissonant posts are more likely to be flamed or attacked.

In fact, Cass Sunstein maintains that the Internet supports the development of "deliberative enclaves" and becomes "a breeding ground for group polarization and extremism" (2001, 67, 71). To assess the contribution and potential of the Internet in encouraging democratic communication would require reconceptualizing the public sphere into an alternative model that can accommodate radical discursive contestation (Dahlberg 2007, 841), and from a state-centric focus to a transnational sphere (G. Yang 2003, 471). Even with the redefinition of the concept of the public sphere, I am still concerned about whether an online public sphere will accelerate the formation of "ethnic enclaves" on the Internet that will deepen the differences and tensions between diasporic communities and their host societies. Given that the voice of a diasporic community may be marginalized and excluded in the dominant discourse of the host society, a diasporic public sphere in cyberspace can allow migrants to gather together and lobby for issues that advance their social and political rights (Dahlberg 2007, 837). But can the diasporic public sphere promote interaction, discussion, and understanding between a diasporic community and the host society, or will it lead to "a hardening and non-engagement (both online and offline) with oppositional identities?" (ibid., 841).

Besides considering the functions of the online Chinese cultural sphere in public expression and civic association, we also need to critically reflect on its potential to mobilize the Chinese diaspora in various forms of collective action. The role of the Internet was rather limited when the virtual communities of Springdale and AutumnLeaves were involved with the Youzi Qing charity concert. While the Internet may have an impact on the configuration of social and political relations on local and global scales, Jayne Rodgers has cautioned against taking a technological-deterministic approach in dealing with the Internet and activism, reminding us that "social movements existed long before the Internet did" (Rodgers 2003, 5).

By the same token, long-distance nationalism preceded the development of the Internet. The overseas Chinese had historically been known to be involved in long-distance nationalism, from the monetary assistance they rendered to Sun Yat-sen in overthrowing the Qing dynasty to the raising of funds in aid of China's war effort against Japan in the 1930s and 1940s. Sandor Vegh (2003, 71–72) also distinguishes between "Internet-based" and "Internet-enhanced" types of activism. In Internet-based strategies, the Internet is used for activities that are only possible online, such as a virtual sit-in, a spamming campaign, and hacking or sabotaging a particular

Web site. In Internet-enhanced strategies, the Internet is used to support or enhance the activities of the social movement, such as increasing levels of awareness or coordinating action more efficiently.

When studying the relationship between the Internet and long-distance nationalism, we should consider whether this particular technology is introducing something new to the ways in which social movements represent themselves, that is, whether the Internet is transforming opportunities for activists by creating new ways of networking and exchanging information. In other words, perhaps we ought to view the Internet "as a *tool* of political activism, rather than the genesis of it" (Rodgers 2003, 3–6).

NOTES

1. G. Yang (2003) also developed and expounded fully upon the idea of the online Chinese cultural sphere.

2. Pseudonyms are used for all the virtual communities mentioned in this chapter. Research on Springdale and AutumnLeaves was carried out for my Ph.D. dissertation. For discussion of the main findings of the research study, see Chan 2005, 336–68; and 2006.

REFERENCES

Appadurai, Arjun. 1996. *Modernity at large: Cultural dimensions of globalization.* Minneapolis: University of Minnesota Press.

BBC. 2003. China invited to G8's table. April 25. Available at http://news.bbc.co.uk/2/hi/asia-pacific/2976641.stm.

Chan, Brenda. 2005. Imagining the homeland: The Internet and diasporic discourse of nationalism. *Journal of Communication Inquiry* 29, no. 4 (May): 336–68.

———. 2006. Virtual communities and Chinese national identity. *Journal of Chinese Overseas* 2, no. 1 (May): 1–32.

Chen, Wenli. 2006. Computer-mediated social support: Internet use and international migrants. Ph.D. diss., Nanyang Technological University, Singapore.

Chia, Dawn. 2004. China students lash out S'pore's been good to you. *The Electric New Paper,* April 12. Available at http://newpaper.asia1.com.sg/top/story/0,4136,57507,00.html.

Chiew, Seen Kong. 1995. The Chinese in Singapore: From colonial times to the present. In *Southeast Asian Chinese: The socio-cultural dimension,* ed. Leo Suryadinata. Singapore: Times Academic Press.

———. 2002. Chinese Singaporeans: Three decades of progress and changes. In *Ethnic Chinese in Singapore and Malaysia: A dialogue between tradition and modernity,* ed. Leo Suryadinata. Singapore: Times Academic Press.

Cohen, Robin. 1997. *Global diasporas: An introduction.* Seattle: University of Washington Press.

Dahlberg, Lincoln. 2007. Rethinking the fragmentation of the cyberpublic: From consensus to contestation. *New Media & Society* 9, no. 5: 827–47.

Goh, Chok Tong. 2001. New Singapore. Speech at National Day Rally, University Cultural Centre, National University of Singapore, August 19. Available at http://app.sprinter.gov.sg/data/pr/2001081903.htm.

Guerassimoff, Carine. 1998. Legal and illegal mainland Chinese emigration during the 1990s. In *New developments in Asian studies: An introduction,* ed. Paul van der Velde and Alex Mckay. London: Kegan Paul.

Habermas, Jürgen. 1989. *The structural transformation of the public sphere.* Cambridge: MIT Press.

Hill, Kevin A., and John E. Hughes. 1997. Computer-mediated political communication: The Usenet and political communities. *Political Communication* 14, no. 1: 3–27.

Holmes, David. 1997. Virtual identity: Communities of broadcast, communities of interactivity. In *Virtual politics: Identity and community in cyberspace,* ed. David Holmes. London: Sage.

Huang, Shirlena, and Brenda S. Yeoh. 2005. Transnational families and their children's education: China's "study mothers" in Singapore. *Global Networks* 5, no. 4: 379–400.

Liu, Dejun. 1999. The Internet as a mode of civic discourse: The Chinese virtual community in North America. In *Civic discourse, civil society, and Chinese communities,* ed. Randy Kluver and John H. Powers. Stamford, Conn.: Ablex Publishing.

Liu, Hong. 2005. New migrants and the revival of overseas Chinese nationalism. *Journal of Contemporary China* 14, no. 43: 291–316.

Loo, Audrey Wei Min. 1997. The construction of mainland Chinese identity in Singapore. Academic exercise, National University of Singapore.

Low, Linda. 2002. Globalisation and the political economy of Singapore's policy on foreign talent and high skills. *Journal of Education and Work* 15, no. 4: 409–25.

Mackay, Hugh, and Tony Powell. 1998. Connecting Wales: The Internet and national identity. In *Cyberspace divide: Equality, agency, and policy in the information society,* ed. Brian D. Loader. New York: Routledge.

Melkote, Srinivas R., and D. J. Liu. 2000. The role of the Internet in forging a pluralistic integration: A study of Chinese intellectuals in the United States. *Gazette* 62, no. 6: 495–504.

Nyíri, Pal. 1999. *New Chinese migrants in Europe: The case of the Chinese in Hungary.* Aldershot, England: Ashgate.

———. 2001. Expatriating is patriotic? The discourse on "new migrants" in the People's Republic of China. In *Asian nationalism in an age of globalization,* ed. Roy Starrs. Surrey, UK: Japan Library.

Pan, Lynn. 1998. *The encyclopedia of the Chinese overseas.* Singapore: Chinese Heritage Centre.

Rheingold, Howard. 1993. *The virtual community: Homesteading on the electronic frontier.* Reading, Mass.: Addison-Wesley.

Rodgers, Jayne. 2003. *Spatializing international politics: Analyzing activism on the Internet.* London and New York: Routledge.

Singapore Department of Statistics. n.d. Singapore's changing population trends. Available at http://www.singstat.gov.sg/pubn/papers/people/cp-poptrends.pdf.

Skeldon, Ronald, and Graeme Hugo. 1999. Conclusion: Of exceptionalisms and generalities. In *Internal and international migration: Chinese perspectives,* ed. Frank N. Pieke and Hein Mallee. Surrey, UK: Curzon Press.

Sun, Wanning. 2002. *Leaving China: Media, migrations, and transnational imagination.* Lanham, Md.: Rowman and Littlefield.

———. 2005. Media and the Chinese diaspora: Community, consumption, and transnational imagination. *Journal of Chinese Overseas* 1, no. 1: 65–86.

Sunstein, Cass. 2001. *Republic.com.* Princeton: Princeton University Press.

Toh, Mavis, and Shuli Sudderudin. 2007. How much is this? And please speak in English. *The Sunday Times* (Singapore), December 9.

Vasil, Raj. 2000. *Governing Singapore: A history of national development and democracy.* St. Leonards, Australia: Allen and Unwin.

Vegh, Sandor. 2003. Classifying forms of online activism: The case of cyberprotests against the World Bank. In *Cyberactivism: Online activism in theory and practice,* ed. Martha McCaughey and Michael D. Ayers. New York and London: Routledge.

Wang, Gungwu. 1991. *China and the Chinese overseas.* Singapore: Times Academic Press.

Wellman, Barry, and Milena Gulia. 1999. Virtual communities as communities: Net surfers don't ride alone. In *Communities in cyberspace,* ed. Marc A. Smith and Peter Kollock. London: Routledge.

Woguo shi zhongguo liuxuesheng remen di [Singapore is a popular destination for overseas Chinese students]. 2006. *Lianhe Zaobao* (Singapore), January 1.

Wong, Loong. 2003. Belonging and diaspora: The Chinese and the Internet. *First Monday* 8, no. 4. Available at http://www.firstmonday.org/issues/issue8_4/wong/.

World Health Organization. 2003. *Severe acute respiratory syndrome (SARS): Status of the outbreak and lessons for the immediate future.* Geneva: World Health Organization. Available at http://www.who.int/csr/media/sars_wha.pdf.

World Health Organization Representative Office in China. 2003. Severe acute respiratory syndrome (SARS) China update. May 2. Available at http://www.wpro.who.int/china/media_centre/press_releases/pr_20030502.htm.

Wu, Wei. 1999. Cyberspace and cultural identity: A case study of cybercommunity of Chinese students in the United States. In *Civic discourse: Intercultural, international, and global media,* ed. Michael H. Prosser and K. S. Sitaram. Stamford, Conn.: Ablex.

Xie, Wenjing. 2005. Virtual space, real identity: Exploring cultural identity of Chinese diaspora in virtual community. *Telematics and Informatics* 22: 395–404.

Yang, Guobin. 2002. Information technology, virtual Chinese diaspora, and transnational public sphere. Paper presented at the "Virtual Diaspora and Global Problem Solving Project Workshop" of the Nautilus Institute, Berkeley, California, April 25. Available at http://www.nautilus.org/gps/virtual-diasporas/paper/Yang.html.

———. 2003. The Internet and the rise of a transnational Chinese cultural sphere. *Media, Culture, and Society* 25, no. 4: 469–90.

Yang, Ruoqian. 2002. Why Chinese workers falling easy prey in Singapore: Analysis. *People's Daily Online,* July 10. Available at http://english.people.com.cn/200207/09/eng20020709_99414.shtml.

Yap, Mui Teng. 1999. The Singapore state's response to migration. *Sojourn* 14, no. 1: 198–211.

14 The Migration of Chinese Professionals and the Development of the Chinese ICT Industry

YU ZHOU

Robin Li (Li Hongyan) graduated from Beijing's Peking University, the "Harvard" of China, in 1991.[1] Like many of the graduates from China's elite universities at the time, he headed to the United States for a graduate-level education in computer science. Before he finished his Ph.D. at the State University of New York at Buffalo, he interned and worked for a number of Japanese and American corporations in the United States. During this time, Li made a breakthrough in Internet search-engine method and was recruited by Infoseek—once the leading Internet company—to be responsible for its search-engine development. In the white-hot Internet bubble of the late 1990s, Li decided to start his own company and was able to raise $1.2 million in venture investments (Barboza 2006).

Li returned to Beijing in 1999 and rented a space in a hotel next to his alma mater, naming his company Baidu.com. (The name comes from a famous phrase in a classical Song dynasty poem, which describes the arduous search for beauty amid a crowd.) Initially, Baidu struggled for a few years because of China's immature Internet industry, but by 2003, it had become China's favorite search engine. In August 2005, it achieved international fame when it issued its initial public offering on NASDAQ. The share price of Baidu, known as the Chinese Google, more than quadrupled on the first day, setting a record for the best first-day performance of all foreign firms ever listed on the U.S. stock market, as well as the best first-day performance among *all* firms in the previous five years.

While Baidu's turn of fortune might have been exceptional, Li's life trajectory is not. He is among the growing number of Chinese youth who, in the mid-1980s, began to go abroad to study or work, and eventually found that returning home could be an attractive option. In other words, Li represents the Chinese digital diaspora that is making a profound mark on the transformation of China's society and economy, by bringing together the countries on their residential itinerary. Their experiences and inspirations demonstrate the tremendous impact of information and communication technology (ICT) on the diaspora population. Yet they have also in turn played an instrumental role in constructing global linkages. This chapter will outline the main patterns of the Chinese diaspora in the United States. I will focus on the stream of Chinese youth who went abroad to study and work, only to return to their homeland. I will also analyze the forces that drive the flows on both sides of the Pacific, and I will outline some of the opportunities and challenges that returnees face back in their homeland.

The information of this chapter has been collected through documentation research and interviews, conducted mostly in Beijing's Zhongguancun Science Park between June 2005 and June 2007. I have interviewed mostly returnee entrepreneurs who have started up their own companies in Beijing as well as local government officials who are in charge of returnee affairs. A few interviews with returnee professionals also took place in Shanghai or over the Internet.

THE CHINESE PROFESSIONAL DIASPORA IN THE UNITED STATES

Chinese migration to the United States has a long history. The significant Chinese migration first started arriving in the United States around the 1850s to work on the railroad and to make a fortune in the California Gold Rush. Composed mostly of manual male labor, the flow reached its peak around the 1870s, with 123,201 Chinese people migrating to the United States in the decade between 1870 and 1880 (U.S. Department of Justice 1991, 1996). However, immigration was sharply curtailed by the Chinese Exclusion Act passed by the U.S. Congress in 1882. The law was the first U.S. federal immigrant law excluding a population solely on the basis of race. However, the Chinese Exclusion Act did not limit the entry of Chinese students and scholars, and they continued to trickle onto the shores of the United States over the next eighty years.

Some of these individuals upon returning to China became key reformers in the Chinese revolution in the late nineteenth century. For example, Charlie Soong returned in 1886 and became the key financial sponsor of Sun Yat-sen, the founder of the Republic of China. His daughters, the Soong sisters, were also educated in the United States; they were arguably the most famous and influential women in China during the twentieth century (Seagrave 1996).[2] By and large, however, the highly educated constituted such a tiny number in war-torn and impoverished China that they could safely be ignored in demographic terms.

The real turning point for Chinese immigration came in 1965, when the U.S. Congress abolished the national-origin immigration quota that had favored European countries. Since then, the number of immigrants from China has grown sharply. During the decade of the 1960s, 102,649 Chinese immigrants were admitted to the United States. This number increased to 261,151 in the 1970s and to 444,962 in the 1980s (U.S. Department of Justice 1991, 1996). Whereas male labor dominated the first wave of Chinese immigration into the United States, the post-1965 immigrants tended to be permanent residents who brought their families with them. Since 1970, the long-skewed sex ratio within Chinese communities of the United States has become almost balanced (Tsai 1986; M. Zhou 1992).

The immigration reform in 1965 also implemented the professional and family preference as two primary channels for immigration. With only a small established population in the United States, the Chinese relied heavily on professional preference for initial entry, and then used family preference to bring in their family members and relatives. Some Chinese also migrated illegally. The result has been a polarized population composed of both highly and poorly educated groups, which is typical of Asian immigration to the United States. In 2000, there were a total of 3.6 million Chinese people in the United States. About 18 percent of adults had less than a high school education, which was slightly higher than the national population (16 percent). However, 50.5 percent of the Chinese had a bachelor's degree or higher, considerably higher than the U.S. population at large (27 percent) (U.S. Census 2000). This shows that the Chinese population in the United States is heavily biased toward professionals, even though the poorly educated population also represents a significant share.

Between 1949 and the 1980s, because of the cold war and the international isolation of mainland China, most of the Chinese immigrants in the

United States actually came from Hong Kong and Taiwan, although many of them were born in mainland China. However, since the Chinese government adopted the Open Door Policy in 1980, a growing share of Chinese immigrants has come directly from mainland China. The next two sections will map out the changing flows of Chinese immigrants to the United States from Taiwan and mainland China.

THE FIRST PHASE: BRAIN DRAIN

Before 1980, Chinese immigration into the United States could be safely characterized as a one-way brain drain. These immigrants originated from Taiwan, Hong Kong, or Southeast Asia—places that had a heavy American dominance in the political, military, ideological, and educational systems (Ong, Bonacich, and Cheng 1994). For the Chinese professional population, the United States represented the ultimate inspiration and path for self-growth and development. For example, in Taiwan during the 1960s, going abroad to the United States was a rite of passage for the most elite students. According to a survey conducted during that time period, 21,248 students left Taiwan for advanced study between 1960 and 1970, but only 1,172 returned, resulting in a mere 5 percent retention rate (NYC 1987). Between 1970 and 1979, 33,165 students went abroad to study, and 5,028, or 15 percent, returned—a marked improvement over the figures from the 1960s, but still quite dismal.

Overall, nonreturnees in the two decades amounted to approximately 88 percent of the student migrants, with the overwhelming majority of the nonreturnees staying on in the United States. Another Taiwanese government manpower study shows that during 1976 and 1981, one out of every five college graduates in the fields of science and technology went abroad, and only 10 percent of those students returned (Liao and Tang 1984). In fact, more than 40 percent of the graduates of the departments of mechanical engineering, electrical engineering, civil engineering, physics, and chemistry from prestigious universities such as National Taiwan University, Tsinghua University, and Chiaotung University went abroad. Almost 95 percent of them went to the United States. For Taiwan, losing so many of its brightest and most educated students to the United States during the postwar recovery period was not a pleasant experience.

For Chinese professionals, however, it was not until the 1980s that going abroad to the West was even considered an option, due to China's isolation. Yet once the gate opened after the mid-1980s, the brain drain took hold in an even more powerful way. When Deng Xiaoping came into power in 1978, he viewed studying in the West as a necessary step in China's modernization drive. Thus, he began to encourage students to pursue advanced degrees abroad, initially through state sponsorship, and then by allowing self-sponsored students—students paid by foreign institutions—to leave. The hope was that these students would return to China after their educations abroad. The outflowing tide started to emerge during the mid-1980s, and it has gone unabated ever since.

The initial flows of mainland students tended to be a highly selected group from the most prestigious Chinese universities, since they had the best chance of being awarded scholarships by foreign universities or by the Chinese government. Until the mid-1990s, an overseas postgraduate education was practically impossible and unimaginable for most Chinese students without those sponsorships.[3] Since then, as China's interactions with the world have expanded vastly, and the average household income has increased, going abroad has become an option for an increasing portion of the population. As a result, the category of mainland Chinese students abroad has become rather diverse.

According to the Chinese Ministry of Education (2005), a total of 815,000 Chinese went abroad between 1978 and 2004, but more than half of these—about 480,000—left *after* 2000, signaling the growing intensity and diversity of the outgoing flow. Most of these students went to the United States and Europe, with science, engineering, and business management as the main fields of study. A survey of college students in Beijing by *Chinese Youth Daily* in December 2007 found that an astonishing 80 percent of current college students were interested in studying abroad (Li, Qin, and Mei 2007). Forty-two percent thought that doing so would help their personal development, and 66 percent believed that those students who returned after studying abroad would have better employment opportunities in the next five years. The top choice, once again, was the United States, according to 42 percent of those surveyed. Barring unforeseen international circumstances, the outgoing flow of Chinese students is likely to continue well into the future.

As with the case of Taiwan, the first ten to fifteen years of outflow from China saw little counterflow. The high pay of jobs abroad and the low salaries and limited opportunities in China discouraged students from returning to their homeland. Given that the economy in mainland China was even less developed than in Taiwan in the 1980s and 1990s, and with an authoritarian and oppressive government, mainland Chinese students had little incentive to return. In fact, research in 1987 and in the 1990s found extremely low interest among Chinese students to return home (Hertling 1997; Zweig and Chen 1995). Hertling reported in 1997 that only 4,000 out of 130,000 self-sponsored students who had left China since 1978 decided to return to China, although returning rates were much higher among government-sponsored students or scholars. Zweig and Chen (1995) reported that in the early 1990s, the leading reasons for students not to return to the mainland were perceived political instability and restrictions on political freedom, lack of economic opportunities, and low standards of living in China.

Furthermore, in 1989, the Chinese government's violent crushing of peaceful student movements in Beijing shocked the world and brought tremendous anger and anguish to Chinese students abroad. For many, it only hardened their resolve not to return home. Subsequent to the crackdown, Chinese students in the United States lobbied intensively to remain in the United States indefinitely. Political sympathy from the U.S. government and the public allowed the rapid pass of the Chinese Student Protection Act in 1992, which gave all Chinese nationals—around 50,000 at the time—permanent residency in the United States. Thus, studying abroad in the 1980s and early 1990s for mainland Chinese was practically a one-way ticket to the West. Going home, in the meantime, was regarded as extraordinary, if not unfortunate.

SECOND PHASE: THE EMERGENCE OF THE RETURNING FLOW

In the case of Taiwan, the returning flow started to pick up speed in the late 1980s, as the development of Taiwan's manufacturing industry reached a stage where high-tech, high-value-added sectors came into focus. The Taiwanese government actively recruited talent from the United States. The development of the ICT industry and the strength of overseas talent—in addition to the personal networks among Taiwanese engineers—further augmented the returning flow until it became the norm in the late 1990s (Hsu 1997; Saxenian and Hsu 2001).

The Chinese government, in contrast, had a much harder time enticing overseas talent to return during the 1980s and 1990s. The fact that Chinese students had lobbied and taken advantage of U.S. government protection made it difficult for the Chinese government to assume their political loyalty. Indeed, Zweig and Chen (1995, 15) cite several surveys after 1989 that reported that an overwhelming majority of Chinese students in the United States professed very strong antigovernment political attitudes. It was not until the late 1990s that the Chinese government could begin systematically seeking out the assistance and advice of the overseas Chinese population. Until then, Chinese professionals abroad largely shied away from direct collaboration with the Chinese government, particularly in the aftermath of 1989, although they continued to maintain strong family and personal ties.

The situation began to change in the late 1990s during the heyday of the dot-com rush—and, at the beginning, with little governmental initiative. The Internet boom in the United States made China look like an untapped gold mine for prospective Internet entrepreneurs. With the memory of the 1989 event gradually fading, a few Chinese professionals returned to the mainland to test the business waters. Some returnee-founded enterprises, such as Sohu.com, AsiaInfo.com, and UTstar.com, successfully drew foreign venture capital and achieved wealth and fame within a very short period (Sheff 2002). Their examples inspired a growing number of Chinese professionals to contemplate the possibility that China might, at long last, be ready for private ventures in knowledge-intensive industries.

The bursting of the Internet bubble on the NASDAQ in 2000 did nothing to moderate the flow of Chinese returnees; in fact, the flow has actually intensified. The most recent study (Jiang and Yang 2007) suggests that a total of 160,000 Chinese students returned from abroad between 1978 and 2003, yet 110,000 of these students returned between 2003 and 2006. In an Internet survey of Chinese students abroad conducted by the All-China Youth Federation, China's official association of young people, 87.7 percent expressed a willingness to return to China after studying or working abroad for a few years (*China Youth Daily* 2005). Since the survey was not by a random sample and covers only 3,097 people, the results have to be taken with a grain of salt. Still, the trend of accelerating returns is unmistakable. More than half of those surveyed agreed that returning has become a trend, and only 9.5 percent thought otherwise. The survey also reveals that many returnees (32.7 percent) have found work in foreign enterprises or organizations, and more than 20 percent have chosen to become entrepreneurs. In 2000, Beijing's Zhongguancun—China's largest science park—began to compile statistics on returnee-founded enterprises. These statistics show a steady increase of roughly five hundred firms every year (see figure 14.1).[4] Half of the returnee enterprises were founded by people who had returned from North America, particularly from the United States (46 percent), reflecting that it has been the most favored place to study abroad in the past twenty years. Europe, Japan, and others divide the remaining half equally.

The flow of returnees has been sustained by several forces. First, the growth of China's economy has reached a point where an increasing array of high-tech job opportunities is available. For instance, before 2000, the only job options for a skilled integrated chips (IC) engineer would have been in government research institutes or universities. Since then, both multina-

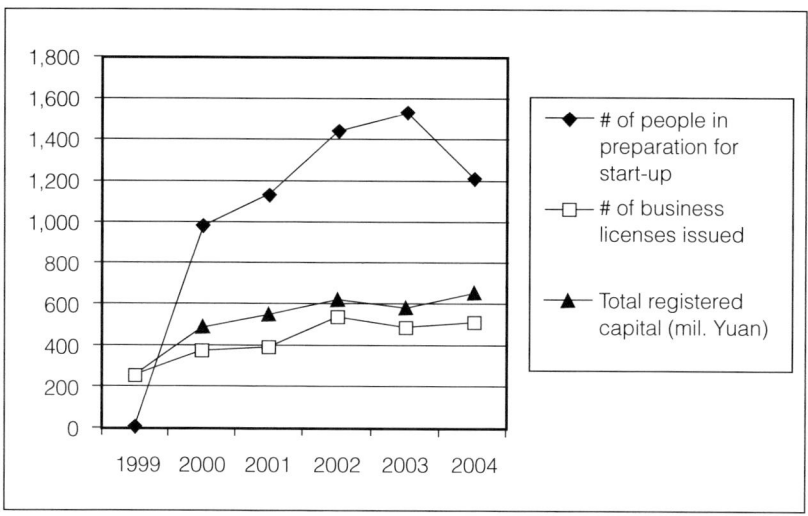

Figure 14.1 Growth of returnee-founded enterprises in Zhongguancun Science Park.
Source: Zhongguancun Administrative Committee; statistics compiled by Headquarters of Returnee Enterprises Services, January 2005.

tional companies (MNCs) and Chinese companies have invested heavily in the semiconductor business, to the extent that China is expected to become one of the world's largest semiconductor manufacturing sites in the near future (Jelinek 2004). As a result, engineering skills in all stages of IC chip production have become highly coveted. Some companies, such as SMIC, are shopping all over the globe for talent. The growing chip-manufacturing industry creates further opportunities for software, IC design, and other specialized products and services. Yet while salaries for highly skilled professionals in China are still low in comparison to the United States, the much lower cost of living in China affords these professionals a comfortable and even luxurious lifestyle that they could not afford in developed countries.

The slow recovery of high-tech sectors in the United States and other advanced countries after the NASDAQ crash and the September 11, 2001, terrorist attacks have also contributed to the increased flow of returnees. Layoffs of personnel in Silicon Valley and elsewhere between 2000 and 2004 made returning more attractive. An entrepreneur explained his choice: "I came back [to Beijing] after 2001. After 9/11, my judgment was that U.S. business would be experiencing several years of lows. It may boom again four to five years later, like the country always does. I figured that I could use these four to five years to develop in China, where market opportunities were expanding. When the U.S. side is growing again and I am strong enough, I might reach back to the U.S. market."

A chief executive officer of another start-up told me that he had worked in California for eight years and had also founded and sold a couple of small firms in Seattle. He had been exploring the China option since 1996. In 2001, after several trips home to China, he felt that the telecommunications sector had generated enough opportunities for his multimedia company and decided it was time to return.

Another reason for the flow home that began in the late 1990s is simply the coming of age of overseas Chinese students. Because the vast majority of Chinese students went abroad from the mid-1980s and onward, many had finished their studies and already spent a decade on their professional careers by the 2000s. Due to racial discrimination and cultural barriers, it has been very rare for Chinese immigrants to reach the higher corporate executive posts in the United States, and even rarer in Japan and Europe. In fact, large numbers of technical professionals face the glass ceiling quite early on, coming to the sad realization that they have reached the peaks of their careers in the West by their late thirties or early forties. Though many do settle on their current track, the more ambitious and restless among them inevitably look for new challenges and excitement. China, as their rapidly developing homeland, holds irresistible appeal. As some returnees started to settle well in China, they drew their friends back, increasing the flow significantly.

Mr. M was among the earliest Chinese students to go to the United States in 1982 with a master's degree in engineering. He earned a Ph.D. in the United States and started to work for IBM in 1987, specializing in semiconductor physics and engineering. After a solid career at IBM, he took an early retirement and returned to Shanghai to work for a major Chinese semiconductor company in Shanghai in 2007. He said that he was responding to the long-standing calls from his graduate school friends, many of whom had already returned to assume the leadership positions in this company. Mr. M was a very active and athletic person in his early fifties. It was clear that he has reached the top of his U.S. career.

Returning to Shanghai—a key hub in China's rapidly growing semiconducting industry—offered Mr. M an entirely new opportunity. Beyond career opportunities, many returnees I interviewed mentioned extended family, friends, cultural affinity, active social lives, and—always—Chinese food as attractions. All these things contrasted with the cultural alienation, isolation, and boredom that many suffered in foreign countries. The China bug hit even the most elite scientists in the United States. Steve Chen, considered one of the United States' most brilliant supercomputer designers, set up a company in China to build a supercomputer after he was unable to secure American venture capital for a similar venture in the United States (Markoff 2004).

To be sure, of all the students who went abroad, only a minority of the Chinese professional population has returned. In 2007, China's own estimate was that roughly more than 70 percent are still abroad (Jiang and Yang 2007). Many professionals are so well established in Western countries that it would be difficult, if not impossible, for them to move back. In particular, it is very difficult for American-educated children who lack fluency in Chinese to be reintegrated back into China. Even for professionals who have returned, it is common for them to leave their families behind in order to accommodate their children's educations or their spouses' careers while maintaining an exit option. This suggests that most returnees are not permanently committed to one side or the other. In many ways, their activities are consistent with those of generations of the Chinese diaspora, who have, over the centuries, woven business networks connecting China, Southeast Asia, North America, and other parts of the world (Ong and Nonini 1997; Weidenbaum and Hughes 1996; Y. Zhou and Tseng 2001).

LOCATIONS OF RETURNEES

Overseas returnees are highly sought after in many Chinese cities, because of their scarcity and because of their perceived technological and managerial superiority. But most returnees prefer large metropolitan centers such as Beijing and Shanghai, as they are rich in job opportunities and offer more comfortable lifestyles. For entrepreneurial returnees, China's national science parks are often the most attractive. Beijing's Zhongguancun Science Park offers RMB100,000 (US$12,500) in grants for quality entrepreneurial projects, providing tax breaks as well as other benefits for start-ups. The local government reports that by 2004 it had issued RMB4.1 million in grants to 509 returnee enterprises (Zhongguancun Administrative Commission 2004). Other cities offer even greater cash incentives. Though the money may not be much in terms of U.S. dollars, it can help small firms significantly, given the low costs of business in China.

However, according to interviews with returnee entrepreneurs, such cash incentives are not the main reason returnees start companies in Beijing. Rather, the most important factors are personal attachments, social networks, and human resources (Y. Zhou 2008). The entrepreneurs tend, disproportionately, to be northerners and people who attended universities in Beijing;

a common sentiment is that "it is only natural to come here, since this is the place I know the best, including the universities and professors." A number of interviewees admitted to looking at other locations such as Shanghai or Shenzhen, which also offer highly developed economies and comfortable lifestyles. Yet they were deterred by the cultural differences between northern and southern China, as well as their lack of personal networks in these places. An Internet entrepreneur explained, "Start-ups need a lot of support and connections. Most of my friends are in Beijing, and they are crucial for me at this stage since many of them are my experimental clients."

It appears that Beijing's concentration of universities has multiple levels of significance for the entrepreneurial returnees. Not only do universities provide all levels of human resources—from faculty to undergraduate students—that firms can use, but, more important, alumni networks also help returnees renew social networks with former schoolmates and professors in Beijing. They thereby provide an entry into the local social network. Shanghai, in contrast, has also become a favored location for professional employment due to its cluster of large Western or Taiwanese transnational corporations in the finance and semiconductor industries. In fact, it has the highest concentration of Taiwanese businessmen in China.

Most start-ups are concentrated in business incubators in Beijing and Shanghai. These incubators are typically affiliated with universities or the local government bodies. They provide discounted office space for one to three years; afterward, the enterprises either move to other locations or stay put and pay the full rent. Zhongguancun (ZGC) International Incubator, for example, is located at the Shangdi Information Industry Base to the northwest of Beijing, which is a modern and well-maintained office building. One wall of its spacious lobby is covered with large displays about firms in the building. The incubator has been managed by a company that uses subsidized rent as venture investment for the firms. The manager of the ZGC International Incubator explained:

> A start-up needs many things. We provide all the services from the initial stages, such as registration, applying for government grants, setting up office space, firm publications, exhibitions, and so on. We also offer consulting services such as marketing and VC [venture capital] training, project evaluation, financial consulting, and human resource recruitment. We are in the process of establishing a business service platform. Since many company founders have technical backgrounds and many were away from China for years, some of them may not

even realize that they need professional help, or they do not know how to approach it here. We can help them throughout the process.

In 2001, there was only one incubator in Zhongguancun Science Park that had the explicit mission of attracting returnee entrepreneurs. In 2005, there were thirteen. In fact, ZGC International Incubator's office building was fully occupied, and while a second building was under construction, its tenant list was already full (interview). Successful incubators like this one have been able to forge a collective identity and community for entrepreneurs. There are social gatherings, salons, lectures, exhibitions, and routine conversations among the tenants through which they share problems and solutions. One entrepreneur told me that his most critical business partners could all be found in the same incubator. The high-profile incubators such as Zhongguancun International Incubator also attract hundreds of visitors each year, ranging from the Chinese president to foreign dignitaries as well as visiting businesspeople and investors, giving the start-ups valuable external exposure.

At another popular incubator affiliated with Tsinghua University, the managers explained some of the unique resources that it had to offer, including a shared laboratory for biotech start-ups that cannot afford expensive equipment and access to the university's labs. The proximity to Tsinghua's campus has meant that faculty and students who work there part-time are only a short bike ride away. Tsinghua's extensive alumni network has been another invaluable resource. As one of China's most prestigious universities, Tsinghua is renowned for graduating not only engineers but also high government officials. For example, China's president since 2002, Hu Jintao, is a Tsinghua alumnus; so was the powerful former Prime Minister Zhu Rongji. Countless Tsinghua alumni now occupy key positions all over China. The director of this incubator told me how she helped arrange a meeting between a start-up entrepreneur at the incubator and a high official of the Bank of China, a Tsinghua alumnus, to discuss possible collaboration between the enterprise and the bank. She also helped connect the start-ups with major government organizations, such as the Beijing Organizing Committee for the 2008 Olympic Games. In a *guanxi*- (connections-) driven society such as China, it is impossible to overstate the value of access to key officials, as they can grant small firms credibility, opportunities, and resources that would otherwise be hard to imagine.

THE BUSINESS STRATEGIES OF RETURNEE ENTERPRISES

Returnee-founded enterprises are hybrids between Chinese and Western businesses, as they are deeply engaged with both the global technology centers and China's innovative resources. Many of the returnee enterprises are miniature transnational corporations, as they maintain foreign branches or headquarters. This is especially true for those founded by returnees from the United States, which are most likely to have their foreign branches in Silicon Valley. Even firms that do not have foreign affiliates usually have close business partners in Silicon Valley. "You have to have a presence in the United States to keep in touch with the most up-to-date trends in the market and technology," one interviewee explained. For business and family reasons, these returnees led an almost astronaut lifestyle, regularly shuttling back and forth between the two sides of the Pacific. The earliest group of returnee enterprises followed the dot-com rush, and mainly sought to replicate the successful U.S. dot-com models. They focused primarily on the Chinese market, using the United States as their capital, technology, and information source.

However, since 2002, returnee enterprises have become more diverse and more transnational in their integration with international technology markets. They now cover a wide range of business areas, but the majority of them in Beijing and Shanghai concentrate on the ICT or biotech sector. The heaviest concentrations are in the Internet, IC design, digital media, and software fields, again reflecting China's fastest-growing fields as well as the traditional human resource strengths of the region. Some returnee enterprises, such as Baidu.com, focus on the Chinese market with little desire to grow abroad. Others are geared more toward serving advanced markets, using China primarily as a research and development (R & D) site while keeping an eye on the potential of the Chinese market. There are no statistics on which types of firms predominate, although my interviews in several incubators suggest that those focused mainly on advanced markets are growing.

BUSINESS OPPORTUNITIES I: CONVERGING AREAS OF CHINA'S MARKET AND EXPORTS

One of the most concentrated fields for returnee enterprises is wireless technology. This is no surprise, as China is the world's largest mobile market. Mobile phones rely on complex technology involving many components, which some returnees have pinpointed as an area where the Chinese market has the potential to lead the world. Vimicro Corporation, established in 1999, is one of the most successful multimedia chip designers in Beijing. The company was publicly listed on NASDAQ in 2005. Its main product is the chip used in desktop computer cameras, for which it held 60 percent of the world market share in 2005. The company is also developing multimedia chips for next-generation mobile phones. Its chief financial officer, who previously worked for several large Western investment banks, explained why he believes that technology from his Chinese firm could have a competitive edge in the global market:

> You need three conditions to compete in the international market:
> 1. Market demand: not only does China have the world's largest mobile-phone market, it is also among the most demanding in designing features for communication and entertainment functions. 2. Supply: the IC chip production chain is already established in China. All major world companies have come into China. All stages of IC production ranging from design, foundry, packaging, testing and others are present here. There is no problem to produce quality chips here. And 3. Human resources: we have U.S.-trained experienced entrepreneurs, professionals, and engineers, including senior engineers from Intel, HP, Lucent, and Kodak. In addition, we have many smart Chinese students from Tsinghua and elsewhere.

His comments underscore the power of synergy between China's domestic demand and export capacity (conditions 1 and 2). The reason a new company such as Vimicro can sharply increase its world market share, he explained, is because most of the world's mainstream PC accessory makers are already located in China. As long as Vimicro can convince manufacturers such as Samsung, HP, Lenovo, Logitech, and others to use its chips in their PC cameras, those chips can quickly become prevalent in the world market.

Yet Vimicro sees its biggest opportunity not in supplying for export but in creating multimedia chips for third-generation (3G) mobile phones in the domestic market. These chips integrate various audio and visual functions. After China started its 3G system in 2008, Vimicro hopes that it will be in a good position to benefit from the extensive growth of this market. Quite a few other entrepreneurs also hope to benefit, and consequently a critical mass of returnee firms collaborating on mobile phone–related technology has already formed, particularly in Beijing and Shenzhen. These firms tend to specialize in designing chips or devising software solutions. They rely on manufacturing facilities in the Yangtze River delta to make their technology into hardware. Most interviewees commented that the capacity for manufacturing IC chips or related devices on the coast has become quite mature: "They can do anything you ask them to do and do it well" was a common sentiment of these IC designers. The convergence of China's market growth with its export manufacturing strength in this area offers promising opportunities for local firms.

On the flip side of the coin, however, returnee firms have also discovered, sometimes rather painfully, that they can run into serious difficulties if they work in areas outside the synchronization zones of the Chinese market or its export specialization. For example, the Chinese market may not even be ready for their advanced products. Many returnee enterprises were originally lured by China's seemingly rapidly expanding information technology market, only to find out that it has not materialized for their products. Considerable adjustments are almost always necessary. And even then, they may not be able to find a market. One entrepreneur, who had returned from Japan, commented, "China has the largest number of home appliance consumers. It is also the world's largest manufacturing base. There should be no better place for firms like us [multimedia-device manufacturers]. But so far, our marketing in China has not been very successful. Most of our products still sell to Japan because the Japanese market is more technologically sophisticated. The Chinese market is more price sensitive, so you have to be cheap to sell in this market. We are working on it."

One returnee, Hu Hui, set up his company in Zhongguancun for $150,000—too small an amount for a venture in Silicon Valley but a significant sum in China. Hu developed a software solution for remote medical diagnoses at Zhongguancun International Incubator, but he could find no buyers in China, nor could he convince Chinese VCs to invest in his firm. He tried to donate the manufactured device with his software to Chinese hospitals during the SARS outbreak in 2003, but the units were never used. Relief finally came from the United States, and an American firm bought his company for the princely sum of $18 million in 2004.

In addition to the lack of a market for advanced products, many returnees have to come to terms with their weak marketing ability. Since these entrepreneurs were typically abroad for an extended period, starting at a young age, they have been unfamiliar with China's technology market and its business culture. Baidu, for example, struggled for some time to gain name recognition, even though it had received U.S. venture capital and Chinese governmental support. It was only after Li Hongyan hired energetic local marketing personnel that the company started to effectively push its brand name and create a profitable business model (interviews). The marriage between the returnees' overseas technology and capital and local marketing and cultural expertise is critical for the success of these firms. But, as with all marriages, fitting partners are hard to come by.

At other times, it has not been the market but the commodity chain that has caused the problem. Since returnees tend to specialize in rather narrow niches, they depend on others to turn their specialization into a product. If China does not have mature and high-quality production partners for a specific product, these returnee enterprises end up running into trouble. One entrepreneur told me that he spent eight months rethinking his business strategies after returning to China: "The commodity chain in China is far from developed, as compared to in America. If you just specialize in your technological niche, it will be impossible to survive here. You have to extend your work up or down the chain. It might be enough for me to just do software in the U.S., but here I have to make it into a piece of hardware, so it is a so-called product. Otherwise, the clients do not recognize the value of your technology. To make these adjustments, I have to have considerably more capital and some business partners."

The experiences of returnee firms suggest that there were some serious constraints on the development of innovative technology in China in the early 2000s. In fact, the Chinese state has put more and more emphasis on autonomous technological innovation since 2003 in an attempt to reduce technological dependency; the encouragement of returnee enterprises is one example. But the state's strategy may not bear fruit if the Chinese market does not support innovative products. Generally speaking, the limited purchasing power of the majority of Chinese consumers means that the market still best supports the inexpensive adaptation of MNCs' mainstream products and has yet to develop stringent demands on quality and features, except in selected areas such as the Internet and mobile phones.

These market characteristics limit the ability of Chinese companies to profit from innovative products in broad areas, thus discouraging investment in them. In China, as elsewhere, the supply of advanced technology per se is not sufficient to create the market. Companies whose products and services are out of sync with China's market and export strengths, whether founded by returnees or not, are going to have a harder time. Thus, it is no wonder that returnee firms tend to thrive in selective areas such as mobile phones and the Internet—areas in which the Chinese market is rapidly growing and the commodity chain is the most well established. These are also the areas where new innovation is most likely to emerge from China.

BUSINESS STRATEGIES II: R&D FOR ADVANCED MARKETS AND WAIT AND SEE FOR THE CHINESE MARKET

An alternative to focusing on the Chinese market is to concentrate on advanced markets and to use China primarily as an R&D outsourcing base. This is a relatively simple operation on the Chinese end of returnee enterprises. Returnees are very familiar with the technical and learning ability of Chinese engineers. Local engineers may lack experience, but they are fast learners and much more willing to put in long hours than their counterparts in America—and at one-tenth of the cost. For Chinese entrepreneurs who are involved in start-ups in advanced countries, it is logical to consider moving part of their R&D operations to China so that their capital can stretch further. Operating in China also gives them opportunities to observe and experiment with the Chinese market.

Here the spatial mobility and networks distinctive to returnee enterprises give them advantages over local firms. First, they are more flexible: they can work on technological accumulation and test their products through serving advanced markets while observing or making adjustments for later developments in the Chinese market. Local firms with advanced technology but no access to advanced markets, on the other hand, are often faced with life-and-death choices if the Chinese market cannot support them. Second, by operating in advanced markets and forming partnerships with large transnational corporations, returnee companies gain credibility that is not usually given to small firms in the Chinese market. "We may be small, but if Sony is using our product, why would you doubt my technology?" said one entrepreneur in the computer-security sector. Third, they have better access to capital and more funding options than local firms do. As mentioned earlier, overseas VC activity in Beijing has grown considerably in recent years, and VC firms find dealing with returnees much easier because they share a similar business language and culture. The *San Jose Mercury News* reported in February 2006 that venture capital from California has surged in China in recent years, and that 70 percent of Silicon Valley investment in China is in firms with executives from the Bay Area (Ha 2006). Many Chinese start-ups hire overseas returnees as managers to communicate better with VC firms.

In short, these returnee-founded firms adopt a strategy of *exporting* China's human resources, but their focus is on skilled labor for R&D, rather than on unskilled labor in China's coastal regions. Generally speaking, China is lagging behind India in exporting services and the software industry. But this does not signify that Chinese professionals lack technical skills. In fact, their much weaker proficiency in English is what makes them less competitive than their counterparts in India. This disadvantage cannot be overcome in a short period, and it is likely that China will never be as competitive as India in English-dominated software or service export markets. However, in technological areas where language communication may not be as crucial—and returnee firms are experts in identifying these areas—Chinese engineers can still be used to serve the global market. Also, operating in China helps these returnee firms observe and adjust to the Chinese market, which is often their ultimate goal. Though most of these firms are currently out of sync with China's market, it won't be long before some of them determine how their core competencies intersect with Chinese consumer demands so that they can enjoy rapid growth.

THE SEARCH FOR CAPITAL

Besides marketing, another critical barrier that returnee companies face in mainland China is capital. Most returnees find that they can count on only limited access to domestic capital. But start-ups in innovative technologies require considerable capital backing, and China's state-dominated financial infrastructure has been inept at providing capital for these companies. While the Chinese government has helped fund a few VC agencies through government financial institutions, these agencies have either a small financial base or little experience in high-tech venture-capital operations. One returnee entrepreneur told me, "It is weird. Chinese VCs do not look at your growth primarily; they look at whether you can survive. If you can survive, they might give you money. American VCs look at the possibility for growth."

The behavior of Chinese VCs is not surprising, given the institutional culture that comes with being state-owned agents. They are very concerned with risk, but not so much with profitability. As long as the firms they are supporting are surviving, it looks good on paper, and good for their supervisors. Whether these firms can generate a high return for the investment is not their immediate concern. Local VCs are also not qualified to make judgments on the prospects of a particular piece of technology in the international market.

Overseas VCs, in contrast, function based on the promise of potential growth and significant share-value increases in the international stock market with initial public offerings. While their influence in China's high-tech parks is growing, it is difficult for them to serve the diverse needs of local firms as these overseas VCs lack local roots. In terms of capital sources, start-ups in China have had a very different experience from Hsinchu Science Park in Taiwan, where local capital was the primary driving force for new enterprises (Saxenian and Hsu 2001). The manager of the Zhongguancun International Incubator told me that international VC firms are very active there, even if they do not yet have a permanent presence: "They have contacted many firms in our park, as the returnees have open communication channels with the international VC community. The homegrown VC firms, on the other hand, are rather conservative. We have made a lot of recommendations to push our companies, but nothing much has come of them."

However, some returnee enterprises have received Chinese state capital. A few select ones are able to receive substantial investment if the founders can successfully develop relations with the government. Vimicro, for example, received RMB10 million from China's Ministry of Information Industry (Xia 2004, 53). In 2005, it also won a top national prize for progress in science and technology. It is often held as an example of the successful synergy between the Chinese government and returnee entrepreneurs. But, to date, it has been a rather unusual case, since government investment still tends to go to state-owned institutions rather than private enterprises founded by returnees. For the majority of returnee enterprises, support from the government is available but limited, mostly in the form of state-sponsored incubators, small start-up grants, bank guarantees for small loans, rent breaks, and other limited subsidies. These are helpful in the beginning stages of the start-ups but are of little use for sustaining further development.

Perhaps the best hope for local enterprises is the growing breed of new local VCs who were previously entrepreneurs or professionals in the ICT fields in China or elsewhere. Lenovo, for example, operates a successful VC subsidiary. Some entrepreneurs, such as Zhou Hongyi, the former Yahoo China executive and longtime ZGC resident, have also become involved in VC investments for local firms. Some of the most active VC investment firms in China were founded by returnees who had been entrepreneurs in Silicon Valley (Ha 2006). For example, Hong Chen, the chairman and CEO of the Hina Group, was the founder of two technology companies in California. The Hina Group now has offices in Beijing, Shanghai, and California and specializes in investments in ICT sectors in China (http://www.hinagroup.com/abouthina.htm). One of the largest and most influential Chinese high-tech business associations, the Hua Yuan Science and Technology Association (HYSTA) in Silicon Valley has more than three thousand members, with chapters in Beijing and Shanghai. The three former chairs of the HYSTA were all entrepreneurs in the valley, and each is now involved in the venture-capital business in China (Du 2007).

This new breed of VCs is more locally rooted than mainstream VCs or investment agencies from abroad. They are also far more professional and commercially aggressive than their predecessors in locally grown VC firms, which had often been moved from state banks to VCs on government assignments. Yet most of these new VCs still have very limited access to locally generated capital. The lagging development of China's capital market has clearly imposed a major constraint on the availability of capital tools for the development of innovative enterprises.

Despite the current interests of global capital, the sustainable development of Chinese technological firms still depends on reforms in China's financial sector, especially in creating ways to encourage the investment of local capital in innovative industries. The situation, however, is changing quickly. China's stock exchange in Shanghai has experienced rapid growth since 2006, and was among the world's best-performing markets in 2007, signaling the potential of domestic capital. A smaller NASDAQ-style stock market was established in Shenzhen in 2009. Although one should not expect the overall situation of capital shortage for start-ups to change in the immediate future, one can safely expect that the capital provision will gradually improve over time.

CONCLUSION

While the Chinese immigration flow to the United States between the 1960s and 1980s was a clear case of brain drain, the returning flow has been established so that many Chinese immigrants return to their homeland after some time. Immigrants from Taiwan were the first to participate in what experts have coined *brain circulation* (Saxenian 2006), with immigrants from mainland China to follow. With their deep linkages with both the American and the Chinese technical worlds, the Chinese digital diaspora is making a monumental impact on Taiwan's and China's technological industries by creating synergetic fields of capital, technology, information, and marketing on both sides of the Pacific. The emergence of high-tech and innovative industry in Taiwan and China is in turn redrawing the map of global industry.

Compared to Taiwan, the out- and inflow of mainland Chinese professional migration occurred with a fifteen- to twenty-year lag, but with a larger population and more sustained timeline. While the returning flow to Taiwan has become well established, and their impact on the global industry clearly demonstrated, the returning flow to mainland China is only at the beginning stage. Although a critical mass of returnees in high-tech industry has been reached in Beijing and Shanghai, their success in creating and sustaining innovative industry in China is yet to be ensured.

Many obstacles continue to plague the returnee entrepreneurs, whether these obstacles are capital, market connections, or making necessary adjustments to the Chinese business culture. Many returnees have been taken aback, at least initially, by the extent of the cultural gulf they have encountered with homegrown entrepreneurs and managers. Some returnees, disappointed with the low business rewards or unable to tolerate the cultural shock or prolonged family separation, have gone back to the United States. This phenomenon suggests that the returnee flow, although robust now, is vulnerable to the shifting political and economic situation in China and other parts of the world. Continued growth of the Chinese economy could encourage returns, but a deteriorating economic situation could inhibit or stop them at any time. In the end, the contribution of returnees to China's development will be determined by China's ability to encourage and sustain innovation and its ability to foster a tolerant but also rewarding political, social, and economic environment for the returnees.

NOTES

1. This article draws from the author's book, *Inside Stories of China's High-Tech Industry: Making Silicon Valley in Beijing* (Lanham, Md.: Rowman and Littlefield, 2008). I would like to thank the National Science Foundation, Chiang Ching-Kuo Foundation, and Vassar College for their financial support for the research.

2. The oldest of the Soong sisters, Soong Ai-Ling married H. H Kung, who was once premier of the Republic of China. Soung Ch'ing-Ling, the second daughter, was the wife of Sun Yat-sen, the founder of the Republic of China. Soong Mey-Ling, the youngest Soong sister, was the wife of Chiang Kai-shek, the successor of Sun Yat-sen.

3. Even after twenty years of sustained income growth in China, in 2005 minimal living expenses for one year in the United States ($12,000) equaled five years' total earnings for an average resident of Beijing. And those living expenses did not include tuition and other college costs. In 1995, the cost of an overseas education could very well have equaled the entire lifetime earnings of an ordinary urban Chinese citizen.

4. In 2004 there was a decline in the number of people said to be preparing to launch start-ups. It is possible that the recovery of the high-tech sector in the United States and elsewhere that started in 2004 might have prompted fewer people to return to China. But other indicators of new start-ups, such as the number of business licenses issued and the amount of registered capital assets of returnee-founded firms, were slightly higher in 2004 than in previous years.

REFERENCES

Barboza, David. 2006. The rise of Baidu (that's Chinese for Google). *New York Times*, Business sec., September 17. Available at http://www.nytimes.com/2006/09/17/business/yourmoney/17baidu.html?pagewanted=1&_r=1.

China Ministry of Education. 2001–2005. *Yearbook of Ministry of Education*. Available at http://www.moe.edu.cn/.

China Youth Daily. 2005. Haiwai Liuxue yu guiguorenyuan xianzhuang dadiaocha jieguo fabu. Press release of the Survey Report of Overseas Students and Returnees, no. 18 (December). Available at http://www.cunews.edu.cn/html2006/jyxw/083803809.htm.

Du, Chen. 2007. Guigu guilai [Returning from the Silicon Valley]. Editorial. Vol. 223. Available at http://www.ceocio.com.cn/store/detail/article.asp?articleId=13356&Columnid=3049&adId=10&view=.

Ha, K. Oanh. 2006. Valley VC firms boost bets on China: Talent, money flow east to huge market. *San Jose Mercury News,* Local news, February 13. Available at http://www.mercurynews.com/mld/mercurynews/news/local/13861269.htm.

Hertling, James. 1997. More Chinese students abroad are deciding not to return home. *Chronicle of Higher Education* 43 (March 28).

Hsu, Jinn-Yuh. 1997. The historical study of Taiwan's integrated circuit industry—high technology, state intervention, and returnee entrepreneurism. *Journal of Geographical Science,* no. 23: 33–48.

Jelinek, Len. 2004. China is catching up to leading-edge technology. *Solid State Technology* 47, no. 2: S20–S22.

Jiang, Dan, and Xiaojing Yang. 2007. Zhongguo Liuxue rencai anquan de xianzhuang yu zhengce fenxi [Security implications of China's overseas talents and policy analysis]. In *Zhongguo Rencai Fazhan Baogao* [The report on the development of Chinese talents], ed. Chenguang Pan. No. 4. Beijing: Social Science Press.

Li, Tao, Zhou Qin, and Han Mei. 2007. Liuxue, Neng wei ziji yingde weilai ma? [Can study abroad win the future for you?] *China Youth Daily,* December 7. Available at http://campus.cyol.com/content/2007-2/07/content_1986475.htm.

Liao, C., and M. Tang. 1984. *Research and analysis on the employment of the returned scholars and students.* Taipei: National Youth Commission.

Markoff, John. 2004. Have supercomputer, will travel. *New York Times,* November 1, C1–C4.

NYC. 1987. *A helping hand to overseas scholars for their service at home.* Taipei: National Youth Commission.

Ong, Aihwa, and Donald Nonini. 1997. *Ungrounded empires: The cultural politics of modern Chinese transnationalism.* New York: Routledge.

Ong, P., E. Bonacich, and L. Cheng. 1994. *The new Asian immigration in Los Angeles and global restructuring.* Philadelphia: Temple University Press.

Saxenian, Anna Lee. 2006. *The new argonauts: Regional advantage in a global economy.* Cambridge: Harvard University Press.

Saxenian, Anna Lee, and Jinn-Yuh Hsu. 2001. The Silicon Valley–Hsinchu connection: Technical communities and industrial upgrading. *Industrial and Corporate Changes* 10, no. 4: 893–920.

Seagrave, Sterling. 1996. *The Soong dynasty.* London: Corgi Books.

Sheff, David. 2002. *China dawn: The story of a technology and business revolution.* Vol. 1. New York: Harper Business.

Tsai, S. H. 1986. *The Chinese experience in America.* Bloomington: Indiana University Press.

U.S. Census Bureau. 2002. *2000 Census of Population.* Washington, D.C.: U.S. Bureau of Census.

U.S. Department of Justice. 1991–1996. *Statistical yearbook of the immigration and naturalization service.* Washington, D.C.: U.S. Government Printing Office.

Weidenbaum, Murray, and Samuel Hughes. 1996. *The bamboo network: How expatriate Chinese entrepreneurs are creating a new economic superpower in Asia.* New York: Martin Kessler Books.

Xia, Y., ed. 2004. *Haigui qiangtan Zhongguancun* [Overseas returnees' innovation in Zhongguancun]. Beijing: China Development Press.

Zhongguancun Administrative Commission. 2004. Annual report of Zhongguancun Science Park. *Touzi Zhongguancun—'Hu Hui xianxiang' yantaohui shilu* [Investment in Zhongguancun—Huhui phenomenon. Discussion panel transcript]. Available at http://gov.finance.sina.com.cn/zsyz/2004-07-20/16782.html.

Zhou, M. 1992. *Chinatown: The socioeconomic potential of an urban enclave.* Philadelphia: Temple University Press.

Zhou, Y. 2008. *Inside story of China's high-tech industry: Making Silicon Valley in Beijing.* Lanham, Md.: Rowman and Littlefield.

Zhou, Y., and Y. Tseng. 2001. Regrounding the "ungrounded empires": Localization as the geographical catalyst for transnationalism. *Global Network* 1, no. 2: 131–54.

Zweig, D., and C. Chen. 1995. China's brain drain to the United States: View of overseas Chinese students and scholars in the 1990s. Berkeley: Institute of East Asia Studies.

15 "Cybernaut" Diaspora
Arab Diaspora in Germany

KHALIL RINNAWI

The proliferation of satellite broadcasting and new transnational media technologies has become accessible to the Arab diaspora, allowing diasporans to reintegrate into their homeland's life and society.[1] This situation raises the important questions of how these new media work and what their implications are for the relationship between the homeland and the diaspora (Morley and Robins 1995). Visits to refugee *(Azulheim)* buildings in Berlin during the early 1990s revealed that one of the most important and essential pieces of electronic equipment owned by refugees was a decoder connected to a satellite dish outside the window of their respective rooms. The refugees told me about this equipment, and how television connected them to their homelands. Despite the low socioeconomic status of most of the refugees, a significant percentage of them owned this electronic equipment. As the refugees moved from the *Azulheim,* they took this equipment with them as a continuous link to their original homelands.

Similar to voluntary migrants, refugee populations (e.g., asylum seekers) desire to maintain links with their countries of origin. Transnational networks play an essential role for both groups of migrants. Nevertheless, this information and its implications are largely ignored by the academic literature (Breidenbach 2001). Since the mid-1980s, transnational Arab media have played a principal role in allowing Arabs in the diaspora (and also Arabs in the Arab world) to engage in a new multidirectional *hiwar* (dialogue) with the Arab world, through cultural, political, news, and current affairs issues.² Transnational Arab media have been more complexly imaginative and effective than the rest of the Arab world in "reengaging" Arabs in the diaspora due to the technical expertise, ideas, and human resources of diaspora Arabs.

Particularly since the beginning of the 1990s, Western-based Arab transnational television has become an integral part of Arab life, inside the Arab world and in the diaspora. One of the most fascinating results of Arab transnational media, and particularly TV, is the extent to which it has allowed for the reintegration of Arab immigrants into Arab life and society (Alterman 1998; J. Anderson 1997). It has the ability to reunite communities scattered by war, exile, and labor migration, while also tapping into the talent they offer. There exists a "virtual" TV community of Arabs based all over the world, receiving, engaging with, and reacting to the same information *simultaneously*.

ARAB SATELLITE TV'S EMERGENCE

In 1990, the Middle East Broadcasting Centre (MBC) (privately owned by Saudi Arabia) created the first Arab satellite broadcasting network (Graieb and Mansour 2000). In October 1993, the creation of two additional privately owned Arab satellite television broadcasting systems, ART and Orbit TV, then both broadcasting from Italy, dramatically accelerated the spread of Arab satellite TV broadcasting. Arab satellite television entered a new phase after 1996, marked by the arrival of the new players from Lebanon: the Lebanese Broadcasting Corporation International (LBCI), Future TV, MTV, and New TV as private stations, and al-Manar, a television station owned by the Lebanese Shia party Hezbollah. In 1996, the al-Jazeera Satellite channel also set a new record, as the first Arab all-news and public affairs satellite channel (Rinnawi 2006; Mellor 2005). At that time in the North African countries of Morocco, Algeria, and Tunisia, national television stations started to broadcast via satellite in order to expedite communication between expatriate labor in Europe and the home country. By the end of 2000 every Arab state in Asia, the Persian Gulf, and North Africa had its own satellite television station.

Presently, there are approximately 370 Arab satellite TV stations. Major channels are directed at national markets such as Saudi Arabia and the Gulf State, as well as the rest of the Arab world (Arab Advisors Group 2007). Beyond them, there are many subnational and ethnic-language channels aimed at specific Arab communities transplanted from their original location to sites across the Arab world or in the diaspora (Rinnawi 2006). Some Pan-Arab satellite channels have specialized in business, entertainment, film, news, religion, educational programming, children's programming, and women's channels.

THEORETICAL FRAMEWORK

The emergence of transnational media has made an important impact on the relationship between local societies and nations and their global diasporas (e.g., see Morley and Robins 1995; and Naficy 1993). Quantitatively, there is an increasing degree of intensification of connections and communication between groups. Qualitatively, one may consider the impact of a new intensive relationship upon diasporas that affects their relationship(s) with hosting societies, and the development of their collective identities and other sociocultural mechanisms (Curran and Park 2000). For example, Korean Americans may watch Korean TV on cable TV in California, just as Indian British viewers may watch Indian films on cable in London. This transnational media offers diaspora communities constant engagement with their original homelands, sustained through daily input. Importantly, these daily inputs do not necessarily originate from specific places, towns, cities, or villages of origin but rely on a greater sense of abstract identity to engage diasporas. Transnational media also encourage diaspora communities to identify with *pan* rather than *local* identities.

Little research has been done on the Arab diaspora and its media consumption. My investigation of the literature revealed only three publications, all concentrating on the Arab diaspora in the United Kingdom. The first is the study of Noureddin Miladi (2006), who investigated how the Arab diaspora in Britain use and interpret Arab satellite TV broadcasts in comparison with their response to the BBC and CNN. The second is the study of Zahera Harb and Ehab Bessaiso on the Arab Muslim diaspora audiences in the UK in regard to TV after September 11th (see Curran and Park 2000). The third is Dina Matar's (2006) study on the Palestinian diaspora in the UK and their reactions to the September 11th attacks and the media reports on the event. The main conclusion of the three studies is that the diaspora's members have heavily consumed Arabic satellite TV since its inception, particularly since they mistrust the local British and Western TV programs, because they perceive them as biased and one-sided against Arabs and Islam.

In the case of Arab diasporas in general, an essential implication of transnational media is the strengthening of ties and relationships to the Arab world on real and virtual levels. These can be viewed within the framework of the Andersonian (B. Anderson 1993) "imagined community," the revival of "Pan-Arabism," the creation of a "New Pan-Arabism" (Alterman 1998; J. Anderson 1997), or, in Rinnawi's (2006) terms, "McArabism." The emergence of sociocultural, religious, and political discourse among Arab diasporas through Arab transnational TV reveals the creation of an Andersonian "Creole" discourse for Arabs (B. Anderson 1993). Where the effects of regionalization initially limited the scale of identification and discourse to an older Andersonian conceptual model of the "imagined community," now the creation of community participation on a transnational level has resulted in a new virtual nationalism, *McArabism*. This McArabism extends beyond the traditional boundaries of the nation-state (or, for most of the Arab world, state-regime) and includes the Arabs in the diaspora.

McArabism

Benjamin Barber (1992) argues that *Jihad* (localization and tribalism) and *McWorld* (globalization) are two major opposing trends at work. In this context *Jihad* does not refer to its more widespread meaning in the West as a "holy war," nor to its literal meaning in Arabic as "struggle" or "effort," which is primarily directed internally against the powers of the ego. The penetration of new media technologies into the Arab world and their expansion via the transnational media have created a confrontation between the localism and tribalism of Jihad and globalization forces, which Barber terms "McWorld." The outcome of this confrontation in the Arab world is McArabism: a kind of instant nationalism quite different from the traditional Pan-Arabism(s) formulated during the 1950s and 1960s in the Arab world. McArabism is a product of interaction between new media technologies and local trends and powers. Without a clear ideology, spokesperson, or political representation, it is vulnerable to external influences and manipulation. Moreover, new media technologies allow it to bypass central or state political, social, or cultural agencies. McArabism is evident in political, sociocultural, and religious spheres (Rinnawi 2006).

The emergence of McArabism is accomplished via four main components or processes, which are complementary to one another: intensification of a shared stance on issues, unification of language, direct engagement, and emotive footage or cultural sensationalism.

"Intensification" is the use and broadcasting of (Pan-Arab) media content expressed on several levels (Rinnawi 2006). In news broadcasts "intensification" refers to the dramatic increase in the frequency and selection of newscasts, which are broadcast several times daily. This is seen in the case of the "normal" TV stations like MBC or LBCI and every hour on the all-news television stations such as al-Jazeera and ANN. "Intensification" also refers to addressing issues on transnational TV, which are of shared concern to a Pan-Arab audience. Such issues include crises between Arab states in the world and non-Arabs, problems facing the Arab world as a whole, internal problems and trends in the Arab world, and political, social, and cultural phenomena in Arab societies.

In Pan-Arab and Islamic programs the process of intensification that has led to McArabism has been achieved through the broadcasting of historical, educational, and political programs that aim to educate specifically Arab audiences regarding Arab and Islamic world history, religious or social development, or other various issues. Virtually simultaneous broadcasting encourages audiences to engage in symbolic events, developments, and milestones of Arab Islamic culture, which motivates the viewer to find relevance in his or her own life and to re-create or emphasize the viewer's worldview based on his or her ethnic or religious identity.

"Intensification" is also achieved through entertainment programs such as movies and dramas that have traditionally been popular on Arab television screens. Alongside the traditional fare of Egyptian movies and Syrian soap operas, it has also encouraged a new Pan-Arab music market, with regular music shows and a new emphasis on video clips.

Part of this component of "intensification" derives from a "shared stance" on Pan-Arab and Islamic issues or crises like the Palestinian Intifada, the Arab-Israeli conflict, the Iraqi crisis with the United States, or Osama bin Laden and al-Qaeda in Afghanistan. Thus, a regional "Pan-Arabist" dialogue among intellectuals has begun to emerge, not only in such regional Arab newspapers as *al-Hayat* and *al-Sharq al-Awsat* but, more important and effectively, through satellite television stations like LBCI, ANN, and al-Jazeera (Alterman 1998).

The second component is "unification of language." Whereas the Arab world has traditionally been divided by a plurality of dialects, transnational TV broadcasts news and serious programs in Modern Standard Arabic. This is based on classical Arabic, with a simplified grammar system and the inclusion of contemporary terms in their Arabicized form.

The third component, "direct engagement," refers to the use of modern styles of news and broadcasting that allow the audience to understand the news with minimal state intervention. This includes using different broadcasting effects and techniques such as live broadcasts, figures and maps, and the use of reporters stationed in both Arab and non-Arab countries. The broadcasts provide in-depth reports on various issues and conduct interviews with individual people, leaders, or groups representing different points of view.

The fourth component is "emotive footage," which is also described as "cultural sensationalism" of the news and documentary programs through the use of various kinds of rhetoric in reporting language as well as pictures, style of presentations, and other effects (Ayish 2002). Importantly, the process of allowing viewers to see (often live) footage is clearly intended to provoke an emotive effect and causes audiences to experience deeper forms of engagement. Beyond a rational level of acknowledgment or sharing issues of concern, the use of emotional footage positions the audience member in an imagined community, and enables him or her to become a participant viewer of a part of his or her community onscreen.

As suggested through this discussion, McArabism requires Arab audiences to be exposed simultaneously to identical content in order to enjoy opportunities for interaction. Just as important, McArabism is achieved not only through media content but also through the medium itself. The medium is a vital element in this process, as it gives Arabs in different locations a greater opportunity to engage with the content. This element helps create a collective discourse that raises issues that have meaning to all Arabs. Consequently, Arab diasporas all over the world have become "virtually united" as a part of this imagined community created by McArabism through exposure to Arab transnational TV channels.

METHODOLOGY

In light of the theoretical background given above, the purpose of this study is to examine the television consumption habits of members of Berlin's Arab diaspora, according to gender and generation. More important, this study also seeks to examine the implications of Arab transnational TV for this community and determine the effect of new media technology and its role in the relationship between the Arab diaspora and the Arab world.

An in-depth qualitative analysis of media use was conducted among members of Berlin's Arab refugee community. Twenty families participated in face-to-face interviews, which were held during the summer of 2002. Participants were randomly chosen. By chance, all participating families were Palestinian or Lebanese refugees. A primary criterion for participation was that the potential families owned a satellite dish that allowed them to view Arab satellite TV channels. I personally carried out all interviews. All participating families were composed of an Arab mother and father with children.

That is, the research did not include families where one parent was non-Arab. Interviews were composed of open-ended questions that I prepared previously. A pilot interview session was held with two families to test the validity and appropriateness of the interview questions.

The interview session was divided into two parts, the first a one-hour interview conducted separately with each member of the family, parents and children. The second part involved a discussion-type interview that involved all members of the family simultaneously. Data obtained during the interview was hand recorded by the research assistant, since most of the participants were suspicious of tape recording the interview.

Interview questions were divided into two sections. Questions from the first section intended to obtain background information about the family, including the circumstances that led to their immigration into Germany and their current status. The second section involved the media consumption habits of the family, especially viewing television. This section itself was divided into several smaller parts, focusing on general TV consumption: consumption of German media and Arab transnational channels in detail, including what was watched, when, and why. This part was extremely broad and expanded.

Background of Participating Families

All participant families sought political asylum in Germany between the early 1980s and the early 1990s following the war in Lebanon and political crises in Syria. As refugees may not legally work, although some do, they usually have a low standard of living and plenty of free time. Consequently, the selected families (all have children) receive social welfare and financial assistance from the German government. Most families interviewed were unfamiliar with the other participants and tended to be scattered throughout Berlin. The majority of participating families are Muslim by faith.

The Arab Community in Berlin

The Arab community in Berlin may be considered a "new" community, especially in comparison to the Turks who have been in Germany since the 1950s and 1960s. The majority of Arabs are of Palestinian or Lebanese origin, and most of them came to Germany as refugees. A further distinguishing feature of Arab communities in Berlin, particularly since the 1990s, is their high degree of politicization, which relates directly to cultural and ideological divisions and struggles within the Arab world. The force of those cultural and ideological dynamics has been reinforced by the consumption of Arab media among this community. Images of daily events in the Arab world have served to strengthen the importance of their world as a cultural and political reference point.

There is no exact number of how many Arabs are in Germany. Unlike the Arab communities in other parts of Europe, there is no real Arab community in terms of local supporting structures dealing with culture, welfare, and so forth. Being of refugee status, they are a social minority that does not truly fit in. In Berlin, it is easy to locate neighborhoods composed of "foreigners." They usually live in poor areas, and the rooftops of their houses are filled with satellite dishes, as if a sea of satellite dishes has overtaken the neighborhood. All Arab homes visited during the interviews were furnished and decorated in typical Arab style.

FINDINGS

Interestingly, the most noticeable feature in the households where the interviews were conducted was a large television located in the family living room. Regardless of the financial and economic situation and living standards of Arabs in Berlin, nearly all the participating families owned several televisions, typically one in the living room as well as one in each bedroom. The television in the living room, just as sometimes the one in the master bedroom, was usually connected to Arab satellites. The remaining televisions in the house tended to be connected only to German TV. From this, it can be suggested that although the usual habit in Arab homes is to view television collectively, here television had generational-cultural cleavages. Through interviews, it was clear that the mother and father usually watched Arab television in the living room. Under normal circumstances the children typically viewed television in their bedrooms.

TV Consumption

Hours of consumption: All families interviewed mentioned that television, not written media, was the sole source of media information. This is due to several reasons. Primarily, there is no local written media in Arabic in Berlin. Even if Arab newspapers were available, in most cases the parents are illiterate or are not accustomed to reading newspapers. The children also do not read Arabic fluently, if at all. While school-aged children mentioned that they sometimes read German newspapers, most are illiterate in Arabic. Participants reported that they were heavy television viewers, especially the parents, who viewed a minimum of six to seven hours of television per day, typically during the afternoon and evening hours when they had nothing else to do.

From VCR to satellite: The Arab families interviewed all purchased and connected to the Arab satellites during the early 1990s (between 1992 and 1994), during the same period that the Arab satellites were established. As new satellite stations were brought into operation, they connected to them. Significantly, the families revealed that in the years before satellite Arab programs became available, the television located in the living room—the main television—was also hooked up to a VCR, and this was their means of connecting to Arab media. During this period, a large number of video stores throughout Berlin carried Arab cassettes. The families interviewed said that at that time they would watch whatever they could obtain in Arabic. In other words, their televisions "spoke" Arabic nearly all the time. Thus, during this period prior to the birth of Arab satellites, most of the viewing was done on videos. Since they spoke no German, the parents watched the German satellite only when there was something they could understand or comprehend, preferably in English. In fact, some of the families revealed in their interviews that at that time not one Arab video available was left unseen. For example, one participant stated, "In the first five years after our arrival to Berlin till our penetration to the Arab satellite TV we have seen all the Arabic video cassettes available in the video cassette stores in Berlin."

Viewing TV: One of the most important findings that emerged in every single interview is that the parents watched Arab satellite TV intensively on a regular basis. This point is clearly made by this participant's statement:

> When I first walk into the house from outside and turn on the Arab satellite channels, preferably the news channels, I feel like I have been transferred from the German world to the Arab world. I watch the Arab news channels and when the news is finished I watch channels that are showing programs related to the news I have just heard, such as talk shows, interviews, or documentaries, even if they are re-runs and I have seen them before. When I watch the news, especially the Palestinian and other Pan-Arab news that concerns me the most, it is a type of compensation for my not being there. The news makes me feel a part of my home again because I at least know what is going on.

This was also evident in the findings of Miladi (2006) in the case of the Arab community members in Britain.

Arabic or German? The parents viewed a majority of Arab channels and only to a much lesser extent viewed German channels, primarily because they did not understand German well, and did not feel the same cultural affiliation. The situation was the reverse for the children, who viewed more German programs and less Arabic. Importantly, the parents also felt that Arab channels had more credibility regarding news on Arab issues, as opposed to the German channels. Regarding this point, Miladi (2006) and also Harb and Bessaiso (2006) made the same findings among the Arab diaspora in Britain. For children, viewing Arabic channels was unattractive, particularly for news and current affairs, as a formal dialect is used, which differs from the local dialect of their parents' country of origin, and makes it difficult for these children to understand and remain interested.

Gender differences: Most women interviewed were housekeepers and spent the majority of their time within the home. When they did make contact outside the home, it was usually with other Arab families. Most women were not fluent in German. This greatly affected their daily lives, including patterns they formed in media consumption. The women interviewed did not view German television, and they relied on Arab TV—at the beginning through videocassettes. One female participant stated, "I watch a lot of television every day, anywhere about eight hours, but a small amount of it is German and the majority of it is Arabic." Her husband said, "Before fifteen years ago when we first arrived in Germany, she would simply sit in front of the television and stare at the screen, not understanding anything being said. I brought her Arabic video cassettes until the Arab satellites became available."

Arab men responded somewhat differently. The men tended to be more integrated into German culture, whether through education or work. They socialized more with Germans, and most speak the language, although many cannot read or write. They were able to watch and enjoy German television. However, like their female counterparts, in addition to the German TV they consumed, they also watched Arab TV, especially news and Arabic talk shows. One participant clearly makes this point: "There is not one newscast or talk show that I will not watch on the Arab channels, even if it is on the hour." Another participant also stressed this point, saying, "I watch a great deal of news; as long as I am at home I watch Arabic news."

Satellite penetration: The majority of Arab families connected to Arab satellites are connected either to one or both of the following satellite broadcasting companies: ArabSat and HotBird2, both digital. These two satellites broadcast the majority of the Arab channels, especially the "big" players—

al-Jazeera, Abu Dhabi, MBC, LBCI, ANN, Al-Manar, Syria, Egypt, and other Lebanese satellite channels.

A Link to Authenticity: TV Viewing Patterns and News Credibility

Preferred Arab channels: Research findings divide the patterns of viewing into three categories: news and current affairs, religious, and entertainment. Differences were found among the participants according to gender.

News: The preferred channel regardless of gender was unanimously al-Jazeera. It was considered to have the most credibility and the most comprehensive news. Matar (2006) reaches the same conclusion among her Arab diaspora interviewees in Britain. There also it was thought to provide news quickly (often virtually simultaneously with real life), in an up-to-date fashion, and reliably. One Palestinian participant from Syria said, "The news I like most is al-Jazeera because it is realistic, quick, and has a lot of scoops." Another Palestinian from Lebanon stated, "I only rely on al-Jazeera's news because it is the only true and accurate news agency that gives news for the sake of news and not for something or someone else. That is, important news pertaining to the people and not only to what the leaders are doing."

Following the al-Jazeera news channel, three other channels were also reported to be viewed by those interviewed: al-Manar (the Lebanese Shia Hezbollah TV channel), ANN, and Abu Dhabi Satellite TV. These three also have updated and accurate news similar to al-Jazeera, especially on the Arab world and the Intifada. One participant from Lebanon said, "After al-Jazeera, I like to watch al-Manar, ANN, and Abu Dhabi, who broadcast the best news about the Arab world, especially the Palestinian Intifada." A Palestinian from Syria said, "I view ANN and al-Manar a great deal because they provide complete, accurate, and up-to-date news about the Arab world and the Palestinian issue, which we cannot find in other news channels."

Religious: One of the most important facts coming out of our interviews is that there is substantial consumption of Islamic religious programs among our interviewees. Most of them declared that Arab TV, like Iqra' and Monajah TVs, was very instrumental in connecting them to Islamic beliefs, because religious information was vital to them in their daily life. Among men the viewing of Islamic religious programs is a part of their daily viewing habits, and they pay special attention to the speeches of religious sheikhs and broadcasts of Friday prayers, as well as historical TV series and other programs that present the Islamic heritage. This is clearly evident from this male participant's statement: "I enjoy watching these TV movies and dramas that present the 'real' heritage and novelty of the Islamic Empire, which strengthen our religious self-esteem, as well as those programs which teach us how to behave as a good Muslim."

Some of the women interviewed also viewed the channel Iqra,' a semi-private channel from Saudi Arabia that broadcasts Islamic religious issues. "I like to watch the shows on the channel Iqra' because it keeps me all the time in the core of Islam as a way of life here in a Christian society. It reminds me of when to pray and creates a religious atmosphere within the home for my children. What's more, it teaches us a great deal about Islam so that I may be able to teach my children."

Entertainment: Women tended to view more entertainment than men. Despite the gender differences in the amount of viewing, the preferred channels for both men and women were Syrian satellite TV and the Lebanon satellite TV stations: LBCI, al-Mustaqbal, MTV, and TeleLiban. A female participant stated, "I like to watch these channels, the Syrian and Lebanese ones, because they have many shows to entertain, such as Arab movies, *telenovelas,* both Egyptian and Syrian, as well as music clips and parties, that let me feel like I am still living in the Arab society. It makes me nostalgic for the Arab culture that I lived in before coming to Berlin." A further insight was expressed by a male participant: "I greatly enjoy the entertainment shows of LBCI and Syria because it lets me forget the stress of the daily news we see all the time." In general, the study revealed that women tended to view more entertainment programs than the men while the men viewed more news broadcasts than women did.

Gender and news: Despite the fact that the Arab women interviewed tended to view relatively more entertainment programs than the men, they nevertheless viewed a great deal of news—up to two or three times a day—especially news related to the Intifada and Palestine. Interestingly, women viewers tended to focus almost exclusively on news broadcasts about Palestinian issues and specific news from their countries of origin. One woman stated, "My favorite channel is al-Jazeera. I watch news on it two or three times a day and especially because I can focus on our issue as Palestinians: what has the Intifada accomplished lately, have they been successful, have they given up, have they taken action against Israel? After that I am only concerned with the news of Jordan, where I came from and where my family is now, and finally the rest of the Arab world if there is any news." Another female participant reported, "When we first arrived in Germany, I mostly watched the channel MBC because they covered the Palestinian issues a great deal. However, recently the MBC has stopped putting so much news about the Palestinians, and al-Jazeera shows much more. Therefore, I have stopped watching MBC and now watch al-Jazeera. Although the MBC does show a lot of Arabic music, plays, movies, and other enjoyable entertainment shows, this does not matter to me. As soon as I turn on the television, I put on al-Jazeera." Once again this is evident from this female participant's statement: "From all the channels I prefer ANN because in recent times it has been showing much more news about the Palestinians and the Arab issue than any other channel." Miladi's findings (2006) based on his Arab diaspora respondents in Britain support this tendency.

When news was not about the Palestinian issue, most women tended to search for entertainment programs on other Arabic-speaking channels. The same applied to talk shows, in the sense that they were interesting only if they discussed Palestine. A female participant stated: "When I turn on my television I always check al-Jazeera and al-Manar first before anything else. If neither of them is showing news about the Palestinians then I simply change to an entertainment channel. If they are showing news about the Palestinians I will keep the program on and watch."

Arab men interviewed explained that the primary reason they installed satellite TV was to watch the news in Arabic, particularly news on Palestine. Other issues, such as the Iraqi crisis, were also important in their eyes. In almost all cases, when we asked, "What do you watch on the Arab satellites?" their reply was "Arab news regarding the Palestinian issue." A sixty-year-old Palestinian father of eight from Lebanon said:

> Prior to the presence of the Arab satellites, we as refugees lived in a state of uncertainty. We did not know what was happening around us and could not understand our own situation since we had no access to any news in Arabic. Now we can be informed and kept up-to-date about the Palestinian issue, and this has in turn given us hope. We are able to follow any developments and know exactly where we stand, and will hope and know when the issue is resolved and we are able to return to our homeland. Because of this I am greatly attached to the Arabic-speaking news channels and will watch news as long as I am home, to the point of, if one channel stops showing the news, I will go from one channel to another and another. And while the most important issue to me is the case of Palestinians, and then the issue of Iraq, I will also remain watching the news until the very end because all the news provided by the Arab news agencies is of interest to me.

Since the introduction of Arab satellites, the men take part in watching the news as a type of daily ritual. In the majority of interviews conducted, when questioned about home television viewing, the men nearly all answered that as soon as they arrived home they turned on the Arab channels and searched for any channel showing news, especially live updates, and that is what they would watch. A male participant stated:

> The first thing I do when I walk in the door is turn on the television to al-Jazeera to watch the news, as I usually know when the news will be on. If there isn't anything or if the news finishes, I simply turn to any another that may also be broadcasting news. I will keep doing this until there is no news, and then I will watch shows that discuss the news, such as talk shows or conferences or interviews that deal with what was in the news. This could sometimes last for up to five hours and may even reach a point where I am so involved and drawn into the news that I forget that I am in Germany, especially when I am joined by my wife and we are talking in Arabic with one another. What awakens me from this dream is when one of my children walks into the room and asks me something in German.

The evening news broadcasts were considered especially important. All the men interviewed stated that they did not miss the main evening news broadcasts, including talk shows and interviews, on al-Jazeera, ANN, or al-Manar. In fact, many of those interviewed would switch from the main news broadcast on al-Jazeera to al-Manar for its array of shows and documentaries regarding the news. At the same time they kept watching German channels for domestic news in Germany and for attempting to perceive the German view of current events in the Arab world. Similarly, one of the British studies stated that among the British Arab diaspora, British channels were watched for domestic news or to get what they called the "British view of incidents taking place in the Middle East" (Harb and Bassaiso 2006, 1069).

When the male participants were asked whether they watched news pertaining to Israel, typically answers were affirmative. The men gave several reasons: Some watched the news to see the many problems within Israel, which gave them hope that the country could be defeated and they could return to their homes. Others wanted to be kept informed of developments in the peace process. Still others wanted to see how the Palestinian Intifada was affecting Israeli society. A male participant commented, "I am concerned with watching the Arab news and want to see the results of the suicide bombings and how it affects the economic situation of Israel, primarily the Jewish people who have fled the country out of fear." Another stated, "I am concerned with hearing the internal Israel news, such as issues of economics, politics, and like to watch the news of the peace process." Most women asked did not watch any news pertaining to Israel.

The study revealed that Arab participants did not watch issues pertaining to Arabs on German-speaking channels. When questioned about this, all the male participants replied similarly. That is, following a comparison between the German and Arabic channels, they found two key points of difference in the reporting. First, the issues surrounding the Palestinian people were dealt with in the Arab satellites in more depth, providing greater details, background, and information, while the German satellites only skimmed through the news concerning Palestine. Second, the Arab men we interviewed all shared the opinion that the German channels were biased against Palestine, especially in instances of suicide bombings or Israeli attacks on Palestinians. The popularity of Arab satellite channels is also due to the perceived inaccurate and biased reporting of Western media after the Gulf War (Miladi 2006). This view was also shared by the women and, even more interestingly, by the children. Not understanding formal Arabic, the children would ask their parents how the same news event was reported on al-Jazeera to fully understand what was happening.

Those interviewed also made a distinction between private Arabic channels, such as al-Jazeera, ANN, Abu Dhabi, and al-Manar, and government-run stations. Participants expressed the opinion that those stations run by governments were instruments of the regime and did not express real news. One participant stated, "The government TV stations did not attract me; they are boring, their news are rulers' news, they did not tell me anything about what's going in the Arab world, unlike al-Jazeera or the other private TV stations." This explains the popularity of the four channels that all of the participants unanimously favored: al-Jazeera, al-Manar (Hezbollah operated), ANN, and Abu Dhabi, the latter two of which, like al-Jazeera, are both privately owned.

Interactive Dialogue (Hiwar)

Among the male participants of this study, the level of engagement with the news extended to active interaction with the news channels. That is to say, they often called the shows and expressed their own ideas. Indeed, during the course of this investigation, a survey of the call-in shows revealed that the majority of those who do in fact participate in the call-in shows are usually Arabs from the diaspora, and not the rest of the Arab world. Because of the developments in satellite technology, Arab audiences have become active participants rather than remaining "passive dupes" (Alterman 1998; Miladi 2006).

There are several reasons for this phenomenon. For more successfully integrated Arabs who have achieved financial stability in their country of exile, their financial status allows them the opportunity to make international calls and participate from their own homes, regardless of how long they have to wait on hold. Interestingly enough, even refugees in Germany living under poor financial conditions were sufficiently motivated to incur these expenses. Second, Arabs in the diaspora obviously feel more secure to make such calls and voice their real opinions than those living within the Arab regimes, as they are free to express themselves without fear of any consequences. More important, however, is the absence in the diaspora of communities where these issues can be discussed in social settings in real life. In the diaspora, the virtual TV community fulfills this social and political need. This is clearly evident from this participant's statement: "I enjoy watching the talk shows following the news, especially those pertaining to the Palestinian issues, and I always try to call in and participate in the discussions."

Back to Roots

Television and Islam: Living in a primarily Christian environment, those interviewed regretted their loss of connection to the typical festive atmosphere of Muslim holidays in their home countries. Prior to the entrance of Arab satellite stations into Germany, Arabs would not sense there was a holiday and would not experience all the cultural aspects associated with such events. In fact, sometimes the holiday would come and go without people's even realizing it. This is evident from this participant's statement: "One of the reasons I connected to the Arab satellites is so that my children would get a sense of their Islam religion, especially during the holidays and Ramadan." Needless to say, the German media agencies did not mention the Muslim holidays. Arab satellite TV, however, bases a large percentage of its programming on holidays. Indeed, particularly during Ramadan, competition is rife to hold viewers' interest during fasting hours through entertainment shows.

Furthermore, Arab satellites not only create a pleasant and comforting environment for the Arabs in Berlin during periods of holidays, especially Ramadan, but also draw the people closer to their religion and their heritage (see also Miladi 2006; and Harb and Bessaiso 2006). The Arab satellites have made it easy for the people in Germany to be able to follow prayers during Ramadan and thus know when to break their fast, whereas prior to the presence of the Arab satellites, they could not fast properly during Ramadan for lack of knowledge of the proper schedule of prayers.

One youth stated, "I watch Arabic only during the period of Ramadan or other holidays because they put on many entertaining programs like *Jameel and Hanaa* or *Abu alhana*." This point is further strengthened by a mother's remark: "Since the entrance of the Arab satellites I have been able to create a festive environment for my children during the holidays and especially Ramadan. I can also teach them the habits and prayers of Ramadan and how to enjoy it. Also, I can now know more punctually and specifically when the prayer times and fasting times are, and thus respect them."

Also, as mentioned earlier, Arab satellites also play a role in teaching the values and traditions involved in Islam. From the majority of interviews, families, primarily the parents, follow a great many of the religious programs. This point is clearly evident in this participant's comment: "I always follow religious programs on the Arab channels, especially on Munajah, al-Manar, and Iqra,' especially the Friday sermon, or other religious programs on al-Jazeera where they argue for the first time religious issues and questions. This way they provide us with answers to our daily problems through religion, which we can't find here because there are no religion authorities." The satellite programs have also helped provide Arabs in Germany with a better understanding of Islam. One participant stated, "I was never very religious, and I rarely prayed before. Now, however, I understand Islam especially through the Arab TV stations, and I have returned to it. I can pray, and I understand its implications and meaning for many things in my daily life."

Arab satellite shows also teach Arab families what is allowed and what is not from a religious perspective. They learn these things from the various religious shows, not only as passive viewers but also as active viewers. The term *active* here refers to some of the programs that invite a Muslim sheikh to the show and the audience has the chance to call in and ask him questions regarding their daily lives. One family stated, "We [the mother and father] follow religious programs that are now teaching us what is right and wrong according to the sharia [Islamic law], and we can now also teach this to our children." Arguably then, Arab satellites are teaching parents to teach their children about Islam (Harb and Bessaiso 2006).

Television, Youth, and Children

Contrary to their parents, the children that arrive in Germany are quickly integrated into German culture through schooling, and they learn the language much more quickly. Immersed in German culture for a large part of the day, they are in continuous contact with the German language. Most of the children could not read or write Arabic. Many could also not speak Arabic well, particularly those between four and eighteen years old.

According to Miladi's (2006) research in Britain, interest in al-Jazeera is high mainly among the first generation of the Arab community regardless of gender, while Western channels are more popular among the young generation. One could assume that this pattern is reflected in Germany as well, and the interviews did reveal that the television viewing habits of the children were mostly German. When they did watch Arab TV, it was mostly for entertainment—music in particular. A thirteen year old stated, "I only watch Arabic music, especially Arabic music video clips that I am familiar with."

However, many children claimed they watch Arabic programs with their parents if previously told the program was interesting. A six-year-old girl revealed, "I watch with my mother sometimes Egyptian films or shows, but I don't always understand what is said and so my mother has to translate or explain for me." Children also watched some news with their parents, particularly in times of crisis in Palestine. They usually asked their parents for explanations and translations, although Arab channels usually show a great deal of pictures and footage, which allow the children to understand visually. Most of the children said that these events helped them to understand their heritage and culture, while also making them feel a part of the Arab world. A seventeen-year-old male stated, "After watching only German for two consecutive days, I feel like watching Arabic in order to understand what is happening in the Arab world and to improve my Arabic language skills while viewing a bit of my heritage."

The studies of the children of Arab refugees in Berlin revealed that, after around the age of eighteen, they began to feel a need to relate more to their Arab heritage, and became more interested in mastering Arabic. This appeared to occur simultaneously with transition into the workplace, where many mentioned they are treated by the Germans as Arabs. Moreover, following the events of September 11th, many members of the Arab community in Berlin observed that they were treated differently or with caution. This led some children to seek answers from Arab transnational TV. An eighteen-year-old participant mentioned, "Prior to the September 11th events we never experienced any problems from the German students and especially my friends in the school as foreigners. Now, during breaks, the German students form one group and the Arab students form another separate group, and they do not want to have contact with us."

The second Intifada (since October 2000) has also played a significant role. As their parents now watch more television, their children do as well. As

Germans obtain their information from German newscasts, which the Arabs consider to be primarily from an Israeli perspective, the Arab youths feel that they must watch from the Arab perspective. A fifteen year old stated, "I am now more interested in and concerned with the Palestinian news and affairs because before anything else, I am a Palestinian. And when the German students accuse us, the Palestinians, of attacking Israel and making terrorism, it is important for me to have the real facts, which are usually the opposite of what the Germans hear, and argue and defend the Palestinians."

Discussion

Some European governments were concerned that cross-border connections could be subversive and prove unsettling to immigrant groups. When Arabic-speaking households in France started to install very large satellite dishes, the French authorities were alarmed, interpreting this as an affront to the country's policy of cultural assimilation. Similar concerns about integration in host communities were expressed in Denmark and Sweden, where researchers charged with assessing the impact of Arab satellite channels on Arab immigrant households likened satellite dishes on rooftops to ears listening out for news of "home" (Sakr 2001).

Intensification of Shared Stance

The findings above reveal that the main implication of the intensive broadcasting of satellite (Pan-Arab) media content for those interviewed in Berlin was to increase and strengthen their sense of instant Arabness—McArabism, converting isolated viewers to members of a virtual community. This media content consists of news of shared concern, entertainment programs, and Pan-Arab and Islamic programs. These effects were achieved through several components:

Political awareness and involvement: The Arab satellites led to an almost complete (virtual) political awareness and involvement among members of Berlin's Arab diaspora that were interviewed. This was accomplished not only through actual broadcasting of news events daily but also through talk shows and background documentaries. Through this means, a greater political understanding was reached (al-Hitti 2000; Graieb and Mansour 2000; Mellor 2005). This was also carried out through the intensive coverage of Arab issues, especially regarding the Palestinian-Israeli conflict, the Intifada, and the Iraqi issue (Alterman 1998). A participant remarked, "Since the presence of the Arab satellite TVs within my home, I no longer feel alone, but rather that I belong somewhere within the Arab world. That I am more involved and active within the Arab world and its issues and conflicts." This was even more influential and even acted as more of a trigger following the second Intifada and the September 11th events in the United States. For instance, as was made clear by Matar (2006) concerning the events of September 11th, in the subsequent days many Arabs in Britain favored Arabic channels such as al-Jazeera, for their more varied analyses and wider range of viewpoints, including people "outside the loop" of the mainstream Western media.

This is also true of the Arab diasporans in Berlin who, through their intensive consumption of Arab satellite television's Pan-Arab programs, became more aware of the various Arab issues around the world as well as the different political situations of the many Arab states. This point is revealed in this participant's comment: "The Arab satellite channels provided us with more information about the Arab issues and the Arabs themselves. We now learned about the Arab politics around the world and what happens in this political situation around the Arab world." This political awareness increased and became more apparent not only within the older generation but also within the younger generation (the children). Matar (2006) also concludes that the preference of transnational Arabic satellite TV during events such as September 11th and events related to the Palestinian-Israeli conflict or the Iraqi situation can be understood not only in terms of the channels' utility and functionality in providing in-depth coverage but also in how their "informed readings" of news stories give audiences discursive powers, where making sense of the news is tied to some form of ideologically resonant position (Lewis 1991).

Arab culture and heritage: The Arab satellite TV stations played a significant role among their viewers in their awareness of, and education in, Arab culture. These programs expose their viewers to Arab arts, music, talents, and other cultural issues within the Arab world (Rinnawi 2006). Through programs discussing the Arab cultural arena itself, cultural awareness was heightened. One participant expressed this opinion: "The Arab satellite TV channels allowed me to view other Arabs that I had never heard of before, such as singers, poets, writers, philosophers, and other literate individuals I had not known before." This important effect was also evident in the case of other Arab diasporas in Europe, as we can see in the research of Miladi (2006), who states that other concerns of the Arab community in Britain are related to preserving their home culture and passing it on to their children, going beyond the national boundaries to encompass an international Arab identity.

Furthermore, our findings show that most of our interviewees interact with news stories on a daily basis, not only to find out what is going on in the homeland and elsewhere but also to negotiate the social meanings of collective and personal memory, belonging, and community. Memory requires a framework or a shared material and social context in which to be reproduced (Halbwachs 1992). Collective memories, as well as historical and familial links, sustain diaspora members' real and imagined relationships with the homeland. These relationships are also sustained by the sense of loss and constant yearning that is to a large extent an integral component of diaspora experiences.

Arab identity: Arab satellite TV stations have acted as a means of distancing the Arabs even further from the environment they currently live in, drawing them even more strongly into their own separate Arab identity. This appears to have halted, or at least slowed down, assimilation into German society. In fact, as a result of the presence of the Arab satellite televisions, people interviewed dismissed the possibility of integration within German society. It was clear from the interviews that all of the participants admitted that there had been some process of integration into German society prior to the presence of the Arab satellites. However, since the entrance of the Arab satellite televisions, this has diminished and in fact the process of segregation has increased.

There is need for further research in particular areas. The study completed by Robins and Aksoy (2000) concerning the Turkish community in Germany brought up the question of how media and cultural practices relate to the process of identity formation, and this needs to be investigated in greater depth. On the one hand, there is a fear that Arabs are taking refugees into an exclusively Arabic cultural space. This provokes anxieties about their dissociation from the cultural and political life of the country of residence. On the other hand, the new media and cultural space clearly provide opportunities for the Arab community to develop extensive forms of transnational associations and solidarity. And, of course, such developments do not necessarily preclude continuing involvement in the country of settlement. Media and cultural practices are absolutely central, then, to the new transnational developments. The problem is that very little is actually known about media use. There is a great need for more solid empirical evidence about new cultural developments in order to respond constructively to the questions posed by the Arab diaspora in Germany, but more important in western Europe and the West.

Arab values and traditions: The Arab satellites played an immense role in the revival of Arab values and traditions among the Arabs in Berlin. This involves programs that directly state proper behavior for the Arabs and other programs with subliminal messages of acceptable behavior. Furthermore, the satellites assist the parents in the socialization and convincing of their children to behave in accordance with Arab tradition and values. This participant's statement clearly indicates this point: "I often seek assistance from the Arab satellites in convincing my children to behave in a manner acceptable to the Arab way of life and not the German one."

Islam and religion: The Arab satellite channels also played the role of virtual mosque via programming. For those already religious, Arab satellite TV developed their religious knowledge and understanding. One participant said, "Arab satellites help me to actively live in my Islamic beliefs, especially the Islamic religious programs that provide me with solutions to daily religious questions and inquiries." Through the many religious programs available via Arab satellite television, parents were taught how to teach their own children issues of religion. Children and secular parents also began to learn more about Islam, some becoming more religious.

Arabic Language

Transnational Arab TV worked to sustain and enrich viewers' fluency in Arabic, particularly for those—the large majority of those interviewed, including all the parents—who watched Arab TV daily, as if someone within the home were speaking Arabic with them. It was stated that "since the entrance of the Arabic TV to our home the Arabic language of my children improved, and even my and my wife's Arabic." In households with young children, a preference for Arabic-language satellite channels was also related to the importance of maintaining the ancestral language in the home (Matar 2006).

Arab TV, primarily "speaking" in formal Arabic, also indirectly provided new phrases and figures of speech, which many did not know before. Arab satellite TV programs also allowed children to improve their Arabic. Most important, the use of a standard formal Arabic allowed transnational TV to appeal to diverse groups of speakers of Arabic dialects and established a common denominator shared by all. This also worked, as I have noted earlier, to actually develop feelings of Pan-Arabism, by allowing refugees to develop from their traditional allegiances to family, village, city, region, and nation-state or state-regime to a wider, more abstract notion of being *Arab*.

Direct Engagement

The intensification of the programs by Arab transnational TV channels introduced concomitantly modern styles of news and broadcasting that allow the audience to understand the news with minimal obvious political or "state" intervention. This includes using different broadcasting effects and techniques such as live broadcasts and figures and maps, having reporters stationed in both Arab and non-Arab countries, providing in-depth reports on various issues, and conducting interviews with people, leaders, or groups representing different points of views. As in the Western media, these techniques produce the impression of "real" and "objective" reporting. The hidden bias lies in the producers' choice of issues, representative interviews, and actual footage—especially the sensational.

Cultural Sensationalism

As was clearly stated by our respondents, Arab satellite TV has adopted the consensus narrative of Arab society as a whole, and this is reflected in the repetition on a daily basis of images of Palestinian and Iraqi suffering, narratives of humiliation, and sensationalistic representations of Palestinians and Iraqis as victims. This has been achieved by a process of allowing viewers to see footage, often live, clearly intended to provoke an emotive effect that causes audiences to experience deeper forms of engagement. This "sensational" presentation is intensified through the use of various kinds of rhetoric in reporting language as well as pictures, style of presentations, and other effects. The respondents of Matar's study describe it in the following words: "We find in al-Jazeera ways to vent our feelings and emotions" (2006, 1035). This component was evident in most of the conversations with our interviewees, and has played an essential role in the creation of the sense of overidentification at the national and cultural levels among this diaspora's members though their intensive consumption of Arab satellite television in their houses in Berlin.

CONCLUSION

For further clarification of the dynamics of the four component factors that result in "McArabism," I present figure 15.1. In this figure we can see that the "intensification of the shared stance" leads directly into "cultural sensationalism," which in turn feeds directly into McArabism. At the same time the other two contributing components also directly influence the development of McArabism and are active factors in "cultural sensationalism." This occurs, on the one hand, through the use of the widely understandable normative "Arabic language" and, on the other hand, through the techniques of "direct engagement"—modern styles of news broadcasting and broadcasting effects and techniques such as live broadcasts and figures and maps that increase the sense of immediacy and reality of the programs. These last

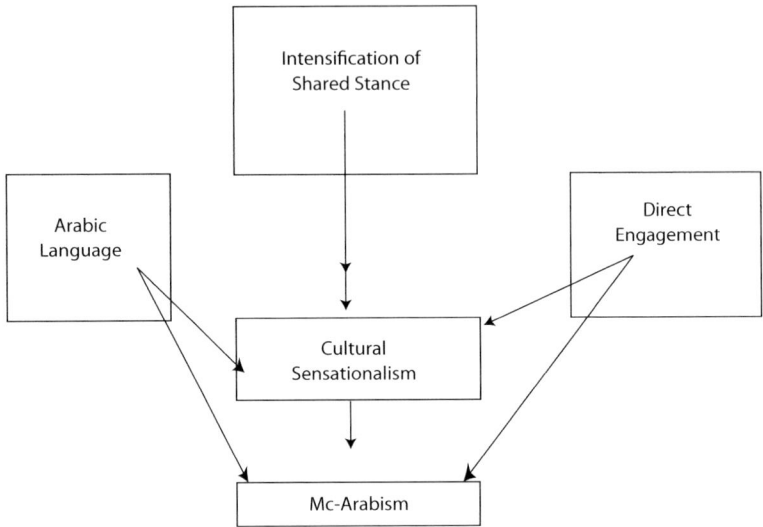

Figure 15.1 Mc-Arabism model

two factors are further intensified by the emotional appeal of sensationalist footage. The final result is the creation of an instant nationalism called "McArabism."

For refugee communities, satellite broadcasting has become an important tool to overcome barriers of distance and time and has considerably broadened the scale of Arab cross-border interaction. News channels like al-Jazeera contribute to fragmenting the onetime national German news audience while, at the same time, creating a Pan-Arabic and Muslim transnational public sphere, where people feel connected to their countries of origin wherever they might be (Harb and Bassaiso 2006; Miladi 2006). Thus, for exiled and refugee communities, satellite TV stations like al-Jazeera open up the possibilities for what Cohen calls "multiple affiliation of associations" and are giving rise to a "diasporic allegiance"—a "proliferation of transnational identities that cannot be contained in the nation state system" (1997, 78).

Finally, these effects and implications indicate the process of virtual Pan-Arabism, which Rinnawi terms *McArabism,* created by the exposure of this community to the Arab transnational media. This effect can be thought of as a kind of instant nationalism, because it creates a new separate and conscious "national" identity that separates its adherents from the German society and the reality in which they live—that is, a physical reality, which is so deeply divergent from the virtual reality that they absorb and internalize through the Arab TV channels at home. Arguably, Berlin's Arab refugee community has become a part of this virtual community of "McArabism."

NOTES

1. This study was accomplished due to the funding of the Deutscher Akademischer Austausch Dienst (German Academic Exchange Service) in Germany.
2. *Arabs in the diaspora* refers to those people of Arab origin beyond the Arab world who culturally and emotionally identify as Arabs, or identify, through a smaller state or regional grouping, as part of the Arab world. Therefore, the term *Arabs in the diaspora* might not apply to children or grandchildren of Arabs in the diaspora.

REFERENCES

Alterman, J. B. 1998. *New media, new politics? From satellite television to the Internet in the Arab world.* Washington, D.C.: Washington Institute for Near East Policy.

Anderson, B. 1993. *Imagined communities: Reflections on the origin and spread of nationalism.* London: Verso.

Anderson, J. 1997. Cybernauts of the Arab diaspora: Electronic mediation in transnational cultural identities. Paper prepared for Couch-Stone symposium "Postmodern Culture, Global Capitalism, and Democratic Action." University of Maryland, April 10–12.

Arab Advisors Group. 2007. Polls and surveys in the Arab world. Amman, Jordan.

Ayish, M. 2002. Political communication on Arab world television: Evolving patterns. *Political Communication* 19, no. 2.

Barber, B. 1992. Jihad vs. McWorld. *Atlantic Monthly* 269, no. 3 (March).

Breidenbach, J. 2001. *Migrants in Germany: Between assimilation and transnationalisation.* Available at http://www.politeia.net/seminar/migration/Breitenbach%20 paper.htm.

Cohen, R. 1997. *Global diasporas: An introduction.* London: UCL Press.

Curran, J., and M. J. Park, eds. 2000. *De-Westernizing media studies.* London and New York: Routledge.

Graieb, A., and Kh Mansour. 2000. Arab media in the 21st century: Between the hammer of the globalization and the anvil of the state. *Bahethat,* no. 6: 119–41.

Halbwachs, Maurice. 1992. *On collective memory.* Chicago: University of Chicago Press.

Harb, Z., and E. Bessaiso. 2006. British Arab Muslim audiences and television after September 11. *Journal of Ethnic and Migration Studies* 32, no. 6: 1063–76.

al-Hitti, H. 2000. The international satellite broadcasting and its possible political effects in the Arab world. *Al-mustaqbal Al-A'rabi* [Arab Future], no. 16.

Lewis, J. 1991. *The ideological octopus.* London: Routledge.

Matar, D. 2006. Diverse diasporas, one meta-narrative: Palestinians in the UK talking about 11 September 2001. *Journal of Ethnic and Migration Studies* 32, no. 6: 1027–40.

Mellor, Noha. 2005. *The making of Arab news.* Lanham, Md.: Rowman and Littlefield.

Miladi, N. 2006. Satellite TV news and the Arab diaspora in Britain: Comparing Al-Jazeera, the BBC, and CNN. *Journal of Ethnic and Migration Studies* 32, no. 6: 947–60.

Morley, D., and K. Robins. 1995. *Spaces of identity: Global media electronic landscapes and cultural boundaries.* London: Routledge.

Naficy, H. 1993. *The making of exile-culture.* Minneapolis: University of Minnesota Press.

Rinnawi, Khalil. 2006. *Instant imaginings: McArabism and al-Jazeera: Transnational media in the Arab world.* Lanham, Md.: University Press of America.

Robins, K., and A. Aksoy. 2000. Thinking across spaces: Transnational television from Turkey. *European Journal of Cultural Studies* 3, no. 3: 343–65.

Sakr, N. 2001. *Satellite realms: Transnational television, globalization, and the Middle East.* London and New York: I. B. Tauris.

16   Net Nationalism
The Digitalization of the Uyghur Diaspora

YITZHAK SHICHOR

Following the deadly riots in Xinjiang in early July 2009, news agencies and other organizations received e-mail messages calling on Tibetans to participate in worldwide protests and demonstrations in front of China's diplomatic missions in support of and solidarity with the suppressed Uyghurs in East Turkestan (Xinjiang). Allegedly sent by the Uyghur American Association and the World Uyghur Congress, these e-mail messages proved to be fake; they had never been sent by the UAA or the WUC. This is not the first time that hackers, most probably Chinese (as no one else has a motive in such provocations), have used the Internet to undermine Uyghur activism by impersonation, delivering viruses and blocking Web sites and occasionally interfering in personal e-mail exchanges (including academic, based on my own personal experience).[1] This incident highlights both the positive and the negative significance of digital communications. On the one hand, Uyghurs have greatly expanded the use of the Internet, achieving a higher degree of visibility than ever before; on the other hand, by doing so they have become exposed to disruptions, malicious penetrations, and cyber-attacks as a part of a long-standing conflict between them and China.

Uyghurs, a Turkic-Muslim nationality mostly located in Northwest China, were hardly known—let alone felt—in the international system until the early

1990s. Claiming an independent homeland from China, Uyghurs have been accused by Beijing of separatism, terrorism, and religious (Islamic) extremism. Suppressed inside China, Uyghurs have become more visible outside China, apparently because of the dynamic and extensive use of technologically advanced media adopted by their growing diaspora communities. Still, these communities are relatively small and spread over many countries (with the majority in central Asia). This has entailed difficulties in communication not only among diaspora Uyghur communities but also between them and their host countries as well as international and nongovernmental organizations (NGOs). Many of these difficulties have apparently been removed by relying on digital means, but the main question is to what extent these means helped to overcome Uyghurs' internal divisions, to improve their organization, to consolidate their collective identity, and, most important, to promote the Uyghur national cause.

Based on a few years of research of Uyghur diasporas, as well as theoretical concepts and comparative studies of transnational diaspora communities and communication, this chapter discusses the role of conventional and digital media in shaping the Uyghur agenda, nationalist identity, and international impact, in the post–cold war environment. Its first part ("Outline") provides the background for Uyghur history, nationalism, and the creation of its diaspora. The second part ("Off-Line") deals with the Uyghurs' use of conventional media that had failed to accomplish their nationalist vision and to create a transnational unified movement. The third part ("Online") concentrates on the emergence of Uyghur digital transnationalism in the 1990s, based primarily on the use of online communication media, first and foremost the Internet. The fourth section ("Bottom Line") evaluates the effectiveness (and vulnerability) of digital media in promoting Uyghur national identity and vision and their impact on China's foreign and domestic policies and the international system at large.[2]

UYGHUR DIASPORA NATIONALISM: OUTLINE

Uyghurs are the largest non-Chinese nationality in the Xinjiang Uyghur Autonomous Region of the People's Republic of China. Of Inner Asian Turkic stock, the Uyghurs trace their origins to the Huns, well before the arrival of the Chinese in Eastern Turkestan in the second century BC, during the Han dynasty. Though there is some debate on their origins (Gladney 1990, 1992, 1996; Geng 1984; Koçaoğlu 1997), the Uyghurs, who are mentioned in Chinese historical records as a tributary of the Chinese Empire, reached their climax in the mid-eighth century, when China's Tang dynasty managed to survive at their mercy. Uyghur kingdoms had begun converting to Islam by the mid-tenth century and maintained their independence for about one thousand years until invaded by the Manchu (Qing dynasty) in the mid-eighteenth century. Failing to regain its independence, in 1884 Eastern Turkestan (as the region is still called by Uyghurs and Turks) was officially incorporated into the Chinese Empire as a province named Xinjiang, or the "New Dominion" (for the best and most recent history, see Millward 2007).

Although there had been inklings of Uyghur collective identity before (Brophy 2005), Uyghur nationalism is a new phenomenon allegedly invented by Soviet scholars in the 1920s when Moscow arbitrarily divided central Asia into several republics based on different nationalities (Mackerras 1972, 1990; Rudelson 1997). Yet while Kazakhs, Uzbeks, Kyrgyz, Tajiks, and Turkmens have been granted a "homeland," Uyghurs, mostly living under Chinese rule, have become a stateless nation. An initial attempt to establish an Islamic Republic of Eastern Turkestan in 1933 was crushed within a few weeks. Launched under Soviet auspices in 1944, an Eastern Turkestan Republic managed to last a few years, only to collapse in late 1949 following the Chinese Communist "peaceful liberation" of Xinjiang (Forbes 1986; Benson 1990; Wang 1999). The Uyghurs have (once again) been incorporated into China. While suppressing the Uyghurs' quest for independence ever since, Beijing declared Xinjiang an autonomous region in 1955, thereby granting the Uyghurs what may be termed an "autonomy with Chinese characteristics," leaving them with very limited political (as well as social, economic, religious, and cultural) breathing space inside China—yet with better opportunities outside.

Probably unaware of their collective "national" identity, though surely well aware of the risks of staying in Xinjiang, Uyghur migrant communities had begun forming in central Asia as early as the nineteenth century, primarily in what was to become Kazakhstan and Kyrgyzstan. This had constituted the first wave of migration that was followed in the 1930s and 1940s by a small number of Uyghurs who reached farther west, settling in Turkey and Saudi Arabia. The third wave began in the late 1940s when hundreds of Uyghurs, including a number of top leaders, had managed to escape Xinjiang overland to Kashmir and, after spending a few years in India, finally ended up in Turkey. It is only since 1949 that the Uyghur diaspora has begun to pursue—consciously though cautiously—its nationalist vision of seeking independence from China.

In the 1960s, the fourth wave of migration had been triggered by internal Chinese hostility and external Soviet hospitality, which underlay the flight of thousands of Kazakhs and Uyghurs from Xinjiang to neighboring Soviet-controlled central Asia. The fifth wave, using Beijing's post-Mao Open Door Policy, had begun in the early 1980s and gathered momentum since the early 1990s following the disintegration of the Soviet Union and the consequent independence assumed by the central Asian republics. Under these circumstances, many Uyghurs have left China for central Asia, both legally and illegally, and some of them later spread all over the world, rather thinly, settling in the Middle East, East Asia, Australia, Russia, Europe, and North America.

As mentioned above, the Uyghur diaspora is relatively small. By the early 2000s, about 93 or 94 percent of the Uyghurs still lived in Xinjiang, China (9,650,629 according to 2007 official figures [Statistics Bureau of Xinjiang Uyghur Autonomous Region 2008, 74]), and only 6 or 7 percent—perhaps 650,000—lived abroad, most of them nearby in central Asia. There are disagreements and inconsistencies as to the dimension of the Uyghur diaspora communities (and even those in China). The claim that there are nearly 1 million Uyghurs just in central Asia, made recently by Dolqun Isa, secretary-general of the World Uyghur Congress, seems to be exaggerated.[3] The Uyghur community of Kazakhstan, the largest outside China, was estimated at 210,365 in 2000, though the unofficial number is 248,000 and some say twice as much, since many Uyghurs are registered as Kazakhs. More recent figures mention 370,000.[4]

The population of Uyghurs in Kyrgyzstan had reportedly reached 44,400 in 2000, though some claim that by now their number has already reached 53,000. The number of Uyghurs in Uzbekistan is about the same, 45,800 in 2000 and 51,000 now, though some claim that because of their similarity, many (perhaps between 500,000 and 600,000) have identified as Uzbeks, or have been forced to do so. In fact, this may be the total unofficial number of Uyghurs in *all* of central Asia. A few hundred Uyghurs may have settled in Turkmenistan, Pakistan, and Afghanistan.

Outside central Asia the number of Uyghurs becomes even more uncertain, not only because of a lack of data but also because their numbers keep changing. Turkey, the historical base and main source of inspiration of Uyghur nationalism since the late nineteenth century, hosts some 4,000 or 5,000 Uyghurs. Since Turkey has served as a junction for Uyghur refugees from China and central Asia, perhaps an additional 1,000 have passed through Turkey. In fact, there are possibly no more than 10,000 Uyghurs outside China and central Asia, of whom no more than 2,000 have settled in Europe and the rest in Saudi Arabia (between 2,000 and 4,000 or even 6,800), Australia (between 1,000 and 1,200), Russia (between 3,000 and

5,000) and North America (about 1,500). Thus, the total Uyghur diaspora is estimated at 300,000 to 600,000.[5] Even where they constitute a relatively large minority (e.g., in Kazakhstan), Uyghurs account for a very small share of the population of their host countries, 1 percent at the most. Most of these Uyghur diaspora communities have been targeted by Beijing over the past fifteen years or so.

Chinese leaders and the government-controlled media have always regarded Uyghur unrest in Xinjiang primarily as a domestic issue. At the same time, they have all along associated Uyghur subversive and "terrorist" activities in Xinjiang (and elsewhere in China) with the influence of external forces, the infiltration of foreign subversive and separatist agents, and the interference of foreign governments and other organizations, never actually named. Yet it is only since the mid-1990s that Beijing has finally realized that its domestic Uyghur crackdown could not succeed without neutralizing the external sources of the Eastern Turkestan Independence Movement. Therefore, China's concern has been expressed not only by the growing attention that Chinese media and leaders have been paying to the movement since the early 1980s. No less important, Beijing began to intensify its efforts, and to adopt drastic measures, to eradicate the movement at home and, for the first time, abroad.

Consequently, the Chinese began to apply pressure on those foreign governments that have enabled Uyghur and Eastern Turkestan organizations to use their territories, and facilities, to promote their independence and so-called separatist activities. Most of Beijing's diplomatic and economic measures, which are beyond the scope of this chapter, have concentrated on the central Asian countries bordering China and on Turkey (Shichor 2009, unpublished manuscript 1). Directly or indirectly affected by these measures, some of these Uyghur organizations have been forced to relocate their activities from central Asia and Turkey to Western Europe, North America, and Australia, away from China's reach. Evidently, Beijing's ability, and willingness, to influence Western democratic governments to suffocate proindependence Uyghur activities is much more limited. Therefore, although, as mentioned above, the number of Uyghurs in Western Europe is small, their impact is substantial.[6]

Apparently, Beijing's powerful domestic control and influence on central Asian governments, on the one hand, and the thin spread of small communities over numerous locations, some inhospitable, on the other hand, should have compromised the Uyghur diaspora collective identity and nationalist aspirations. Indeed, from the 1950s to the 1970s, exiled Uyghur leaders had barely managed to keep the quest for Eastern Turkestan independence alive by relying on conventional media and survival strategies. Still, since the 1980s, organized Uyghur transnationalist activism began to grow, gathering momentum in the 1990s. This has been an outcome of Beijing's far-reaching reforms and Open Door Policy undertaken since early 1979. While Beijing's hold on Xinjiang looks as firm as ever, if not firmer, the region has become exposed to unprecedented international attention and involvement, all the more so following the disintegration of the Soviet Union and the independence acquired by the central Asian republics. These domestic and regional developments have been converging with the emergence of globalization processes, the increased quest for human rights, and the pervasiveness of sophisticated digital communication, information, and media technologies to enable Uyghur communities to reassert their commitment to securing a free homeland in Eastern Turkestan on a worldwide scale. This has been done in two stages, first off-line and then online.

UYGHUR DIASPORA NATIONALISM: OFF-LINE

Until the 1990s, Uyghur diaspora communities relied on conventional media for the promotion of their nationalist vision. In central Asia this was undertaken under Soviet auspices and exhortation. Moscow, though reluctant to support minority nationalism, mobilized Uyghur grievances against Beijing as ammunition in the Sino-Soviet conflict to undermine China's claim for sovereignty over Xinjiang. Motivated primarily by Soviet interests yet also converging with the cause of Eastern Turkestan independence, Uyghur newspapers, journals, theater, books, and radio transmissions undoubtedly helped to preserve Uyghur identity (already partly Russified) but failed to make any difference as far as the Uyghur nationalist vision was concerned. Enjoying much more freedom and sympathy under Turkish auspices, Uyghur nationalism appeared to thrive there but, continuing to use conventional media, ultimately did not fare better.

Led by Mehmet Emin Buğra and, after his death in 1965, by Isa Yusuf Alptekin (both of whom had fled China in 1949 to India and then to Turkey), the Uyghur community in Turkey has always aimed at two interconnected targets and at two different audiences simultaneously. For one, they acted to enlist external support, win international recognition, and promote solidarity with other stateless nations for the cause of Eastern Turkestan independence. For another, they wished to preserve Uyghur collective identity, revive the memory of the two defunct Eastern Turkestan republics of the 1930s and 1940s, and sustain Uyghur culture and language. To these ends they used, among other things, the conventional media available in those times.

Launched in Turkey as early as the 1920s, Uyghur publications were given a boost only after 1953, when a new group of Turkestani émigrés from Xinjiang became politically active. Dealing mainly with cultural, social, and scholarly issues, these publications had a definite political and Pan-Turkic character, and included attacks on China's hostile policies in Xinjiang.[7] Printed mostly in Turkish and Uyghur, though occasionally in English, these journals were aimed at enlisting Muslims and other anticommunist sympathizers, including in Taiwan, to support Eastern Turkestani independence. Newsletters were regularly published starting in the early 1960s (Landau 1981, 122; Koçaoğlu 1997), as well as numerous books dealing with Eastern Turkestan's history and politics (Buğra 1952, 1954; Gayretullah 1965; Alptekin 1973). This policy continued in the 1980s and 1990s with new publications, such as *Doğu Türkistan'in Sesi* (Voice of Eastern Turkestan), a quarterly journal of cultural studies founded in 1984 by Isa Yusuf Alptekin, and issued in Turkish, English, and Uyghur.

Established in Munich in 1991, the East Turkestan Information Center (ETIC; in Uyghur, Sherqi Turkistan Axbarat Merkezi) issued the bimonthly *Eastern Turkestan Information Bulletin* "to disseminate objective current information on the people, culture and civilization of Eastern Turkestan and to provide a forum for discussion on a wide range of topics and complex issues." ETIC is the official information and news agency of the Eastern Turkestan National Center, an umbrella organization of the Uyghur diaspora, since its foundation in 1992 and until the establishment of the World Uyghur Congress in 2004.[8] *Doğu Türkistan* (Eastern Turkestan), the Union of East Turkestani Youth's newsletter, has been published in Munich since 1993 in Turkish, with separate English, German, and Uyghur issues. Another journal with the same name has been published in Istanbul in Turkish as a journal from 1979 to 1987, as a daily from 1987 to 1994, and again as a journal since 1994. Supposed to be aimed at English readers, the first—and last—issue of the *Uighur Affairs Survey* was published in Stockholm in September 2001.

Yet printed matter, especially in non-Western languages, had narrow circulation and could reach only a small audience. Still, the Uyghurs' acts to promote their national cause had been ineffective not only because of the conventional media they had employed but also because their performance and agenda had been affected by political and international constraints. Inside, the so-called Uyghur national movement was divided over strategic, tactical, and personal issues that precluded an effective united action against China. Outside, in those years China, excluded from much of the global community, was inaccessible and, therefore, practically immune to external pressure. The cold war against the Soviet Union had sapped all the attention and energy at the expense of other international issues. It was also difficult to portray China as a colonialist power in the age of the struggle against Western colonialism, in which China claimed to play a significant, if not a leading, role.

Many of these predicaments have been gradually, and sometimes dramatically, removed since the late 1970s. China has withdrawn from the struggle against colonialism and has become more accessible and thereby more vulnerable to external (and internal) pressure. Under these circumstances, the struggle for Eastern Turkestan independence that had been suppressed for thirty years could now be revived both inside China and abroad. Consequently, the Chinese, who heretofore had considered Uyghur nationalism primarily a *domestic* issue, have begun to blame *external* agents for fomenting and exacerbating unrest in Xinjiang (Shichor 1994). Nearly impossible during the Sino-Soviet conflict from the 1960s to the 1980s when the borders between the two countries had been sealed, Uyghur communities all over the world could now resume and expand their relationships in promoting the vision of Eastern Turkestan independence and applying pressure on China to stop Uyghur persecution. By that time, the international environment had also undergone some dramatic changes.

In addition to the unprecedented reform in China and the coming collapse of the Soviet Union, the West has for the first time begun to encourage democracy, human rights, and self-determination in non-Western countries, up to a limit. Public opinion, governments, and NGOs for the first time have begun to show interest in the plight of Uyghurs and in their nationalist agenda. Consequently, the new Uyghur organizations—though traditional in structure and goals—have become nontraditional in terms of means and audience (e.g., by consolidating solidarity among themselves and with other stateless nations). Joining hands on July 14, 1985 (a symbolic day), they launched the Allied Committee of the Peoples of Eastern Turkestan, Inner Mongolia, Manchuria, and Tibet in Zurich, Switzerland (*Eastern Turkestan Information Bulletin* 1993).[9] Since then the committee—which is no longer active—has been consistently used to promote the vision of Eastern Turkestan independence, while condemning Chinese persecution and execution of Uyghurs. Underscoring this solidarity, two volumes of the Allied Committee's publication, *Common Voice,* were issued (in 1988 and 1992). Despite the international change, much of this effort has still been undertaken by conventional means, such as founding new organizations and publications.

The effectiveness of such conventional communication media was quite limited. Until the 1980s the Uyghur cause was largely ignored not only by Western governments and NGOs but even by Middle Eastern and other Muslim governments and organizations. However, by the 1990s China's transformation and the redesign of the international system have coincided with the emergence of sophisticated communication media to create unprecedented opportunities for promoting the Uyghur cause. Some of these possibilities were beyond the reach of Uyghur diaspora communities, as most of them are too small to benefit from tailor-made satellite communication programs, movies, or television broadcasts (Husband 1994; Riggins 1992). However, by far the most widespread—and probably the least-expensive—means for the promotion of the Uyghur national quest for independence or greater autonomy has been the proliferation of Internet networks and Web sites since the mid-1990s (Dahan and Sheffer 2001). Uyghur diaspora nationalism ended its off-line phase and has become digitalized.

UYGHUR DIASPORA NATIONALISM: ONLINE

With postmodern and sophisticated computer-mediated communication (CMC), dispersed ethnic migrant communities can not only be easily and quickly linked together but also, and much more effectively, articulate their nationalist (as well as cultural, social, religious, linguistic, and other) interests, overrunning geographical borders, traditional space, as well as political constraints (Karim 1998). Apparently, by using CMC separate and weak local or national communities could be transformed into more powerful and influential international "online" constituencies.

This is evident among Uyghur diaspora communities that have begun using CMC to promote their nationalist mission of resurrecting the Eastern Turkestan Republic and restoring its independence. To this end, a variety of links have been created, including links to different Uyghur communities, to non-Uyghur communities, to the country of origin, to the host country or countries, and to governments and NGOs. Impossible to launch in China, most of these links, or Web sites, if not all, are based and maintained in the West, and for good reasons.

There is already substantial evidence that unequal development and different political environments produce separatism not only between Uyghur communities and other communities but also *within* the different Uyghur diaspora organizations, notably between central Asian organizations and those in the West. More militant in their pursuit of Eastern Turkestan independence to begin with, central Asia's Uyghur nationalist activities have been handicapped by two disadvantages. For one, they are much more exposed to persecution by their own authoritarian regimes, which in turn have become more susceptible to Beijing's economic, political, and military pressure. For another, due to the region's authoritarian political orientation, its backward technological infrastructure, and its low standard of living, central Asian digital communication is underdeveloped and, where it exists, often inefficient (Johnson 2001, 11–12).[10] That much cannot be said about Uyghur communities in many other countries such as the United States, Canada, Australia, Germany and other European countries, and even Turkey. Though much smaller, these Uyghur communities are wealthier, enjoy unlimited access to sophisticated communication technology, and, most important, have been sheltered by liberal democracies and granted freedom of speech, religion, and association. Consequently, they are much less vulnerable and sensitive to Chinese harassment, and therefore their contribution to the promotion of the Eastern Turkestan cause is substantial and out of proportion to their actual size. This is why most Uyghur Web sites originate in the West.

One early Web site that had links to a number of East Turkestani activities all over the world was hosted by Geocities and does not exist anymore.[11] It was maintained by Jack Churchward, an American who ran a Web site called "Free Eastern Turkestan" as part of an American organization called Citizens Against Communist Chinese Propaganda. Another early Web site was hosted by Euronet,[12] which has been affiliated with SOTA, a Dutch acronym for the Foundation for the Research of Turkestan, Azerbaijan, Crimea, Caucasus, and Siberia. Established in 1991 by Mehmet Tütüncü, SOTA is dedicated to the study of, and provides information about, the Turkic peoples of the former Soviet Union and the promotion of human rights, democratic governance, and just peace. Located in Haarlem in the Netherlands, SOTA has so far launched four Web projects, one of which was the *Turkistan Newsletter*, now discontinued.

Launched on May 9, 1997, the *Turkistan Newsletter* was an electronic distribution list and newsletter whose purpose was to report on all the Turkic peoples worldwide. While the Uyghurs and developments in Xinjiang related to Eastern Turkestan were occasionally covered in the regular issues, there was a special subseries called *Uyghur Perspectives,* seven issues of which had been published by mid-December 2000 (it too is now discontinued). Using Turkistan-Net,[13] nearly 250 issues of the *Turkistan Newsletter* were published every year. The number of subscribers increased four and a half times, from 650 in June 1997 to 2,940 (in sixty-five countries) by December 2000.[14]

There are additional general Web sites that provide occasional information on Uyghurs, Xinjiang, and Eastern Turkestan. EurasiaNet,[15] created by the Open Society Institute's Central Eurasia Project in New York, provides a critical analytical viewpoint on political, economic, and social developments in the countries of central Asia and the Caucasus, as well as Afghanistan, Iran, and Turkey. This Web site emphasizes in-depth coverage of issues that are usually not addressed by other data resources. Within two years after its launch in 2000, EurasiaNet already received more than 300,000 visits per month. Another Web site is called Central Asia–Caucasus Analyst,[16] created by the Central Asia–Caucasus Institute's School of Advanced International Studies at Johns Hopkins University. In 2002 it combined with the Silk Road Studies Program to create the joint Transatlantic Research and Policy Center with offices in Washington and Stockholm. Also available on the Net are Radio Free Europe and Radio Liberty reports and analyses covering Uyghurs and Eastern Turkestan (Xinjiang).[17] Based in The Hague, the Web site of the Unrepresented Nations and Peoples Organization also provides occasional information, documents, and reports on the Uyghurs.[18]

Most important, however, were Web sites exclusively devoted to the Uyghurs, Eastern Turkestan, and Xinjiang. I say "were" because most, if not all, of them have been blocked or overwritten, probably by Chinese hackers, most likely acting on official orders. One example is the International Taklamakan Human Rights Association (ITHRA), formed in the United States in 1996.[19] It operated and maintained a number of mailing lists to provide information on Eastern Turkestan, using its Web site also on behalf of other organizations. Thus, ITHRA maintained a mailing list on the activities of the Allied Committee of the Peoples of Eastern Turkestan, Inner Mongolia, Manchuria, and Tibet, including its publication, *Common Voice*.[20] ITHRA also maintained Uighur-L, an open mailing list created in March 1997 to serve as a communication tool for the Uyghur–Eastern Turkestan community, the only mailing list dedicated to discussions on Eastern Turkestan, the Uyghur people, and related issues. It covered a variety of topics, including culture and arts, history, religion, economics, politics, humanitarian aid and refugees, education, literature, science and research, "and, of course, the topic of high importance: organizing and coordinating Eastern Turkestani/Uighur support activities."[21] Also, ITHRA provided online access to all issues of the *Eastern Turkestan Information Bulletin* (1991–1996), which, unfortunately, are no longer available online.[22]

The East Turkestan Information Center offers additional major Web sites under the title "The World Uyghur Network." These include an electronic newsletter, the *World Uighur Network News,* which was first published on June 23, 1996. All issues are accessible through the World Uyghur Congress Web site.[23] ETIC's *Reports 2000* are published online under the title *Spark* (Turkish Press on Eastern Turkestan). They also cover Uyghur organizations, issues of history, human rights, daily world news, and so on. Many Uyghur Web sites frequently change their URLs to avoid Chinese hacking. The Uyghur Human Rights Coalition in Washington, which had become very active in 2000, used to have its own home page.[24] Since 1998, the Eastern Turkestan National Freedom Center (also in Washington) maintained a nearly identical Web site, displaying two newsletters so far.[25]

The Washington Uyghur Information Agency launched its Web site on September 10, 2000.[26] Active to this very day is the International Uyghur Human Rights and Democracy Foundation and the Uyghur Human Rights Project, associated with the Uyghur American Association, both sharing the same address in Washington.[27] The Eastern Turkestan National Center now uses its own Web site[28] to run other Web sites.[29] Last but not least, the Australian National University's Asian Studies WWW VL (World Wide Web Virtual Library) maintains the Eastern Turkestan WWW VL, providing elaborate bibliographic data on Eastern Turkestan.[30]

A number of these Web sites that existed in 2002 are no longer updated, or accessible, because of routine neglect but probably due to Chinese hacking. For example, Uyghur.com used to be one of the most diversified and extensive Uyghur Web sites that dealt with all aspects of Eastern Turkestan and its quest for freedom, independence, and democracy.[31] It provided links for donations, Eastern Turkestan maps, history, music, geography, civilization, oil, literature, and information about religious training, and addresses of international Uyghur organizations and access to their leaders. All has been overwritten.

Uyghurs, and anyone interested, could become members of a discussion group called the Uyghur Yari community.[32] Messages posted mention the glorious history and culture of Uyghurs, and the abundant mineral resources of Eastern Turkestan (Xinjiang) that have been "appropriated" by Beijing. Posted messages complain that the Chinese treat Uyghurs like slaves who are despised and exploited. "We have to be ACTIVE—to show the world that we don't fear the Chinese government! We have to show the world that it's not possible for us to accept the situation of assimilation and discrimination." The message ends with a militant call: "We Turkic brothers have to fight for our right to live in freedom, peace and harmony." Another message says, "The only thing the Uighurs can do is to make their problems PUBLIC!!

TABLE 16.1 | POLLS ON THE DESIRABLE CHINESE ATTITUDE TO XINJIANG

ANSWERS	NUMBER	PERCENT
Strike hard against Uyghur separatists	303	2.54
Give Uyghuristan independence	11,356	95.25
Open negotiations with moderate Uyghur leaders	57	0.48
Allow greater local autonomy in Xinjiang	206	1.72
Total	11,922	100.00

SOURCE: http://www.theuygur.com/uyghurindepen1e.html

And to make this problem the problem of all Turkic states—only in this way the pressure on China can be risen [sic]." (Capital letters, spelling, and exclamation marks are in the original.) The last posting is dated December 12, 2002 (ibid., May 19, 2003).

To keep the Eastern Turkestan cause alive, these Web sites created a network that not only distributed extensive information on the situation in Xinjiang but also organized forums for an exchange of opinions and disseminated news about meetings, speeches, appeals, demonstrations, and publications. The Eastern Turkestan Movement has been using CMC both internally (to preserve cultural and linguistic legacies and to enhance historical continuity) and externally (to become more visible, salient, and efficient in its lobbying efforts, in mobilizing international support, and in twisting the Chinese arm). One example was the attempt to pressure BP (British Petroleum) Amoco to forfeit its venture with PetroChina to build a gas pipeline from Xinjiang to Shanghai. Another one was the polls, the first of its kind, allegedly conducted by the *Times of Central Asia* (an English online newspaper) on the question, "How should the Chinese government address the Uyghur issue in Xinjiang?" (see table 16.1). According to the Uyghur Web site (now overwritten) that recycled the dispatch, "Even though the results are not necessarily scientific and respondents may not represent a wide range of ethnic groups, they do however express the general feelings of the Uyghur people for their future as a nation." But how effective is the use of digital means in promoting the Uyghur cause?

UYGHUR DIASPORA NATIONALISM: THE BOTTOM LINE

To begin with, following the earlier somewhat enthusiastic studies on the social and political advantages of CMC published in the 1990s, by 2008 it had become evident that CMC communities in general and digital diasporas in particular suffer constraints and disadvantages. While most of these constraints are universal and are common to all digital diasporas, including the Uyghur, the latter appears to suffer from a particular disadvantage as a "targeted diaspora." Put differently, in addition to all other constraints, to be discussed below, the Uyghur digital diaspora is subject to a constant, uncompromising, and ongoing Chinese brutal cyber offensive.

Resentful about the increased Uyghur international visibility, Beijing has directed its outrage less against any *actual* harmful and "subversive" consequences, marginal at best, of the Uyghurs' actions and more against their symbolic context. Beijing has always been more concerned with *potential* consequences over the long run, all the more so since the message and mission of Eastern Turkestan independence no longer depend on circumscribed conventional means like conferences, meetings, or printed matter. The proliferation of Uyghur CMC Internet Web sites has theoretically expanded its transmitting horizons almost endlessly, with a capacity to effortlessly penetrate the Great Wall of China, far beyond Beijing's reach and apparently with little risk. Aware that the Internet could be easily utilized to coordinate and encourage dissidence, since the late 1990s Beijing has occasionally blocked numerous foreign Web sites dealing with controversial issues that "harm national security" or "the interests of the State" (Kalathil 2001, 74–75).

Undoubtedly acting officially and professionally on behalf of their government, since the early 2000s Chinese hackers have apparently used a variety of means to sabotage Uyghur and East Turkestan CMC, blocking many Uyghur Web sites, creating threats to prevent or reduce access to these Web sites, installing viruses in the computers of those who try to use these Web sites, and overwriting some of these Web sites to replace their nationalist, occasionally anti-Chinese, content with sterile and harmless information. This policy has succeeded in undermining the effectiveness of the digitalized Uyghur diaspora since, unlike conventional media that are impossible to kill completely, CMC and digital media can be easily and effectively interfered with and killed altogether. This was evident in several Uyghur Web sites checked between August and October 2002.

Of twelve Uyghur Web sites blocked in 2002, either partly or totally, no more than three are active today. Two are not accessible at all, and the rest have been "overwritten," most probably by the Chinese, and now display information on economics, culture, tourism, and the like. Beijing avoided blocking twenty-four Uyghur Web sites that are clearly associated with officially recognized Uyghur associations in various countries (mainly in the United States, the United Kingdom, Russia, Japan, Turkey, and elsewhere),[33] official Web sites run by foreign governments (such as Radio Free Asia and Radio Free Europe–Radio Liberty), and Web sites dealing with purely cul-

TABLE 16.2 | ACCESSES TO ENGLISH-UYGHUR DICTIONARIES

COUNTRY OF ORIGIN	NUMBER	PERCENTAGE
China	2,352	47.0
United States	559	11.2
Japan	477	9.5
Israel	338	6.8
Turkey	164	3.3
Germany	161	3.2
Canada	137	2.7
New Zealand	91	1.8
United Kingdom	70	1.4
Sweden	54	1.1
Unknown	159	3.2
All others	439	8.8
Total	5,001	100.0

SOURCE: http://freeud.tripod.com/index.html.

tural topics, as can be seen in the number of accesses to http://freeud.tripod.com, an English-Uyghur, Uyghur-English dictionary, between December 2, 2001, and May 5, 2003. Visits from China are nearly 50 percent (see table 16.2). Therefore, in spite of all its efforts, Beijing cannot completely block external or internal access to Uyghur Web sites.

Indeed, Beijing has always considered the Internet an essential component of its modernization and development drive. All provinces, Tibet and Xinjiang included, have been provided with access to the Internet, extended to hundreds of cities. Also, it is relatively easy to access a blocked site from within China by using a proxy server located outside China (Foster and Goodman 2000, 6, 29, 43). Given the sophisticated technology involved, "there was no way the Party-state could have total control over activities conducted via the Internet" (Yang 2001, 67). Even if Beijing has failed to block Uyghur Web sites—an unlikely possibility—we have to assume that domestic access to these Web sites in Xinjiang, and therefore their effectiveness, must be limited, as the record shows.

By the end of 2008 Xinjiang's share in Web addresses was 0.7 percent, ranking twenty-fourth out of thirty-one provinces and administrative units. There were 6.25 million netizens (rank 18) with a penetration rate of 27.1 percent (rank 7) and a 2.1 percent share of the total compared with the end of 2007 when Xinjiang had 3.63 million netizens (rank 22) with a penetration rate of 17.7 percent (rank 7); growth rate from 2007 to 2008 reached 72.1 percent, ranking fifth of all provinces. By the end of 2008 Xinjiang's share in the geographical distribution of Web sites was 0.3 percent (compared to 0.2 percent a year earlier), ranking 26 (28 a year earlier). This means that the use of the Internet in Xinjiang is developing faster than in most other provinces (China Internet Network Information Center 2009, 17–18, 61, 63). Thus, we may assume that students, officials, and intellectuals have access to the Internet in Xinjiang. To be sure, the number of Internet users in China has been growing dramatically within the past few years, from 2.1 million in January 1999 to 33.7 million in January 2002, reaching 162 million in June 2007, and 298 million in December 2008, overtaking the United States and reaching number 1 in the world (see table 16.3). As the numbers of Internet users increase in China, and especially in Xinjiang, the impact of Uyghur Web sites is expected to grow, despite Beijing's efforts to prevent it and although China's Internet penetration rates are still comparatively low (see table 16.4). Yet the Uyghur digital diaspora has to face additional problems.

It has been suggested that globalized universalistic processes facilitated by extensive migration, easier and quicker transportation, and communication media, especially the World Wide Web, would by necessity contribute to the fading of particularistic, nationalist, ethnic, and linguistic identities (e.g., see Hobsbawn 1991, 9; and Appadurai 1996).[34] Yet by now there are indications of the opposite: in some cases, primarily where and when a scattered ethnicity shares a concrete and common *political agenda* and an *active* commitment to a nationalist mission, these processes of globalization expedite uniqueness, underscore unity, and consolidate collective identity. Unexpectedly, one outcome of these enhanced contacts has been the accentuation of a particularistic and unique *Uyghur* identity not only vis-à-vis the Chinese

TABLE 16.3 | GROWTH AND PENETRATION RATES OF CHINESE INTERNET USE

YEARS	NUMBER OF USERS (MILLIONS)	GROWTH (%)	PENETRATION RATE (%)
June 2002	45.8	72.8	3.6
December 2002	59.1	75.4	4.6
June 2003	68.0	48.5	5.3
December 2003	79.5	34.5	6.2
June 2004	87.0	27.9	6.7
December 2004	94.0	18.2	7.3
June 2005	103.0	18.4	7.9
December 2005	111.0	18.1	8.5
June 2006	123.0	19.4	9.4
December 2006	137.0	23.4	10.5
June 2007	162.0	31.7	12.3
December 2007	210.0	53.3	15.9
June 2008	253.0	56.2	19.1
December 2008	298.0	41.9	22.6

SOURCE: China Internet Network Information Center, Statistical survey report on the Internet development in China (various years).

TABLE 16.4 | INTERNET PENETRATION RATES, 2009: A COMPARATIVE PERSPECTIVE

COUNTRY/REGION	PENETRATION RATE (%)
Iceland	90.0
United States	73.2
Japan	73.8
South Korea	76.1
Europe	48.9
Russia	27.0
China	22.6
India	7.1
World total	23.8

(which is obvious) but also at the expense of an Uyghur "East Turkestani," "Turkic," "Pan-Turkic," or, all the more so, Turkish identity.

In a sense, this underlined Uyghur digital identity challenges not only Beijing but also Ankara, something that could resurrect tribalism. Thus, universal and easy communication could underline and prolong particularistic identities that would otherwise be lost. Under these circumstances, the emerging universal globalism and cosmopolitanism could be converted into a breeding ground for contemporary fragile parochialism (Schiffauer 1999, 2). At first glance, this is not too bad, as it helps to preserve the Uyghur diaspora collective identity and cultural legacy. Yet, upon second glance, by accentuating their distinctiveness and divergence, Uyghur diaspora nationalists stand to lose traditional supporters in Turkey as well as in central Asia and actually increase their isolation in the international community.

Still, even if Turkey and central Asia would like to help their Uyghur kin in their national cause, there is very little they can do. On the other hand, Western Europe and North America could help Uyghur nationalism but—for a variety of reasons, including their reluctance to support separatism and mainly because of their growing interests in China—will not. Uyghur strategy, therefore, has targeted North America and Western Europe since the late 1990s in order to win their goodwill and support by relying, among other things, on transnational digital media. Put differently, CMC is not just a passive opportunity created by advanced technology. It becomes meaningful and useful only when accompanied by professional political activism. As Portes said, "It is preferable to reserve the term 'transnationalism' for *activities* of an economic, political, and cultural sort that require the involvement of participants on a regular basis as a major part of their occupation" (1997, 17).[35] Online activism, much like revolutionary action, cannot become a sideline employment but should be a frontline, full-time (if not lifetime) commitment.

In reality, however, this process might still be problematic. To begin with, the earlier generation of Uyghur diaspora nationalists did not, and perhaps could not, devote all their time to the promotion of the Eastern Turkestan cause. Many of them did not know foreign languages, lacked higher education, and had to make a living, often with great difficulty. The new generation of Uyghur diaspora leadership is better educated, knows foreign languages, and spends most of its time promoting Uyghur nationalism, using CMC. Yet it appears that CMC is by no means a substitute for more conventional activities such as meetings, petitions, demonstrations, hearings, and lobbying. The main problem is not CMC in itself, which clearly has its own merits, but the link between virtual life and actual life, between cyberspace and human space, between the abstract world and the real world. In other words, the main question is to what extent the use of CMC could, and did, contribute to the promotion of the Uyghur nationalist cause compared to the use of more conventional means: what is the added value of digital media for Uyghur diaspora activism?

Sharing a vague and abstract vision that is not grounded in geographical proximity, personal familiarity, work association, and institutionalized interests, an "online" community can become fluid, flexible, and unstable. The Internet's anonymity could end in reducing accountability. Thus, "Whereas it is true that the Internet overcomes distance, in some ways it also overcomes proximity" (Jones 1998b, xiii). As easy to sign out as to sign in, CMC could end in atomization or solipsism and undermine the very sense of "community," not to mention the great amount of distraction that the Internet offers. In other words, we have just begun to understand the *negative* impact of CMC on personal involvement and commitment to the cause. Obviously, people tend to change much more slowly than technologies. Indeed, there has been so far no explicit connection between information, political activism, and outcomes: an enormous increase in information and proliferation has led to a moderate increase in political activism and to marginal political outcome. Therefore, "the anticipated effects of expanded communication are limited by the willingness and capacity of humans to engage in a complex political life. While the Net will certainly change the informational environment of individuals, it will likely not alter their overall interest in public affairs or their ability to assimilate and act on political information" (Bimber 1998, 135).

In addition, the potential audience of these digital appeals—leaders, parliamentary senators and committees, government ministers, NGO officials, journalists, as well as businessmen and companies—may show sympathy, moral support, and understanding in words but remain unaffected as far as political action and deeds are concerned. Nearly unlimited technologically, the virtual possibilities are, in fact, limited by, and subordinated to, existing realities, actual interests, and political feasibilities. In a ten-year perspective, the added value of CMC to the promotion of the Uyghur diaspora national ends appears to be marginal. There is no substitute for human action.

Moreover, since not all community members can access computers, CMC could become an elitist and oligarchic middle-class instrument that leads to hierarchical paternalism while undermining democratic processes (Jones 1998a). In fact, many Uyghurs who have left China, illegally or, and especially, legally, are well educated and belong to the middle class who have and can use computers. Nonetheless, despite the apparent limitlessness of the Internet, CMC is not unlimited. "Not all immigrants are involved in transnational activities, nor everyone in the countries of origin is affected by them" (Portes 1997, 16). Also, Uyghur migrants, and even more so refugees, usually spend the first few years after their arrival in their host countries trying to settle down, learn the language, find employment, and do whatever they can to be granted citizenship, as quickly as possible.

For these reasons, and especially the latter, Uyghur migrants try to avoid any political activity that might jeopardize their prospects. Many, but by no means all, will remain politically indifferent by choice even after settling down—or may be forbidden to engage in political activism by their employers (e.g., agencies of their host government). New life in a new country with new comfort and access to CMC could become harmful to personal national—and thereby collective—identity. The Internet facilitates mobility without moving and exposure without visibility, and, most dangerous for a deterritorialized nationalist diaspora movement, it offers a (virtual) "site" without or as a substitute for an (actual) territory. In the long run, when diaspora Uyghurs integrate into their new "homeland," this deterritorialization property might undermine the Eastern Turkestan independence movement.

CONCLUSION

Having relied on conventional and ineffective media before, since the mid-1990s relatively small and dispersed Uyghur diaspora communities have been digitalized through the use of CMC. By facilitating instant communication in volume and over a long distance, CMC was expected to turn an unproductive reality of disunity, heterogeneity, weakness, and hesitation into a productive reality of unity, homogeneity, power, and determination. To a great extent, the outcome has been an illusion, as the Uyghur diaspora is still divided, heterogeneous, and weak. This is not necessarily the fault of using CMC as much as the fault of given objective constraints related to China's growing prominence and the Western unwillingness and inability to challenge it.

The use of CMC may have been more effective *inside* the community, in maintaining and upgrading collective identity, although those diaspora Uyghurs who use the Internet do not have to be reminded about Eastern Turkestan, its history, culture, literature, music, and nationalism. For a committed Uyghur activist, the Internet is useful as a means of communicating information and ideas, yet it preaches to the converted and the use of a telephone or a fax is probably safer, though less efficient. It is possible that CMC has enabled an unknown number of idle diaspora Uyghurs, who otherwise might have been cut off from the mainstream, to actively join the cause, but this is sheer speculation. CMC may have contributed to the consolidation of Uyghur diaspora communities but not so much to the accomplishment of their nationalist goals.

Viewed retrospectively, the greatest achievement of the Uyghur digital diaspora is an unprecedented and widespread visibility toward the *outside*. Unlike the Tibetan issue that has been on international agendas since the early 1950s, if not before, the Uyghur predicament had been barely known to outsiders, let alone recognized, before the 1990s. Coinciding with the passing away of traditional leaders, the increased and sophisticated use of CMC has advertised the quest of Eastern Turkestan independence and the plight of Uyghurs in China throughout the world. Becoming aware of Uyghur nationalism, foreign governments, parliaments, leaders, politicians, statesmen, NGOs, and international organizations have been asked to intervene on behalf of the Eastern Turkestan independence movement. Some did, but primarily in *words*.

However, notwithstanding the extensive use of CMC, it should be admitted that the results so far are quite poor as far as *deeds* are concerned. Apart from some inconsequential resolutions, practically nothing has been really done to pressure China to improve the Uyghurs' conditions in Xinjiang, much less to promote their claims for greater autonomy, not to say independence. This is an outcome of a number of factors. For one, it is a tragic irony of history that while these transnational communities are sheltered by democratic governments, have better access to sophisticated communication media, and are free to pursue their vision of Eastern Turkestan independence, they lack an acceptable organized framework and, even more so, an acceptable and recognized world leader.

It was only in 2004 that the World Uyghur Congress was established, but it has yet to be universally recognized and politically effective. Its first leader,

Erkin Alptekin, the son of the legendary Uyghur leader Isa Yusuf Alptekin, quit after less than three years. His successor, Rebiya Kadeer, a Nobel Peace Prize nominee who had been jailed in China and left for the United States, is not yet universally known and recognized like the Dalai Lama, but she stands a good chance of success in winning the respect, if not the support, of the international community. Perhaps less articulate than Alptekin (especially in foreign languages), she is more authentic, intimately familiar with the situation in Xinjiang, and militant. Still, even without a recognized leader, Uyghurs have managed to win greater international attention than ever before but have a long, almost endless, road before they will reach a homeland of their own. Beijing's increased self-confidence, arrogant nationalism, and economic prosperity at home, and its regional and global impact, make the Uyghur mission much more difficult, if not impossible, to accomplish in the future, or ever—with or without CMC.

Even actual, and definitely virtual, transnationalism has its limits. Digitalization could be meaningless as an effective tool for promoting self-determination unless a national liberation movement is recognized as such by other nations, primarily the big powers. These nations and powers, however, are reluctant to support Uyghur national aspirations not simply because they are careful to avoid antagonizing the Chinese government and undermine their various interests in China but, furthermore, because they suffer from potential and also actual separatist threats themselves. Consequently, it seems highly unlikely that Uyghur nationalism and their quest for Eastern Turkestan independence will be accepted as *practically* (to distinguish from *theoretically*) legitimate in the foreseeable future. And no digital communication technology could ever change that.

NOTES

1. Press release by the Uyghur American Association, July 16, 2009. See also Szabo 2009.

2. This paper is part of a larger study titled "Uyghur Expatriate Communities: Domestic, Regional, and International Challenges," supported by a MacArthur Foundation grant, No. 02-76170-000-GSS, to which I am very grateful. Support for my ongoing research on Uyghur collective identity and ethnonationalism in Xinjiang has been provided by the Hebrew University through its Research and Development Authority, the Harry S. Truman Research Institute for the Advancement of Peace, and the Minerva Center for Human Rights. I would also like to thank my research assistants, Ofer Ben-Zvi, Itamar Livni, Ran Shauli, Zhang Hongbo, Gulhan Kariali,

and Shimon Sharbaf. This chapter is a revised, updated, and expanded version of Yitzhak Shichor, "Virtual Transnationalism: Uyghur Communities in Europe and the Quest for Eastern Turkestan Independence," in *Muslim Networks and Transnational Communities in and Across Europe,* ed. Jørgen S. Nielsen and Stefano Allievi (Leiden: Brill, 2003).

3. http://www.uyghurcongress.org/Uy/home/asp?ItemID=1278821765.

4. http://www.bethany.com/profiles/p_code1/1613.html; http://www.joshuaproject.net/peoples.php?rop3=110469, last updated June 2009.

5. These figures are based on a number of interviews with Uyghur activists held between 2002 and 2009. For somewhat different figures, see Federation of Atomic Scientists, Intelligence Resource Program, Uighur Militants Committee for Eastern Turkestan, http://www.fas.org/irp/world/para/uighur.htm. For a claim that 1 million Uyghurs live outside China, see Besson 1998, 162. The figure of 25 million Eastern Turkestanis, given on June 4, 1999, by Anwar Yusuf, at that time president of the Eastern Turkestan National Freedom Center in Washington, is inflated without proportion. Some estimates put the number of East Turkestani exiles in Kazakhstan at 1.5 million and the number of Uyghurs in Xinjiang at 13.5 million (in the early 1990s).

6. By 2002 there were about 500 Uyghurs in Germany (400 in Munich alone), 500 in Belgium (mostly from central Asia), 200 in Sweden (85 percent from Kazakhstan), 40 in England, 35 in Switzerland, 30 in Holland, and 10 in Norway. I am grateful to Mr. Enver Can, president of the Eastern Turkestan National Center in Munich, for providing this information.

7. For example, M. Ruhi Uyghur, "Doğu Türkistan ve Çin Tarihinde Bir Mühim Nokta" (An Important Point on the History of East Turkestan and China), in Landau 1981, 121, 139n142.

8. More details available at http://www.eastturkistan.com.

9. Manchuria was added in 1988.

10. Although the number of Internet users in central Asia has grown quite dramatically since 2000, the Internet penetration rate is still rather low: 12.4 percent in Kazakhstan, 8.8 percent in Uzbekistan, and 14.0 percent in Kyrgyzstan, compared to Japan (73.8 percent), Singapore (67.4 percent), and Malaysia (62.8 percent) (2008 figures). See http://www.internetworldstats.com/asia.htm.

11. http://www.geocities.com/CapitolHill/1730/index/html.

12. http://www.euronet.nl/users/sota/turkistan.html.

13. http://www.turkiye.net/sota/sota.html.

14. See http://www.euronet.nl/users/turkistan.htm for back issues.

15. http://eurasianet.org.

16. http://www.cacianalyst.org.

17. http://www/rferl.org.

18. http://unpo.org/member/eturk.html.

19. http://www.taklamakan.org. Named after the Takla Makan Desert in southern Xinjiang.

20. http://www.taklamakan.org/allied_com.

21. http://www/taklamakan.org/uighur-l/index.html.

22. http://www.taklamakan.org/etib. It is also available at http://www.geocities.com/CapitolHill/1730/etib, mentioned above.

23. http://www.Uyghur.org/wunn and http://www.uyghurcongress.org.

24. http://www.uyghurs.org.

25. http://www/uyghur.org.

26. http://uyghurinfo.com.

27. http://iuhrdf.org and http://www.uhrp.org.

28. http://www.eastturkistan.com.

29. Such as http://www.uighurlanguage.com. Other Web sites include http://www.kivilcim.com, http://www.Uyghur.net, http://www.turpan.com, and http://www.doguturkistan.net. http://www.Uyghur.com runs the *Uyghur Awazı Radiosı* (Voice of the Uyghur Radio), no. 16 of which was released on December 14, 2000. Though completely blocked from time to time, most likely by the Chinese, this Web site, probably based in Turkey, is still active as of the time of the writing of this chapter.

30. http://www.ccs.uky.edu/~rakhim/et.html.

31. http://www.theuygur.com.

32. http://et.4t.com/uyguryari.html. MSN group (under a Chinese name, from Hamburg).

33. For Uyghur Web sites reportedly blocked by China, see Jonathan Zittrain and Benkamin Edelman, "Empirical Analysis of Internet Filtering in China" (available at http://cyber.law.harvard.edu/filtering/china/China-E.html). Blocked Uyghur Web sites included the Eastern Turkestan National Center (http://www.eastturkestan.com, overwritten probably by the Chinese) and Uyghur Information Agency (http://www.uyghurinfo.com, no longer accessible), both partly blocked; the Uyghur American Association (http://www.uyghuramerican.org, still active); Radio Asia (http://www.uygurs.com, overwritten); the International Taklamakan Human Rights Association (http://www.taklamakan.org, no longer accessible); as well as http://www.uyghurs.org, http://www.uygur.com, http://www.turpan.net, http://www.uygur.net, and http://www.theuyghur.com (all overwritten); http://www.uygur.org (still active); and the *Gök Bayrak* (Heavenly [Blue] Flag) Association (http://www.gokbayrak.com, still active)—all totally blocked.

34. On the possible "evaporation" of the nation-state under the influence of new technologies, see Negroponte 1995, 165. On the possible decline of political parties and civic associations, see Grossman 1995, 16.

35. Emphasis added. For an extensive discussion of transnationalism, see Vertovec 1999, 2009.

REFERENCES

Alptekin, Isa Yusuf. 1973. *Doğu Türkistan Davası*. Istanbul: Marifet Yayınları.
Appadurai, Arjun. 1996. *Modernity at large: Cultural dimensions of globalization*. Minneapolis: University of Minnesota Press.
Benson, Linda. 1990. *The Ili rebellion. The Moslem challenge to Chinese authority in Xinjiang, 1944–1949*. New York: M. E. Sharpe.

Besson, Frédérique-Jeanne. 1998. Les Ouïgours hors du Turkestan oriental: De l'exil à la formation d'une diaspora. *Cahiers d'étude sur la Méditerranée orientale et le monde turco-iranien,* no. 25 (January–June).

Bimber, Bruce. 1998. The Internet and political transformation: Populism, community, and accelerated pluralism. *Polity* 31, no. 1.

Brophy, David. 2005. Taranchis, Kashgaris, and the "Uyghur question" in Soviet central Asia. *Inner Asia* 7, no. 2: 163–84.

Buğra, Mehmet Emin. 1952. *Doğu Türkistan: Tarihi, Çoğrafi ve Śimdiki Durumu.* Eastern Turkestan, Istanbul: History, Geography, and the Present Situation.

———. 1954. *Doğu Türkistan'in Hürriyet Davası ve Çin Siyaseti.* Istanbul: Freedom of East Turkestanis and China's Policy.

Chase, Michael S., and James C. Mulvenon. 2002. *You've got dissent! Chinese dissident use of the Internet and Beijing's counter-strategies.* MR 1543. Santa Monica: RAND.

China Internet Network Information Center. 2009. *The 23rd statistical survey report on the Internet development in China.* January. PDF.

China's Xinjiang Posts 45.6 percent Jump in New Internet Users. 2004. *AsiaPulse News,* August 17.

Dahan, Michael, and Gabriel Sheffer. 2001. Ethnic groups and distance shrinking communication technologies. *Nationalism and Ethnic Politics* 7, no. 1: 85–107.

Eastern Turkestan Information Bulletin. 1993. Vol. 3, no. 6 (December).

Forbes, Andrew D. W. 1986. *Warlords and Muslims in Chinese central Asia: A political history of republican Xinjiang, 1911–1949.* Cambridge: Cambridge University Press.

Foster, William, and Seymour F. Goodman. 2000. *The diffusion of the Internet in China.* Stanford: Center for International Security and Cooperation, Institute for International Studies.

Gayretullah, H. B. 1965. *Osman Batur.* Istanbul: Orna Yayınları.

Geng, Shimin. 1984. On the fusion of nationalities in the Tarim basin and the formation of the modern Uighur nationality. *Central Asian Survey* 3, no. 4: 1–14.

Gladney, Dru C. 1990. The ethnogenesis of the Uighurs. *Central Asian Survey* 9, no. 1: 1–28.

———. 1992. Transnational Islam and Uighur national identity: Salman Rushdie, Sino-Muslim missile deals, and the trans-Eurasian railway. *Central Asian Survey* 11, no. 3: 1–21.

———. 1996. Relational alterity: Constructing Dungan (Hui), Uyghur, and Kazakh identities across China, central Asia, and Turkey. *History and Anthropology* 9, no. 4: 445–77.

Grossman, Lawrence K. 1995. *The electronic republic: Reshaping democracy in the information age.* New York: Viking.

Hobsbawn, Eric. 1991. *Nations and nationalism since 1780.* Cambridge: Cambridge University Press.

Husband, Charles, ed. 1994. *A richer vision: The development of ethnic minority media in Western democracies.* Paris: UNESCO.

Johnson, E. 2001. Left behind in the rush to go online. *Turkistan Newsletter* 5, no. 109 (July): 11–12.

Jones, Steven G. 1998a. Information, Internet, and community: Notes toward understanding of community in the information age. In *Cybersociety 2.0: Revisiting Computer-Mediated Communication and Community*, ed. Steven G. Jones. Thousand Oaks, Calif.: Sage.

———. 1998b. Introduction to *Cybersociety 2.0: Revisiting computer-mediated communication and community*, ed. Steven G. Jones. Thousand Oaks, Calif.: Sage.

Kalathil, Shanthi. 2001. China's dot-communism. *Foreign Policy*, no. 122: 74–75.

Karim, H. Karim. 1998. *From ethnic media to global media: Transnational communication networks among diasporic communities*. Paper WPTC-99-02. International Comparative Research Group, Strategic Research and Analysis, Canadian Heritage.

Koçaoğlu, Timur. 1997. A national identity abroad: The Turkistani emigree press (1927–1997). *Central Asia Monitor*, no. 6.

Landau, Jacob M. 1981. *Pan-Turkism: A study in irredentism*. London: C. Hurst.

Mackerras, Colin. 1972. *The Uighur empire: According to the T'ang dynastic histories*. Columbia: University of South Carolina Press.

———. 1990. The Uighurs. In *The Cambridge History of Early Inner Asia*, ed. Denis Sinor. Cambridge: Cambridge University Press.

Millward, James A. 2007. *Eurasian crossroads: A history of Xinjiang*. New York: Columbia University Press.

Negroponte, Nicholas. 1995. *Being digital*. New York: Vintage.

Portes, Alejandro. 1997. *Globalization from below: The rise of transnational communities*. Paper WPTC-98-01. Princeton: Princeton University.

Riggins, Stephen Harold, ed. 1992. *Ethnic minority media: An international perspective*. Newbury Park, Calif.: Sage.

Rudelson, Justin Jon. 1997. *Oasis identities: Uyghur nationalism along China's silk road*. New York: Columbia University Press.

Schiffauer, Wener. 1999. *Islamism in the diaspora: The fascination of political Islam among second generation German Turks*. Paper WPTC-99-06. Frankfurt and Oder: Lehrstuhl Vergleichende Kultur und Sozialanthropologie.

Shichor, Yitzhak. 1994. Separatism: Sino-Muslim conflict in Xinjiang. *Pacifica Review* 6, no. 2: 71–82.

———. 2003. Virtual transnationalism: Uyghur communities in Europe and the quest for eastern Turkestan independence. In *Muslim networks and transnational communities in and across Europe*, ed. Jørgen S. Nielsen and Stefano Allievi. Leiden: Brill.

———. 2009. *Ethno-diplomacy: The Uyghur hitch in Sino-Turkish relations*. Policy Paper 53. Washington, D.C.: East-West Center.

———. n.d. Thorn in the flesh: Uyghurs in China's relations with central Asia. Unpublished manuscript 1.

Statistics Bureau of Xinjiang Uygur Autonomous Region. 2008. *Xinjiang statistical yearbook, 2008*. Beijing: China Statistics Press.

Szabo, Christopher. 2009. The case of the Uyghur email attacks. *Digital Journal,* July 18. http://www.digitaljournal.com/article/276144.

Vertovec, Steven. 1999. Conceiving and researching transnationalism. *Ethnic and Racial Studies* 22, no. 2 (March): 447–62.

———. 2009. *Transnationalism.* London: Routledge.

Wang, David D. 1999. *Under the Soviet shadow: The Yining incident, ethnic conflict, and international rivalry in Xinjiang, 1944–1949.* Hong Kong: Chinese University Press.

Yang, Dali L. 2001. The great net of China: Information technology and governance in China. *Harvard International Review* (Winter).

17 Migrate Like a Galician
The Graphic Identity of the Galician Diaspora on the Internet

XABIER CID AND IOLANDA OGANDO

In autumn 2006 an advertising campaign featuring the slogan "Vivamos como galegos" (Live like the Galicians) was broadcast across all Galician TV stations and cinemas. This campaign had a great impact not only on the physical region of Galicia but also through Web sites and virtual communities. The commercial,[1] commissioned by a Galician supermarket chain, makes use of a series of clichés, which Galicians should—supposedly—be proud of when inciting people to consume local products. The commercial therefore commences in an airport car park, where a family picks up a traveler in his thirties: the very first cliché presented on the idea of being a Galician is that of emigration.

GALICIA AND EMIGRATION

Galicia is a region on the Atlantic European coast, which has belonged to the Kingdom of Spain and its predecessors, the Kingdoms of Castile and León, for the past eight hundred years at least. Galician is widely spoken there (Monteagudo 1999, 2001), a language that is different from Spanish, and shares a common root with Portuguese. It is in the language and in other more or less endemic features that Galicians have grounded their feeling of identity, a feeling that has had clear and steady political translation since the latter years of the nineteenth century (Beramendi and Núñez Seixas 1996; Beramendi 2007). Today, Galicia is an autonomous region in Spain enjoying a certain degree of self-government. Most Galicians identify themselves with both Galicia and Spain (70 percent), even though 30 percent consider themselves more Galician than Spanish.[2]

As suggested by the cliché in "Vivamos como galegos," Galicia has been traditionally an emigrant country. The extensive bibliography written on this particular subject alternates between the negative aspects of emigration—especially from the point of view of its effects on people but also on the economy[3]—and a more balanced view, aware of the fact that Galician emigration has contributed to the economical, political, social, and cultural modernization of Galicia.[4] Nevertheless, the migratory phenomenon has been so wide and intense that it has become impossible to understand Galician identity without its being mentioned.

Galician emigration, the "Great Galician Emigration" to Latin America, took place only from 1870 onward, and it replaced other communities of emigrants, such as Italians, who had arrived earlier. It can therefore be considered a relatively late phenomenon in the wider European context. However, from 1870 on, emigration remains constant, and between 1860 and 1960 almost 2 million Galicians left their country temporarily or definitively. More than half of these 2 million emigrated between 1900 and 1930.[5] Just to provide a point of reference, the population of Galicia in 1900 was nearly 2 million habitants, and it currently does not come up to 2.8 million.[6] In fact, it is purported that a third of Uruguayans have Galician blood running through their veins or that Buenos Aires is the fifth Galician province (apart from the other four existing within the metropolitan territory). Forty percent of all Spaniards who have left the Peninsula during the past two centuries were Galicians (Eiras Roel 1991; Emigración 2002), even though Galicians made up only between 9 and 10 percent of the total population of Spain (Villares 1997a; see figure 17.1).

Galician emigrants have been mostly men, particularly during the emigration to Latin America at the beginning of the twentieth century. Women remained behind, taking charge of the domestic economy and supporting large families. The women's situation was so desperate that they were denominated *viúvas dos vivos* (widows of the living) by Rosalía de Castro[7] (1837–1885)—the enigmatic poet and indisputable reference for an axial identity within Galicia. In addition, it is interesting to point out that when these men left the country they usually settled in much bigger cities than those in Galicia—La Havana, Buenos Aires, Montevideo, or Mexico City

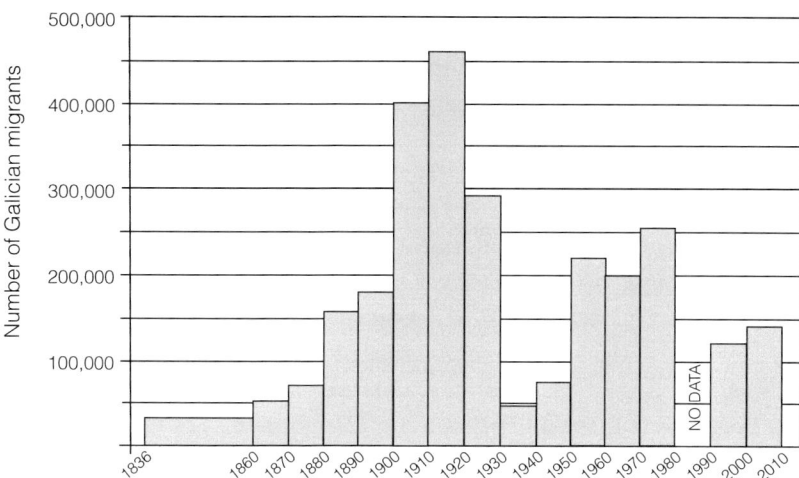

Figure 17.1 Galician migrants, 1836–2010
Source: Ogando and Cid, from data of Eiras Roel, Sixirei Paredes, CIG and IGE.

and later Frankfurt, Zurich, or Paris. As a result, they learned how to move around a Big City and inside the associative, cultural, and economic tissue of the metropolis—that is, for the first time, an urban Galician community was formed.

However, the separation of the families—among other factors—favored communication with Galicia, as well as the high rate of emigrants who decided to return temporarily or definitely. This separation also helps to explain the huge economic and cultural investment that they made in their areas of origin, even though these investments were seldom in the form of industry and more often than not were in the form of property and education. The hand of the emigrant has taken airfields and satellite Internet connections to poor, small, isolated villages in inland Galicia,[8] but at the same time it has bled this territory of its workforce for more than a century (Villares 1996, 1997b; Núñez Seixas 1992, 2001; Sixirei Paredes 2001; Calo Lourido et al. 1997). Galicia can be said to have gone from subsistence agriculture to the information society without having experienced an industrial revolution. After all this, it is easy to see why the migratory phenomenon is one of the main axes of Galicians' identity.

THE GALICIAN EMIGRANT: A PORTRAIT

Galicians emigrated from both the coast and the interior of the country, and most of them were peasants. They were widely known during the Ancien

Régime due to their industriousness in temporary migrations to Spain and Portugal (Leira 2007; Sixirei Paredes 1995a; Villares 1996, 1997a; Núñez Seixas 2001). In 1883 French utopian Paul Lafargue, in his essay on sloth, even considered Galicians to be one of the five damned races on the planet because of their vocation for hard work.

However, and despite the fact that this image persisted in Latin America and Europe during the nineteenth and twentieth centuries, the Galician emigrant was also regarded as a miser, stubborn, stupid, and poor, the butt of jokes and parodies in numerous plays and films. A clear example can be found in the Mafalda comic strips by Argentinean cartoonist Quino, published between 1964 and 1973. One of the most remarkable secondary characters is Manuel Goreiro, a.k.a. Manolito, son of a brutish, selfish, unsympathetic, and tightfisted Galician shopkeeper. Just by googling the word *gallego*—"Galician" in Spanish—we find the amount of jokes Argentineans dedicate to the Galicians:[9] their lack of intelligence is taken to the realm of the absurd in this type of Latin humor.

Nevertheless, since it is almost impossible to find a Galician who does not have family that has not experienced the diaspora (Núñez Seixas 2001), Galicians have also created a vivid image of their own emigrants (Souto 1999). It is a portrait in which tearful mourning for those who leave—again, as presented by the national poet Rosalía de Castro[10] but also as seen in the extremely well-known photography[11] of the emigrant taken by Manuel Ferrol in 1953—alternates with the joy of leaving behind underdevelopment and scarce future prospects, as shown by a popular folk song: "Aí vos quedades, aí vos quedades, entre curas, frades e militares" (There you remain, among priests, monks and soldiers) (ibid.). This feeling of melancholic nostalgia for the land left behind has become a key determinant for the way in which Galicians identify themselves and are identified by others.

In fact, the Galician word for this feeling, *morriña* (homesickness), is one of the very few, along with a half-dozen fish and names of seafood, that the Spanish language has borrowed from Galician. Also, the relations between emigrants ("people from outside") and their relatives in Galicia ("people from the country") have been the source of countless petty conflicts in recent history. To this the recent controversy on political participation of children and grandchildren of emigrated Galicians should be added. Although some consider the vote a symbol of belonging[12] to a community (and a way to access health-care services), others do not want to have their taxes decided by people who do not live—and will probably never live—in their town. In some parts of Galicia the electoral register of emigrants is higher than that of actual residents, and emigrants have already had the opportunity to determine many electoral majorities in the Galician government.

STAGES AND DESTINATIONS OF GALICIAN EMIGRATION

In broad terms, studies on Galician emigration have proposed a periodization in three basic stages: emigration during the Ancien Régime (until 1836), modern emigration (1836–1890), and contemporary emigration (1950–1980). Each stage corresponds to a preferred destination: during the Ancien Régime emigrants headed especially for other parts of Spain and Portugal, while the favorite destination for modern emigration was the American continent; contemporary emigration sees Galicians leaving for affluent countries in western Europe, as well as Venezuela, and more industrialized areas in Spain—Madrid, Catalonia, and the Basque Country.[13]

Ancien Régime emigration will not be discussed in this article: owing to economic and political factors (to escape from military recruitment) young Galicians left to seek temporary employment such as casual laboring, shoe shining, or water carrying, but they often returned after a few weeks or months. The rate of emigrants in relation to the inner population is not significant yet, and all that has remained of it are a few features of the way Galicians are perceived and, again, some combative lines penned by Rosalía de Castro.[14]

Modern Emigration: Galicia in America

From 1870 onward Central America (Mexico and Cuba) and South America (Argentina, Uruguay, Brazil) of the American continent received a large number of Galician emigrants, with figures reaching their peak in the decades of 1910 and 1920. Ramón Villares (1996, 1997b) has pointed out the role played by migratory chains in this process. Indeed, the accounts of those who returned (or their correspondence) created a very beautiful image of economic and career prospects in Latin America, a temptation reinforced by the opening up of immigration policies launched by those countries at the time. This chain effect favored the gradual move of the Galician rural organization to these new countries: relatives and neighbors were still relatives and neighbors in Buenos Aires, despite Buenos Aires being ten thousand times bigger than their village.

In a way, some of the causes of Galician emigration in this period were greatly tied up with the lack of prospects in Galicia, an excess of population in relation to land property, occasional agricultural crises, and obviously increasingly lower prices and improvements in the means of transport. Also, according to Núñez Seixas (2001) and Gabriel 2006, emigrants belonged to those sections of the population with the highest literacy levels, and to those with a better chance of finding a job at their destinations.

In other words, this particular emigration was not a result of famines or peasants embracing proletarian ideas. It arose from the contrast between a world that was going through a process of modernization outside and the old agricultural structures that still remained inside, hindering the economic and cultural development of the individual (Villares 1996, 1997b). Emigrants reproduced in the new cities the same relations that they had in Galicia: they found associations of people from the same parish, and they coordinated their activities in order to create societies of Galicians independent from the Spanish government.

These were the most booming associative structures in the history of Galicia (Núñez Seixas 1992, 1993, 2001): banks were financed, health-care services were created, districts of houses and cemeteries were built, money was invested in the educational development of Galicia, a relevant part of the Galician literature published up to 1950 was produced (the Galician Language Academy and the Galician anthem were all launched in La Havana), and—last but not least—parties, dinners, and balls were organized. Furthermore, after the Spanish civil war (1936–1939), these societies gave refuge to most of the exiles fleeing Franco's dictatorship. The possibility of a return to the motherland justified this effort to form societies. Villares (1996, 1997a) points to the fact that the Galicians who lived in America would not have maintained such strong ties with their emigrated relatives and neighbors if return had not been a real possibility in their minds.

On the whole, Galician societies founded in Latin America reproduced the ideology of the Galicia that they had left behind, in the moment they left: low involvement in politics and a regionalist view of Galician identity. In other words, the preservation of the customs and traditions was paramount, rather than the fight for political autonomy.[15] These were the dominant trends in late-nineteenth-century Galicia, reproduced by Galician associations in Latin America at least until the exiles transformed, up to a certain point, this situation from 1936 onward.

From 1930 on, the descent in the number of emigrants in the American continent was unstoppable. The arrival of exiles in the period between 1936 and 1939 represented a more symbolical than significant increase in these numbers. The aging of the group, dispersion, the integration of children and grandchildren in destination societies, along with returns resulted in Galician societies being less and less important, their functions blurred, and some of them eventually disappearing (Sixirei Paredes 1995b, 2001; Casal Lodeiro 2006).

Contemporary Emigration: Galicia in Europe

As in other parts of the planet, the world economic crisis of 1929 and global military conflict brought migratory movements to a halt for economic reasons. Nevertheless, from 1950 onward the rapid industrialization and development of certain areas in Europe and Spain favored the opening up of emigration policies to receive more workers.

The years 1950 and especially 1960 marked a substantial shift in the direction of Galician emigration, which now focused on European countries such as Germany, Switzerland, or France and other parts of Spain, like the more industrialized areas of Madrid, Barcelona, and the Basque Country (Sixirei Paredes 1995a; Núñez Seixas 2001; Villares 1996, 1997a). In addition, the Venezuelan oil boom (Sixirei Paredes 1995a; Núñez Seixas 2001) led to emigration to that country, a movement that also corresponded to the aforementioned model despite the fact that the destination was Latin America.

Even though the sociohistorical conditions in which the shift to Europe took place were very different from those of the emigration to the Americas, there were some common characteristics. For instance, emigrants were again primarily men, although women often left the country to join their husbands or, if they were single, ended up marrying other emigrants. Now it was the children who were left behind in Galicia and looked after by their grandparents.[16]

However, the integration of these new emigrants was significantly harder, especially as regards European destinations. The cultural and linguistic chasm existing between a German factory manager and a peasant from Galicia in Franco's time was much larger than the one that had existed fifty years before in Argentina or Mexico. Therefore, the socialization among Galician emigrants was lower.

On the one hand, it was now technically much easier for the emigrants to keep in touch with their real families—every year they returned to Galicia to spend their holidays there, and the use of the telephone became widespread. On the other hand, their identity became stigmatized by their low social status. These emigrants had been brought up in Franco's dictatorship, and very few of them were familiar with how democracy works. That is to say, it was difficult for them to create and maintain social networks, and for the migrants within Spanish territory this may have been politically compromising. Furthermore, societies no longer offered financial, medical, or social assistance; they did not stimulate collective work in Galicia. Only very few members gathered (Sixirei Paredes 2001) for traditional celebrations or gastronomic parties. Latin American emigration had exported regionalism and clichés about Galicia that began to circulate at the beginning of the century. In contrast, for those who immigrated into Europe, the image of Galicia merged with the old regionalist patterns of identity, became blurred with Spanish identity, or simply did not exist, especially if it could entail a politically prejudicial issue.

The Present Emigration

Democracy, the improvement of general economic conditions in Galicia, demographic cancellation—Galicians have had close to zero vegetative population growth in the past few years[17]—and the arrival of cheap labor in Europe (Turks and Arabs in Germany or Switzerland, Pakistanis, Latinos, and Arabs in Spain) brought Galician emigration to a halt during the 1980s. However, figures recovered from 1990 onward, as economic differences between Galicia and Europe became more and more obvious. Since the true number of emigrants has been concealed, it is very difficult to apply any methodology to obtain reliable data. In fact, in *Galicia 2000*—a booklet with information about Galicia and propaganda of the Galician government's policies—the following statement is made: "Nowadays, Galicians do not have to emigrate." Nevertheless, other sources[18] maintain that 175,000 people—that is, 7 percent of the population, and 28 percent of those aged between twenty and thirty-five years old[19]—left Galicia between 1990 and 2003.

Destinations vary and jobs are diverse. For instance, the Canary Islands[20] and Andorra attract those who want to work as bricklayers or in the hotel and catering businesses, Madrid and Barcelona offer employment in the service sector, and those who want to practice medicine opt for the United Kingdom or Portugal. For the first time emigrants are mostly skilled workers (Yáñez 2007) who cannot find career prospects within the stagnant Galician economy. These emigrants spread out around the world without fixed destinations, from Japan to California, from Australia to Sweden.

Nevertheless, at the end of the twentieth and the beginning of the twenty-first centuries, more accessible means of transportation have allowed emigrants to return to Galicia not once a year but once a month. Associations and societies in those areas are reduced or have disappeared, especially among those Galicians living in countries where the density of the Galician population is not very high. Even in places where the ratio of Galicians is higher (e.g., in the Canary island of Fuerteventura 15 percent of the population is from Galicia),[21] the associations lack practical functions. After all, it takes only two hours to get to Galicia. However, the perception of Galician identity is much stronger in this third stage of emigration than in the previous ones, as significant sectors of the emigrant population are politically active and support a clearly political nationalism reproduced in those identity symbols employed by the new groups.

EMIGRATION ON THE INTERNET

It was not easy to fit 150 years and two million people in this short and accelerated history of Galician emigration, and obviously there are still many aspects to clarify and details that have been omitted. However, it was essential to explain who the Galicians are and when and how they spread out around the world before dealing with the real objective of this article: to examine how these different types of emigration have produced a series of images of Galician identity, and especially how the model of community and iconographic identity representation are conditioned by the use of the Internet in relation to emigration.

The following conclusions emerge from an analysis of the representation of Galician emigrants' identity on the Internet:

1. The graphic resources used refer not to a different conception of what Galicia is nowadays but to the vision of Galicia at the time of emigration. As far as earlier emigration is concerned, this iconography has become fossilized and has been handed down practically unchanged from parents to children and from children to grandchildren.
2. The relative importance of iconographic representation varies in the Internet products along the different migratory stages.
3. The systems used to create networks of Galicians on the Internet depend strongly on the moment and context in which the act of migration occurred, before or after the use of the World Wide Web became widespread.

PRE-WEB GALICIAN EMIGRANTS ON THE INTERNET

A superficial look at the Web sites of Galician centers and associations in Europe and South America (currently more than four hundred) reveals a clear identity checklist (Thiesse 2001), which reveals how Galician emigrants see Galicia. Iconographic motifs are surprisingly homogeneous and—despite differing nuances—similar on both sides of the Atlantic, both in centers founded in the nineteenth century run now by grandchildren or great-grandchildren of the original pioneers and in the more recent ones in Switzerland or Germany where first-generation emigrants still gather together.

The identity checklist has as its common denominator an obvious fossilization in the choice or reinforcement of old symbols. That is, all these Web sites present symbols that refer not so much to Galicia but to the Galicia they left behind. There is the *pote,* a special three-legged pot used to cook the *caldo* (or Galician broth) in the fireplace. Bagpipers, dressed in their traditional Galician costume, and the *hórreos,* a typical maize grainer unique to Galicia and to the nearby areas, are also featured. However, nowadays Galicians use modern kitchens and appliances, they play bagpipes in jeans, and, of course, they—the very few who still grow it—do not store corn in a construction built on stone pillars to prevent mice from eating the grain. In the same way, the cross of Santiago (González Millán 1993)—a heraldic symbol commonly used at the end of the nineteenth century to represent Galicia—has disappeared in present-day Galicia due to its religious meaning. It remains only in the emblems of some soccer clubs (founded in the 1920s) and in the Web sites of Galician emigration centers.

Another interesting development in Galician identity construction is the employment of Celtic iconography. The very first theories claiming that Galicia—unlike the rest of Spain—had a Celtic past date back to the mid-nineteenth century and were elaborated by historians and pioneers of the development of regionalism or Galician nationalism such as Manuel Murguía (1833–1923). Regardless of the fact that those theories have been harshly refuted by science, triskelions[22] (round swastikas) are found everywhere, although it is important to point out that these Celtic motifs are void of any political or autonomy-orientated claim, and they do not come into conflict with Spanish identity.[23]

This is the Galician checklist, but it coexists with a shorter one—that is, the Spanish checklist. In fact, many of the Galician centers and societies have such pompous names as Paraná Galician Club–Spanish Centre or Spanish Charity Society, whose founding ceremonies mixed both nationalities and a joint sense of belonging. Symbols that are undoubtedly perceived as alien within inner Galicia (such as flamenco) have their own place on emigration Web sites.[24] Web pages showing successive images of key places in the Galician capital advertise Spanish lessons in Brazil as well. The colors sky blue and white on the Galician flag are therefore often combined with the red and yellow in the Spanish flag in a nongarish way.

Another feature of this regionalist vision revolving around popular customs is the way in which language is treated. At the end of the nineteenth century and throughout most of the twentieth century, Galician was spoken by the vast majority of the population; it was, however, banned in other more formal spheres. Present in all the streets and villages, it had no place whatsoever either in offices or in public communication. The political demands levied toward the end of the twentieth century have transformed this situation, and Galician has started the twenty-first century as the most widely spoken language—but closely followed by Spanish (González González 2004)—and this time firmly settled in more formal fields. For instance, all public communication of the Galician government's administration is transmitted solely in Galician. However, emigration Web sites are written mostly in Spanish, and only a few of them include a bilingual version.

It could be argued that users of these Web sites cannot speak Galician anymore, and this may be true: the number of Galician speakers has dropped inside and outside Galicia. Nonetheless, the point of our analysis is not the communicative value of the language but its symbolic value. As far as emigration Web sites are concerned, Galician is not a symbol of identity, even though it plays a central role in the configuration of internal Herderian-based nationalism.[25]

These Web sites have been mentioned because they represent the main support of pre-Internet emigrants on the Net. Such Web sites, the substitute of physical notice boards in the associations' headquarters, are constituted by the basic structure of history plus committee plus calendar. They are Web sites 1.0, with very few forums and very reduced interactivity. Blogs are scarce, and there is only one group on Facebook—the London Galician Centre[26]—with a bilingual English-Spanish version and featuring pictures of their typical dances along with the Santiago cross as its avatar. More important, these scarce communication tools operate independently of metropolitan Galicia. Galicians who live in the inner parts of the region do not participate in these networks.

Also, a high percentage of pre-Internet emigrants have communicated with Galicia only to find out more about their family name or to look for information about the village where their grandparents once lived. Pre-Internet emigrants do not take part in the main Galician social networks, and the Internet has brought very little to them. Contrary to the predictions of Internet pioneers (Rodríguez and Cid 2001; Romero and Vaquero 2001), the Internet has not connected the old diaspora with Galicia. It does not seem to have enabled contact between the members of this diaspora with each other, either. For example, the Galician associations in Brazil remain as far away from those in Switzerland as they were forty years ago.

THE POST-INTERNET EMIGRANTS

We denominate a post-Internet emigrant as somebody who left Galicia after the popularization of the commercial Internet. By this we refer to a not inconsiderable percentage of young people with different professional profiles who have been emigrating all over the world since 1990, and especially from 1996 onward. They left Galicia when the Internet already existed. The first Galician private domain, http://www.vieiros.com, was registered in 1995, and the media that became a referent for the Galician community appeared in February 1996.[27] However, at the same time the Internet has helped emigrants to leave Galicia. It has given them the ability to book plane tickets, make phone calls, read the news in Galician papers, follow institutional announcements, watch videos from Galician TV on YouTube, listen to Galician radio stations, and keep their messenger accounts active.

Emigrants in the past ten years have not founded associations hardly anywhere, and even when they have these societies have been given a very different profile. For instance, the Alexandre Bóveda Association of Fuerteventura (Canary Islands) is named after a politician who defended nationalist ideas and was assassinated in August 1936. Likewise, these emigrants have remained distant from already existing societies, as is the case in Madrid and—as a general rule—in Catalonia as well.

However, this does not necessarily mean that these migrants are not on the Net. They use their previous messenger accounts and social networks, they participate in networks of news syndication such as Chuza! (Digg's Galician clone),[28] or they write blogs and photoblogs read by their old friends (not the Galicians living close to them but the friends that they left behind in their homeland). Because of the Internet, new emigrants keep their community connection with Galicia alive, but they do not establish networks with other Galicians inhabiting the diaspora. One of the reasons for this behavior may lie in the regionalist character of the associations. In other words, Galician centers in Europe or Spain refer in their images or activities not to the Galicia of the new emigrants but to the Galicia of their parents or grandparents. Diasporic Galicia of pre-Internet communities seems too distant in terms of time and space to those who have left recently.

On the other hand, iconographic identity is completely different from that of the pre-Internet era. Web sites have disappeared, and so have the spaces in which to display identity icons—the pictures uploaded in photoblogs refer not to the identity past but to the present moment. The message they send to their peers is not "we are all Galicians here" but "here's how I live in this country." Broadly speaking, gone are the *potes,* the *hórreos,* and the triskelions.

Those who are active members in a nationalist party will probably include a flag, not as an identity symbol of origin but as a sign of embracing a certain ideology. Furthermore, the flag will not be just a Galician flag but will be completed with the symbols of their party.[29] Similarly, there is a change in the use of the Galician language, which is used in the posts of many blogs written in Sweden, Lanzarote, Berlin, or Barcelona, at the level of both communication and symbolic representation, such that present-day Galician language is the principal symbol of identity.

REALITY AND ITS INTENTION: DIASPORA AND THE WILL TO BE A COMMUNITY

However, this general description would not be valid if there were no exceptions, exceptions that in some cases arise from a specific will to break away from the described model, while in others they have entered through the peripheries of the system.

A Pre-Internet Network of Galicians: Fillos de Galicia

When in 1997 Manuel Casal launched the Web site Fillos de Galicia (Children of Galicia),[30] he had very limited resources at his disposal but a clear will to create a diaspora network on the Internet. Manuel Casal was a student of computer science, and his parents are both emigrants living in the Basque Country. He has had two types of experience regarding migration and settlement. On the one hand, Casal has participated in Galician associations and become disappointed with them. On the other hand, in an environment marked by the presence of Basque nationalism, he has evolved toward Galician nationalist stances. He saw in this type of Web site an economic opportunity and has developed an interesting business aimed at providing theoretical support for the creation of online community Web sites in the Galician diaspora (Casal 2006; Yáñez and Yáñez 2008).

Nonetheless, this online tool was not economically feasible. It is true that in ten years he has managed to obtain more than five thousand registered users, but only about fifty (1 percent) helped him to finance the project. In addition, he did not obtain the support that he expected from the Galician administration, either. Moreover, the number of registered users is much higher than the number of real users. That is, a short perusal of the user list reveals that many users are repeated again and again (through spam and so forth), and this leads to the conclusion that there are more than three thousand real users, half of them in Argentina and 10 percent in Galicia. However, there is not one in the United Kingdom, for instance, a place where many Galicians have emigrated in the past forty years.

Fillos de Galicia displays connections with regionalism, especially at an iconographic level. Indeed, the checklist includes a fusion of regionalist elements with some nationalist features, resulting in an identity overload. For example, the face of Rosalía de Castro appears at the front of the Web site. In the background, we find the image and works of Alfonso Daniel Rodriguez, "Castelao"—the second most important identity literary referent in Galicia, the first nationalist member of the Spanish Parliament (1931), and the figure largely responsible for the symbolic ratification of the Galician Autonomy Statute (1936–1938). In addition, *hórreos,* flags, Romanesque architecture, the green landscape of the valleys, rubrics of rural inspiration for each section or articles on ethnography, popular architecture, and typical foods are also featured in the different thematic appendixes.

However, and despite the fact that the Web site is quite participation oriented—it has forums and a chat room—it has created little sense of community. Users enter this Web site not to contact other members but to satisfy a particular need, and mostly to locate an ancestor or to find out the meaning of their family name. Therefore, the communication established is individual and unidirectional instead of multidirectional or bidirectional. Casal has certainly tried to create a community. He has carried out publicity campaigns, started a profiles Web site so that emigrants could contact each other, has set up an online shop,[31] and has even promoted other parallel projects of a clearly nationalist and politically engaged nature.[32] Yet an associated community has not grown from this endeavor, and it probably never will.

A POST-INTERNET GALICIAN NETWORK: GALICIA GLOBAL

Galicia Global[33] was launched in May 2007 as a government initiative on the occasion of both Internet Day and Galician Literature Day[34] (both coinciding, by chance, on May 17). This initiative came from a sector of the administration under nationalist tutelage, and was clearly in tune with one of the two leading views on emigration of Galician nationalism. One perspective assumes that over the past two centuries Galicians have established a Global Galicia, and therefore the union and activation of that network will have a positive impact on Galicia at both economic and social levels. The other perspective is, paradoxically, the opposite. That is, emigrants should not be allowed to vote in the elections in Galicia, and they should be erased from "our map."

In contrast with Fillos de Galicia, Galicia Global is not an initiative coming from emigrants; instead, it is aimed toward emigrants with a fairly limited purpose: "to enable all men and women from Galicia to complete a brief personal file and to place it in a world map."[35] It is devoid of interactive tools. In January 2008 Galicia Global had reached one thousand users, and even though these figures are quite reliable, half of these users reside in Galicia. This Web site was publicized by the Galician media and social networks. As a result, its main recipients were post-Internet Galicians. In fact, there are more registered users in London than in the whole of Argentina.

It is worth mentioning the way in which avatars are selected to be later distributed all over a Google map. In this sense, users can choose from a wide range of Playmobil figures wearing either present-day clothes or regional costumes. Nonetheless, this Playmobil game is not regionalist but parodic. Fillos de Galicia features avatars in regional costumes as well, but there is a basic difference between these two Web sites. Where Fillos de Galicia aims for realism, Galicia Global is funny, childish, and amusing. Obviously, this initiative has not been successful, either. Practically useless interrelation tools have hindered the creation of communities. At most, it was useful for some Galicians living in the South Island of New Zealand to find out that there is another Galician who lives on the North Island and tracks whales in the Pacific Ocean. But this does not qualify as a community.

Mafia Gallega

At the edges of the system appear some examples of small networks, such as the Mafia Gallega.[36] It has nothing to do with organized crime. Instead, it showcases hip-hop music performed by a group of young people from Switzerland. Their parents are Galician, and they return every year to spend their summer holidays there, therefore fitting the pre-Internet emigration pattern in terms of their destination and image of Galicia. In fact, they sing mainly in Spanish—using neither the German nor the French language of their adopted country nor Galician—in accordance with a language model that does not relate to identity. Nonetheless, when they do sing in Galician, it is just in one song ("Pardeconde," rapped by GranPurismo), the biggest regionalist poem[37] in decades in the history of Galician literature, featuring *hórreos*, totemic trees, popular traditions, and so on.

Mafia Gallega occupies a peripheral position in the system, in number (a mere dozen people) as well as in cultural relevance (they sing hip-hop music, not pop). In spite of this, and even though they live in different cities across Switzerland and therefore come from different adopted communities, they get together to promote their records, produce their remixes, and also—we suppose—to enjoy themselves. More than other genres, the hip-hop scene has learned how to make the most of the dynamics of creation and exhibition on the Net, and the Internet has enabled this group to get in touch with the small hip-hop scene in Galicia. Although only on a small scale, children of pre-Internet emigrants and people living in inner Galicia have joined together and actually formed a community.

Facebook

If there were an operational wide-range network, it should be in Facebook. Nevertheless, the reality is that Galicia has just arrived on Facebook, and it does not even have a geographic entity of its own; inner Galicia users belong to the extremely broad and quite vague network of "Spain."[38] However, Facebook easily permits the creation of interest groups, and the media surprise us on a daily basis with strange examples of them. Therefore, if there was any interest at all for Galicians in the diaspora and those from inner Galicia to join together, Facebook would be the forum in which this might happen, but at the moment this is not the case. There is no global network whatsoever, and although some small, very specific networks can be found, we will have to wait a few months to see if they survive.

This is the case of "Ourense the Best,"[39] which sprang from the initiative of two Panamanian girls on April 2007. For a long time it was orientated toward the (unsuccessful) promotion of one of them to win a beauty contest. In recent months its members have increased due to the registration of users coming from France and Switzerland. Despite the fact that no activity is perceived in the common space, perhaps—and only perhaps—there is some activity taking place between its more than two hundred members, none of whom come from Galicia. The images included as galleries and avatars are of a clearly regionalist nature: monuments, food dishes, pictures from the visit to their grandparents, and the like.

Other groups, such as Viva Galicia (Long Live Galicia),[40] Galicia—Fogar de Breogán (Galicia—the Home of Breogán),[41] or Facebook en Galego (Facebook in Galician),[42] are quite the opposite of the communities mentioned earlier. These groups share some characteristics. They combine regionalist elements such as monument images with other more nationalistic iconic proposals, and also confer more symbolic importance on language (in fact, they demanded a Facebook software translation into Galician). In addition, they bring together four different members' profiles: pre-Internet emigrants, post-Internet emigrants, Galicians who still live in Galicia, and, last but not least, people from all over the world with an interest in Galician culture. The number of users is not very high—Viva Galicia reached seven hundred in January 2008—and the activity of their interrelation tools on display is scarce. It is, nevertheless, still too soon to know if Facebook communities will leave their place in the periphery to become a space in which to build relationships for a community of people scattered all over Argentina, Switzerland, Japan, or Galicia.

BEFORE OR AFTER, IN OR OUT?

We have presented in as succinct a manner the most relevant issues concerning Galician emigration, a phenomenon that has impacted crucially upon the Galician economy, politics, and society from 1870 to the present. Three main conclusions can be drawn from our overview. On the one hand, this research allows us to translate into an iconographic point of view—and therefore an identitary point of view—the different steps taken by historians and sociologists to chart this important diaspora. On the other, the present research allows us to corroborate our theory that, with the exception of some personal or marginal attempts, a community of Galician emigrants does not exist on the Internet, in spite of what many scholars had predicted by the end of the 1990s. Finally, this Internet analysis serves to reinforce previous theories about the representation of Galicia by Galician emigrants. We assert that in the future it should be necessary to classify Galician emigrants not according to country of arrival or professional skills but by their Internet usage as a community tool. Perhaps in future years we will contrast not pre- and post-Internet emigrants but in- and out-Internet Galician emigrants, with the logical corollaries noted for the fields of economics, politics, culture, and health care.

NOTES

1. "Vivamos como galegos" (2007), available at http://www.vivamoscomogalegos.com/.

2. Data compiled by sociologist Carlos Neira, available at http://calidonia.blogaliza.org/2007/01/08/enquisas-e-sentimento-de-nacion-en-galiza-e-iii/.

3. That is the case with the works by Xosé Manuel Beiras, published in 1972 and revised in 1995, or by Carlos Sixirei Paredes, particularly *A Emigración Galega,* published in 1996.

4. In this sense, we can cite the interesting research carried out by historians such as Ramón Villares (1996, 1997a, 1997b) and Xosé Manoel Núñez Seixas in 2001

and in 2002 concerning Galician emigration. As for its political and identitary bias, there are also interesting essays by Beramendi and Núñez Seixas.

5. The lack of files with details of this emigration in the past sixty years and, moreover, the fact that emigrants left without any governmental supervision prevent us from considering undisputed numbers (Villares 1996; Sixirei Paredes 1995a). Instead, we base our findings on the data given in Eiras Roel 1991.

6. The Galician census is available at http://www.ige.eu/igebdt/verPpalesResultados.jsp?OP=1&B=1&M=&COD=1373&R=2[all]&C=1[all]&F=0:1;9912:12&S=.

7. This is a poem belonging to *Follas Novas* (New Leaves), a book of poetry published in 1880. This work represented the definitive reputation of Rosalía de Castro as maximum poet in the Galician language. See also Castro 1991.

8. This news item appeared in *La Voz de Galicia,* the newspaper with the largest reported circulation in Galicia. Available at http://www.lavozdegalicia.es/hemeroteca/2004/06/12/2764229.shtml.

9. Some examples can be found at http://bepop.com.ar/humor/Gallegos.html. On the other hand, most Argentineans and Uruguayans use to name *gallegos* to everybody coming from Spain, regardless of their actual Galician origin.

10. This very well-known poem is entitled "Adios ríos, adios fontes" (Farewell Rivers and Springs), and it is from Rosalía de Castro's first book, *Cantares Gallegos* (Galician Ballads).

11. This portrait can be viewed at http://xornal.vigo.org/medi/xornal/galeria/015 Da_marco/006Ferrol.jpg.

12. See http://grupotraballogalego.uk.net/ep04no31.shtml.

13. We draw a concise summary following Sixirei Paredes (1995a, 1995b), Villares (1997a), and Núñez Seixas (2001).

14. This poem was entitled "Castellanos de Castilla" (Castilians of Castile).

15. Concerning these issues, works by Núñez Seixas about the Galician emigration in America should be of particular interest. Núñez Seixas pays close attention to the process by which Galician associations were formed and to their identity background.

16. The phrase *orfos de vivos* (orphans of the living) was created by mimicking the pattern of the aforementioned poem by Rosalía de Castro (Sixirei Paredes 1995a).

17. See http://www.xunta.es/galicia2003/gl/01_02.htm.

18. CIG (Confederación Intersindical Galega). *Non máis Emigración Galega* was published in 2005. It is available at http://www.galizacig.com/actualidade/200503/entre_1990_2003_emigraron_175000_galegos.htm.

19. Data combined with the official census (Instituto Galego de Estatística).

20. *La Voz de Galicia,* "La otra emigración hacia las islas." Available at http://www.lavozdegalicia.es/hemeroteca/2004/11/30/3254738.shtml.

21. *La Voz de Galicia,* "Galicia anida en Fuerteventura." Available at http://www.lavozdegalicia.es/hemeroteca/2004/11/27/3246300.shtml.

22. In order to see the form and meaning of triskelions in Galician culture, a quick overview can be found at http://gl.wikipedia.org/wiki/Trisquel. More meanings and forms are at http://en.wikipedia.org/wiki/Triskelion.

23. This is the case of Web sites such as A Chaiva da Ponte (http://ar.geocities.com/achaivadapontefolkcelta/principal), one of the few music groups formed by emigrants with their own MySpace profile, (http://www.myspace.com/achaivadaponte), though the community is more interested in music than in specifically Galician issues.

24. Web sites such as the Centro Gallego de Tandil (Argentina) (http://personales.ciudad.com.ar/centrogallegotandil/links0021.htm) or Caballeros de Santiago (Brazil) (http://personales.ciudad.com.ar/centrogallegotandil/links0021.htm).

25. The only exception seems to be the Web site of the Centro Galego de México (http://centrogallegomexico.galeon.com/), where most of the text is written in Galician and contains considerable material relating to identity, mainly centered on regionalistic notions and addressed solely to the Galician community.

26. See http://www.facebook.com/group.php?gid=19749751912.

27. See http://www.vieiros.com.

28. See http://chuza.org/. Technically, Chuza! is not a clone of Digg but a clone of Menéame (a Spanish open-source Web site that functions exactly in the same way as Digg).

29. The Galician flag has only two colors, white and blue. The field is white, and a blue band crosses the flag from the top-right corner to the bottom left. The Galician nationalist flag has a red five-point star on the blue stripe.

30. See http://www.fillos.org/galicia/index.php.

31. We would like to underline the interesting case of "parcelaria" (http://www.parcelaria.com), a moneymaking service that provides Internet domains. When these domains are sold, they are displayed as part of a map of Galicia. The "logoization" of national boundaries was widely explained by Benedict Anderson (1996), among others. However, in this case, we have to remark that "parcelaria" was also a common concept underpinning the political measure taken during the twentieth century to redistribute properties in rural Galicia in order to create larger tracts of land. The "parcelaria," widely regarded as a sign of economic development, was never completed and, at the dawn of the twenty-first century, is likely to remain so.

32. See http://www.degalicia.org/planeta-galego/ or http://www.nacionmundialgalega.info/.

33. See http://galiciaglobal.com/.

34. *Día das Letras Galegas,* the Galician Literature Day, is a tribute to the Galician language, writers, and culture. It was inaugurated as a ceremony in 1963 to mark the centenary of the publication of *Cantares Gallegos,* the first book of Rosalía de Castro.

35. Press note of the Galician Ministry of Innovation and Industry. See "+ Internet + Galego + Futuro. Galicia Global," available at http://maisinternetmaisgalego.org/actividade.php?estxt=actividade07_9.

36. See http://www.mafiagallega.com/.

37. GranPurismo, "Pardeconde." Lyrics available at http://pontelouco.blogspot.com/2007/12/pardeconde.html.

38. See Campaign to Get Facebook to Add GALICIA as a Network Region at http://www.facebook.com/group.php?gid=9215100545.

39. See Ourense the Best, http://www.facebook.com/group.php?gid=2283275113.
40. Viva Galicia, http://www.facebook.com/group.php?gid=2317649849.
41. Galicia—Fogar de Breogán, http://www.facebook.com/group.php?gid=2265727732.
42. Facebook en Galego, http://www.facebook.com/group.php?gid=6195880907.

REFERENCES

Anderson, Benedict. 1996. *Imagined communities.* 2d ed. New York and London: Verso.

Beiras, Xosé M. 1995. *O atraso económico de Galicia.* Santiago de Compostela: Laiovento.

Beramendi, Justo G. 2007. *De provincia a nación: Historia do Galeguismo político.* Vigo: Xerais.

Beramendi, Xusto, and Xosé M. Núñez Seixas. 1996. *O nacionalismo Galego.* 2d ed. Vigo: A Nosa Terra.

Calo Lourido, Francisco, et al. 1997. *Historia xeral de Galicia.* Vigo: A Nosa Terra.

Casal Lodeiro, Manuel. 2006. *O papel de Internet na conservación da cultura e identidade galegas entre os descendentes dos emigrantes.* Bilbao: Fillos de Galicia and Xunta de Galicia. First published in 2003. Available at http://www.casdeiro.info/arquivos/ensaio-manuel-casal-lodeiro.pdf.

Castro, Rosalía de. 1991. *Poems.* Ed. and trans. Anna-Marie Aldaz, Barbara N. Gantt, and Anne C. Bromley. Albany: State University of New York Press.

Eiras Roel, Antonio. 1991. *La emigración española a Ultramar, 1492–1914.* Madrid: Tabapress.

Emigración. 2002. In *Enciclopedia Galega universal.* Vol. 8, *Edu-Fad.* Vigo: Ir Indo.

Gabriel, Narciso de. 2006. *Ler e escribir en Galicia: A alfabetización dos galegos e das galegas nos séculos XIX e XX.* A Coruña: Universidade da Coruña, Servizo de Publicacións.

Galicia 2000. 2000. Santiago de Compostela: Xunta de Galicia.

González González, Manuel, dir. 2004. *Mapa sociolingüístico de Galicia.* A Coruña: Real Academia Galega.

González Millán, Antonio Jesús. 1993. La Cruz de Santiago: Una donación del Rey Alfonso III al Apóstol y a su sede de Compostela en el año 874. *Compostellanum* 38, nos. 3–4: 303–35.

Lafargue, Paul. 1883. *Le Droit à la paresse (Réfutation du « Droit au travail » de 1848).* Paris: Henry Oriol Éditeur. First English edition: *The Right to Be Lazy.* Chicago: Charles H. Kerr.

Leira, Xan. 2007. *Historias dunha emigración difusa: A emigración galega a Lisboa.* Lisboa: Centro Galego de Lisboa. Available at http://www.galiciaaberta.com/portal/binary/com.epicentric.contentmanagement.servlet.ContentDeliveryServlet/Documentos/2007/HistoriasDunhaEmigracionDifusa.pdf.

Monteagudo, Henrique. 1999. *Historia social da lingua galega: Idioma, sociedade e cultura a través do tempo.* Vigo: Galaxia.

———. 2001. O idioma. In *Galicia, unha luz no Atlántico*, ed. Victor F. Freixanes. Vigo: Xerais.

Núñez Seixas, Xosé M. 1992. *O galeguismo en América, 1879–1936*. Sada (A Coruña): Ediciós do Castro.

———. 1993. Compromiso politico e galeguismo na diáspora, 1879–1950. In *Galicia-América: relacións históricas e retos de futuro*, ed. Peña Saavedra. Santiago de Compostela: Xunta de Galicia.

———. 2001. La emigración: Galicia en el mundo. In *Galicia, unha luz no Atlántico*, ed. Víctor F. Freixanes. Vigo: Xerais.

———. 2002. *O inmigrante imaxinario: Estereotipos, representacións e identidades dos galegos na Arxentina (1880–1940)*. Santiago de Compostela: Universidade de Santiago de Compostela.

Quino. 1974. *10 años con Mafalda*. Selected by Esteban Busquets. Buenos Aires: Ediciones de la Flor.

Rodríguez, Lois, and Xabier Cid. 2001. Temos un soño? A comunicación na sociedade da información. *Terra e Tempo*, no. 16: 52–57.

Romero, D., and I. Vaquero. 2001. *Da periferia á Rede: Internet en Galicia, lingua e contidos*. Vigo: Xerais.

Sixirei Paredes, Carlos. 1995a. *A emigración*. Vigo: Galaxia.

———. 1995b. *Galeguidade e cultura no exterior*. Santiago de Compostela: Xunta de Galicia.

Sixirei Paredes, Carlos, et al. 2001. *Asociacionismo galego no exterior*. 2 vols. Santiago de Compostela: Xunta de Galicia.

Souto, Xurxo. 1999. *O retorno dos homes mariños*. Vigo: Xerais.

Thiesse, Anne-Marie. 2001. *La création des idéntités nationales*. 2d ed. Paris: Seuil.

Villares, Ramón. 1996. *Historia da emigración galega a América*. Santiago de Compostela: Xunta de Galicia.

———. 1997a. *A Historia*. 7th ed. Vigo: Xerais.

———. 1997b. *Figuras de nación*. Vigo: Xerais.

Yáñez, María. 2007. Ahí os quedáis. *El País*, January 23. Available at http://www.elpais.com/articulo/Galicia/Ahi/os/quedais/elpepiautgal/20070123elpgal_10/Tes?print=1.

Yáñez, María, and Berto Yáñez. 2008. *O uso de internet nas redes sociais de Galiza*. Santiago de Compostela: Observatorio galego da sociedade da información.

18   Basque Diaspora Digital Nationalism
Designing "Banal" Identity

PEDRO J. OIARZABAL

The Basque diaspora Webscape encompasses Web sites authored or commissioned by Basques in the homeland as well as in the diaspora. My interest lies within the framework of the Web landscape established by the Basque diaspora institutional Web sites, which create a common networked set of online discourses across geographical, political, and linguistic barriers. The study presented here draws on previous larger research, where I conducted quantitative, qualitative, and comparative analyses on the online and off-line dimensions of the Basque institutional diaspora in order to better understand its discourses on Basque identity, culture, and nation (see Oiarzabal 2006, forthcoming; and Oiarzabal and Oiarzabal 2005). In relation to the present work, I applied a discursive or rhetorical analysis to ninety Basque diaspora sites from sixteen countries as of July–August 2005, which I complemented by studying new Web sites that have been created since then. Particularly, I focused on the sites' textual, graphic, and multimedia content (575 graphics, thousands of texts, and dozens of songs), while also taking into account the structure (i.e., the number of pages, navigation options, or hierarchical order of pages).

Therefore, in this essay I present the results of the analysis carried out on the multimedia character of the Basque institutional diaspora online. I address the Basque diaspora's construction of a Basque nationalist discourse and its own self-representations by examining its institutional sites' audio-visual and graphic content. In what ways does the Basque institutional diaspora's utilization of the Internet help to foster an online digital nationalism? What is the image that the Basque diaspora portrays online? And what does the Basque diaspora attempt to express and promote?

As of December 2005, the Basque diaspora had engendered 189 associations—club federations, *euskal etxeak* (community-based social clubs), and cultural, educational, political, and business organizations[1]—in twenty-two countries.[2] More than half of those associations (98) were online in sixteen countries with a Basque institutional presence as of November 2005.[3] By June 2007, the Basque institutional diaspora increased by 10 new associations in two new countries, China and Cuba. At that time, 120, or nearly 61 percent of diaspora associations, had a presence on the Internet in nineteen, or nearly 80 percent, of the countries.[4] The Basque diaspora presence on the Internet and particularly on the Web is a phenomenon that is still undoubtedly unfolding. Although it is not wise to prognosticate about any future trends of the Basque diaspora presence on the Web, evidence shows an increasing tendency for articulating an online presence. For example, from the beginning of 2004 to the end of 2005, 13 new diaspora institutional Web sites were created, and from October 2005 to June 2007 another 30 (23 sites and 7 blogs or photoblogs, mostly built by Basque-Argentinean dance groups), the majority of which are Basque clubs, or *euskal etxeak,* and mainly from Argentina.

These associations, each explicitly self-defined as Basque, materialize strong group self-awareness, sustained over a considerable period of time (Douglass and Bilbao 1975; Molina and Oiarzabal 2009). Diaspora communities are formed by emigrants who share a collective identity in their homeland, where socioeconomic or political conditions or both forced them to leave, or who for other reasons chose to settle in another country. Collectively and associatively, some of them attempt to preserve or develop cultural, religious, and even political expressions of their identity, reflecting different degrees of assimilation into their host societies. Diaspora associations create transnational networks that maintain varying degrees of personal, institutional, cultural, social, economic, political, and business ties with the homeland and with other countries where there is a Basque presence: a globe-spanning network of attachments and allegiances.

The Basque Country is a region situated at the Spanish-Franco border of the western Pyrenees. The Basque historical homeland territories are divided into three main political administrative areas—the Basque Autonomous

Community (BAC), or Euskadi; the Foral Community of Navarre in the Spanish state; and three Basque provinces, or Iparralde, in the French state—with a total combined population of nearly 3 million people. The quantification of the Basque diaspora population is nearly impossible to determine, as it depends on the operational definition of "being Basque" as well as a complete database. Nevertheless, the Basque government estimates that the diaspora population consists of 4.5 million people (Gobierno Vasco 1996, 47). However, that figure is extremely difficult to corroborate.

DIGITAL "BANAL" NATIONALISM: DIASPORIC SYMBOLISM

Billig's seminal study on national identity introduced the concept of "banal nationalism" as the "ideological habits which enable the established nations of the West to be reproduced." He states, "National identity in established nations is remembered because it is embedded in routines of life, which constantly remind, or 'flag,' nationhood" (1995, 6, 38). That is to say, banal nationalism refers to mundane daily and nearly subliminal—not insignificant or meaningless—events that promote one's identity without people's really thinking about it.

Banal nationalism also applies to diaspora communities. For example, everyday objects and commodities from "back home," or from the "old country," re-create familiar nostalgic images while helping to evoke memories of belonging and identity (see Cohen 2004; Davis 1999; and Geisler 2005). They are "unconscious" and "unnoticed" reminders of nationhood, "preventing the danger of collective amnesia" (Billig 2003, 133). Symbols, particularly national symbols (*the* flag, *the* anthem, *the* language, monuments, clothing, or food), play a role in creating, maintaining, and expressing an individual and collective identity, as well as providing a sentimental or emotional attachment and identification of individuals within the group. In other words, as Smith argues:

> National symbols, customs, and ceremonies are the most potent and durable aspects of nationalism . . . that evoke instant emotional responses from all strata of the community. . . . By means of ceremonies, customs and symbols every member of a community participates in the life, emotions and virtues of that community and through them, re-dedicates him or herself to its destiny. By articulating and making tangible the ideology of nationalism and the concepts of the nation ceremonial and symbolism help to assure the continuity of history and destiny. . . . Through ceremonies and symbols the individual identity is bound up with the collective identity. (1992, 77–78, 160)

According to scholars such as Alba (1985), Gans (1979, 1994), and Kivisto and Nefzger (1993), symbols are the last vestiges of ethnic identity among assimilated groups in their host lands. For example, Gans describes the phenomenon of consumption of ethnic symbols by third-generation individuals born in the United States as "symbolic ethnicity" or "leisure-time ethnicity." According to the author, they look for "easy and intermittent ways of expressing their identity, ways that do not conflict with other ways of life," such as the celebration of festivals, commemorations, cuisine, and parades (1979, 6).

Regarding diaspora Basques and particularly the descendants of the immigrant generation, Corcostegui argues that diaspora Basques "increasingly rely on symbols to formulate and bolster their ethnic identity" (1999, 250). Focusing on Basque dance as one of the key elements of Basque identity in the diaspora, the author argues that dance "as physical movement is itself indicative of vitality, an image of Basque cultural vitality is conveyed through dance. Images of Basque dancers are often used to signal Basqueness. The web site of the Ventura County Basque Club [Thousand Oaks, California] is an example of this. Pictures of dancers figure prominently on the page, although the club does not have a dance group" (ibid., 251).

Symbols help diaspora Basques to express or to externalize publicly (as in festivals) and privately (for example, homeland decoration, such as Pablo Picasso's *Guernica* painting, or personal adornments, such as Basque-themed tattoos) their own identities, showing individuality but referring to a collectivity. For example, Basque and Basque American identities are represented by the use of a large inventory of symbols. In the Basque American diaspora one can observe *ikurriñas* (Basque flags, which resemble the British flag's double cross, but in red, green, and white); *lauburus* (four-headed crosses);[5] bumper stickers with messages such as "Thank God I'm Basque," "Living with a Basque builds character," "Basque-Euskalduna," or "Proud to be Basque"; and Basque custom license plates.

Banal nationalism is also present on the Basque diaspora Webscape. Although the main format of the Web is textual, the text cannot be taken in isolation while exploring and examining the discourses of the Basque diaspora. Consequently, I have examined the textual, audiovisual (i.e., songs that can be listened to online), and graphic (symbols, photos, maps) aspects of the sites, including their "decorations"—a sort of digital banal nationalism that portrays and reinforces certain Basque national representations of identity, nationhood, and homeland. I found that 95 percent of Basque symbols displayed on the sites (maps, *lauburus,* and *ikurriñas*) are banal assertions of identity. In a sense, the Basque diaspora online is built by a landscape of image-based narratives—an ideoscape, in the words of Appadurai (1991)—that help to reconstruct a Basque imagined community by constituting narratives of "themselves."

Basque nationalist iconography, historical symbols, mythological imagery, and banal ideology, once exclusively identified with a particular homeland party, such as the PNV (Basque Nationalist Party in its Spanish acronym), passed into the Basque people's heritage, at home and abroad. The Basque flag and anthem, for instance, which were created in the homeland during the late nineteenth century by Basque nationalists, were made the official flag and anthem of the first Basque government in 1936. And they were once again ratified as the official flag and anthem of the government of the BAC in the early 1980s.[6] Similarly, other iconographies forged by the Basque Nationalist Party, such as Aberri Eguna (Day of the Homeland), also passed into the wider Basque nationalist heritage and to some extent to the general Basque people's heritage, both in the homeland and in the diaspora.

In the case of the Basque diaspora online, the multimedia character of the Web is exemplified by home pages' flash presentations, which combine richly designed, high-quality graphics, music, and text and serve to introduce the various associations' Web sites. For example, the main upper banner of the Basque club of Valencia's home page (Carabobo, Venezuela) is represented by, in the left corner, the name of the club and the Web address and the *ikurriña* and the Venezuelan flag with poles crossing and is flanked by the coats of arms of the Basque Country, Venezuela, and the Carabobo province, as well as a postal image of the Guggenheim Bilbao Museum and a *lauburu*. In the right corner, there is a photo of San Juan de Gaztelugatxe—a church on a small island that is joined to the Bizkaian coast by a bridge—and a superimposed Basque word that reads *etxean* (at home). On its upper right side are another *ikurriña* and a map of the BAC, and at the bottom of the banner there is an image of the Basque clubhouse as well as a Spanish legend that reads *Bienvenidos al Centro Vasco* (Welcome to the Basque Club).

The digital nationalist displays are easily grasped by users, as they can visualize diverse representations of identity. In fact, words, photos, and sounds emphasize each other in the construction of identity discourses throughout the Basque diaspora Webscape. The use of particular fonts, including peculiar Basque typefaces; music, images, and colors, especially the green, red, and white of the Basque flag; ornaments, backgrounds, and wallpapers;[7] functional link buttons in, for example, the shape of *etxeas* (homes) or *baserris* (traditional farmhouses),[8] *lauburus,* and *ikurriñas* and Navarran flags; layout styles; and even Basque emoticons or smiley faces wearing *txapelas* (berets) all form part of expressing a specific Basque collective identity online—and by extension a collective identity off-line. They embed a Basque particularity and uniqueness into the banal design or decoration of their associations' sites, which set them apart from other communities.

The past and the idyllic rural homeland symbolized by its *baserri* are central to diaspora identity discourse, as evidenced in 80 percent of the Web sites I studied. There are examples of images that evoke a much romanticized view of Basque identity by depicting scenes of the past related to an almost vanished agrarian world. In addition, 80 percent of the total Basque diaspora sites offer the visitor a section dedicated to the Basque Country, and a gallery of photos is usually available. Ninety percent of those photos reconstruct a bucolic Basque Country, far from the most modernized and urbanite aspects of Basque society. At the same time, the sites also offer images of Basques in "traditional" costumes, particularly in dance garments, which evoke a past Basque culture. However, contemporary assertions of identification are also found but in a very small proportion—that is, only in 5 percent of all diaspora sites. The image of the Guggenheim Bilbao Museum monopolizes the contemporariness of Basque identity in the Basque diaspora online. No other contemporary images seem to represent Basque identity as well as Frank Gehry's 1997 museum does.

Diaspora Basques are symbolically connected not only to the homeland but also to their respective host lands' specific dates, places, names, activities, and events. Eighty percent of the Web sites across different countries and types of sites portray the "in-betweenness" of the diasporic condition of Basques abroad. They reflect on diaspora Basques' dual and noncompeting allegiances—homeland and host lands—and on the amalgamation of Basque symbols with the host lands' symbols. For example, in 2005, the Mexico D. F. Basque club created a new "mascot" called "Coatlitxu" (the Little Snake) in order to merge both millenarian symbols: "Coatl" (Snake) from the Aztec culture and "txu," a Basque-language diminutive. Also, the logo of the first Basque association in China integrates both Basque (a red *lauburu*) and Chinese characters and symbols.

In addition, as of June 2007 forty-three diaspora Web sites (i.e., 36 percent of the total number of Web sites) from eleven countries (i.e., 58 percent of the total number of countries) are identified by their country of origin using national- or primary-level domain names,[9] such as: ".fr" for France, ".it" for Italy, ".ch" for Switzerland, ".ar" for Argentina, ".us" for the United States, and ".au" for Australia. For example, the FEVA site's domain name, http://www.fevaonline.org.ar/, acknowledges the Argentinean national domain as an integral part of its online and off-line discursive identity, as an integral discursive space. That is, there is an identification with a geographic location and political identity in cyberspace that could reflect diaspora institutions' desire to be identified in dual terms—Basque as their ancestral origin and Argentinean as their adopted host identity. Significantly, no Basque diaspora association Web site in Spain is identified by the Spanish national identity domain name ".es." Could this mean a problematic coexistence of both identities, Basque and Spanish, on the Internet, as a reflection of the off-line reality, or is it simply a preferential choice for other domain names, such as ".com" or ".net"? In this sense, although the Web allows us to reimagine our allegiances in different ways, they are still closely related to a national imaginary.

GERNIKA

If any symbol is "universally" recognized as Basque it is the Gernika Oak Tree. Gernika, in the province of Bizkaia, has become a symbol of multiple meanings from a local to a global context. It hosts the "sacred" Oak Tree and the Assembly House of Gernika, where the representatives of the Bizkaia territory met under the rule of consuetudinary laws, or *fueros,* for centuries until their abolition in 1876. Raento and Watson argue that "by the late nineteenth century, then, Gernika not only came to represent 'Basqueness' or a Basque ethnic identity, but also its political expression, Basque nationalism" (2000, 714).

During the Spanish civil war, it became the target of the first intentional aerial bombings of a civilian population in Europe, achieving substantial human loss and urban destruction. The bombing of Gernika became the single most important event in Basques' contemporary history. The event is recognizable on a global scale not only by Basques but particularly by non-Basques. The destruction of Gernika as well as other villages and small towns, such as Durango by Nazi Germany in 1937, is constantly used as an identity marker that defines Basqueness and Basque recent history within the context of Franco's victory, dictatorship, and subsequent political and cultural repression against republicans and nationalists. It is estimated that 150,000 Basques, including 25,000 children, went into exile, while an estimated 100,000 were imprisoned and 50,000 died. In 1936, the total population of the Basque provinces in Spain was 1.3 million.

Sixteen diaspora associations online re-create Gernika as an unrelenting historical testament of contemporary collective victimhood as well as the space that embodies the historical continuity of Basque liberties, autonomy, political determination, and nationalism. The diaspora discourse on Gernika follows to some extent nationalist romanticized myths on the assumed protodemocratic and egalitarian Basque society symbolized by the Gernika Assembly House and the Gernika Oak Tree, where all socioeconomic statements' representatives met to decide their own future. For example,

> On the evening of April 26, 1937, the Nazi-fascist air force bombarded during three long hours the defenseless town of Gernika; the heart of the oldest democracy in the world. But the symbol of the Basque nation, perhaps, because of the eternal Oak Tree's divine fate, kept standing amongst the ruins. The "blessed tree" and the Assembly House were the only things that were saved from the disaster. . . . And the spirit, with more strength than ever, screamed from the eternal Oak Tree that the Basques were immortal. (at the Centro Vasco del Chaco's site, http://www.ecomchaco.com.ar/centrovasco/default.htm)

In 1937, Picasso painted what would be a famous work called *Guernica* in order to immortalize the horrors of the war, while universalizing the name of the Basque town as the Basque people's collective suffering (see Cava Mesa 1996). The *Guernica* painting has become an expression of collective memory, identity, and belonging in both the homeland and the diaspora. For example, there are several oak trees planted abroad from seedlings taken from the Gernika Oak Tree.[10] They represent the old traditions transplanted into new soil, as roots of identity and alleged historical continuity between the past and the present. The Basques from Elko, Nevada, commissioned a mural of a young man in front of the Gernika Oak Tree. The mural is on the exterior wall of the Basque handball court in a park in Elko. A commemorative plaque of the mural reads as follows:

> A young man stands one last time before the Tree of Gernika to say good-bye before leaving his homeland to start a new life in America. He will carry on the Basque tradition to be free, responsible only to himself, his family, and God. . . . [W]e Basques who are American born dedicate this mural to all those Basques living and dead who left their homeland to carry on their traditions. We honor them with our belief that all Basques, living and dead, are one. . . . As we salute them let us cry *"Ama, aita, euzkaldunak, inoiz ez dugu ahaztuko"* . . . mother, father, Basques everywhere, we shall not forget! Our roots run deep.[11]

In addition, political homeland and diaspora associations related to the defense of ETA (Basque Country and Freedom in its Basque acronym) political prisoners and refugees (for instance, the Josu Lariz Askatu Campaign in Argentina [http://www.josu-askatu.org] and 6 de México [http://www.6demexico.org]) and the defense of human rights (such as the International Basque Organization of Human Rights in the United States [http://www.euskojustice.org/]) have deconstructed or fragmented the painting into individual icons, appropriating them as powerful logos.[12] By using those logos the associations attempt to claim the message embedded in the original painting itself as well as the historical legitimacy of the Basque struggle against fascism.

Furthermore, according to the contents displayed on 60 percent of diaspora associations' Web sites, Gernika constitutes a central reference for the Basque diaspora identity political discourse. This discourse not only pertains to political associations but is also widely accepted by different types of associations across geographical and linguistic barriers. Gernika is undeniably a focal discourse on the reconstruction and maintenance of the Basque politics of identity in the diaspora. Discourses on the Spanish civil war, Franco's dictatorship, and mostly the Gernika bombardment are emotionally charged discourses that bring significance to "the other" as a constitutive part of the diaspora's own identity construction.

Diaspora associations interpret Gernika as a rhetorical construction of Basqueness not being the same as Spanish, which has been politically articulated since the inception of Sabino Arana's Basque nationalist ideology (Arana was the founder of the PNV). For example, the Basque Collectivity of Concordia (in Argentina), created in 1982, recalls that during the local celebration of the five hundredth anniversary of Christopher Columbus's voyage to the New World, the organizing committee decided that the Basque association should parade under the Spanish flag. However, "The organizers did not take into account the inflamed reaction of the recently created Basque association. . . . [Basques] paraded through the streets of Concordia with the 'ikurriña,' pride of an ancestral tradition and the heritage of the Basque blood" (the Concordia Basque club's site, http://www.concordia.com.ar/Vascos).

CONCLUSION

I have examined how the Internet and particularly the Web help to foster a digital banal nationalism in the Basque diaspora Webscape. The Web allows diaspora Basques and their associations to portray and reimagine, on a global scale, certain Basque national representations about identity, culture, nation, and homeland, while constructing a specific Basque imagined transnational and diasporic community. That is to say, the Internet allows the Basque diaspora to crystallize a shared image of itself that it distributes to the entire planet. At the same time, diaspora Basques' self-images challenge to some degree the notions of identity and ethnicity in the homeland as they incorporate into their own identity discourse migration and resettlement experiences. Estrangement also becomes an intrinsic characteristic of the relationship between homeland and diaspora.

The texts and graphics displayed on the Basque diaspora sites are similar across the different types of sites and across geopolitical and linguistic spaces. *Lauburus, ikurriñas, baserris,* and the Gernika Oak Tree are easily recognized by Basques and non-Basques as Basque symbols common to the homeland and diaspora. These images and symbols are imbued with history and are part of the collective memory and cultural traditions of specific periods. They convey distinct messages as collective representations of community and identity and are used profusely as representations of Basque identity by both homeland and diaspora Basques. Particularly, Basque homeland nationalist images and symbols are embedded, for more than a century in some cases, in the landscape of diaspora communities' imagery and ideology. That is to say, they have been reconfigured over time according to different sociohistorical and geographical contexts within and outside of the homeland. They are indeed reminders of the existence of a Basque nation.

NOTES

1. The result of my database on the Basque diaspora Webscape is a Web site, http://euskaldiaspora.com, which has become a useful tool to access diaspora sites as well as an experiment in hypertextuality.

2. Andorra, Argentina, Australia, Brazil, Canada, Chile, Colombia, the Dominican Republic, El Salvador, France, Germany, Italy, Mexico, Paraguay, Peru, Puerto Rico, Spain, Switzerland, the United Kingdom, the United States, Uruguay, and Venezuela. Basque associations, registered with the government of the Basque Autonomous Community (hereafter Basque government), from Colombia, Brazil, El Salvador, Paraguay, Puerto Rico, and the Dominican Republic had no presence on the Web as of November 2005.

3. The first Basque Web site was created in the diaspora by Blas Uberuaga in 1994 (http://www.buber.net). (Uberuaga is also the Webmaster of the New Mexico Basque Club's site, http://www.buber.net/NMEE.) The first Basque institutional diaspora Web site was created in 1996 by the political association Asociación Venezolana de Amigos de Euskal Herria (Venezuelan Association of Friends of the Basque Country, Caracas, Venezuela, http://earth.prohosting.com/avaeh/). Prior to this, the Basque presence on the Internet was related to two mailing lists: Basque-L (December 1993) and soc.culture.basque (July 1996).

4. New associations from Brazil, China—the first Basque diaspora association in Asia—and Colombia joined the Basque diaspora Webscape. As of June 2007, Basque associations, registered with the Basque government, from Cuba, El Salvador, Paraguay, Puerto Rico, and the Dominican Republic still had no presence on the Web.

5. The *lauburus*—four-headed crosses or solar swastikas—are not exclusive Basque symbols but are shared by other cultures. However, they are considered one of the most popular and universal Basque symbols. In addition, the *lauburu* symbolizes, according to current Basque nationalists, the unity of the four Basque provinces that lie within the Spanish state.

6. Eighteen percent of the sites reproduce the BAC anthem, "Eusko Abendaren Ereserkia."

7. For example, the background of the Web pages of the Valladolid club's site (from Spain, http://www.geocities.com/guretxoko/) is made up of *ikurriñas* and *lauburus*. Similarly, the wallpaper of the Parisian choir association's site, Anaiki (http://www.anaiki.com/), is a concatenation of *lauburus*.

8. *Etxea* or *baserri* symbols are used as the home-page or index-page buttons in 15 percent of the total Basque diaspora Web sites.

9. As of June 2007, 64 percent of Basque diaspora Web sites are identified by generic domain names, such as NABO's site, http://www.nabasque.org.

10. There are seedlings planted in several countries, including Argentina, Chile, the United States, and Uruguay.

11. Significantly, there is no image of the mural on the Elko Basque club's site, despite its being a symbolic cornerstone of the Basque community's social identification.

12. The advocate sites mentioned, Josu Lariz and 6 de México, were taken off the Web once their campaigns ended.

REFERENCES

Alba, Richard D. 1985. *Italian-Americans.* Englewood Cliffs, N.J.: Prentice-Hall.

Appadurai, Arjun. 1991. Global ethnoscapes: Notes and queries for a transnational anthropology. In *Recapturing anthropology: Working in the present,* ed. Richard Fox. Santa Fe: School of American Press.

Billig, Michael. 1995. *Banal nationalism.* London: Sage Publications.

———. 2003. Banal nationalism. In *The language, ethnicity, and race reader,* ed. Roxy Harris and Ben Rampton. London: Routledge.

Cava Mesa, María Jesús. 1996. *Memoria colectiva del bombardeo de Gernika.* Gernika-Lumo: Gernika Gogoratuz.

Cohen, Erik H. 2004. Components and symbols of ethnic identity: A case study in informal education and identity formation in diaspora. *Applied Psychology: An International Review* 53, no. 1: 87–112.

Corcostegui, Lisa M. 1999. Moving emblems: Basque dance and symbolic ethnicity. In *The Basque diaspora/La diáspora vasca,* ed. William A. Douglass et al. Basque

Studies Program Occasional Papers no. 7. Reno: Basque Studies Program, University of Nevada.
Davis, Thomas C. 1999. Revisiting group attachment: Ethnic and national identity. *Political Psychology* 20, no. 1: 25–47.
Douglass, William A., and Jon Bilbao. 1975. *Amerikanuak: Basques in the New World.* Reno: University of Nevada Press.
Gans, Herbert J. 1979. Symbolic ethnicity: The future of ethnic groups and cultures in America. *Ethnic and Racial Studies* 2 (January): 1–20.
———. 1994. Symbolic ethnicity and symbolic religiosity: Towards a comparison of ethnic and religious acculturation. *Ethnic and Racial Studies* 17 (October): 577–92.
Geisler, Michael E., ed. 2005. *National symbols, fractured identities: Contesting the national narrative.* Middlebury Bicentennial Series in International Studies. Lebanon, N.H.: University Press of New England.
Gobierno Vasco. 1996. *Euskaldunak munduan* [Building the future]. Vitoria-Gasteiz: Servicio Editorial de Publicaciones del Gobierno Vasco.
Kivisto, P., and B. Nefzger. 1993. Symbolic ethnicity and American Jews: The relationship of ethnic identity to behavior and group affiliation. *Social Science Journal* 30: 1–12.
Molina, Fernando, and Pedro J. Oiarzabal. 2009. Basque-Atlantic shores: Ethnicity, the nation-state, and the diaspora in Europe and America (1808–1898). *Ethnic and Racial Studies* 32, no. 4 (May): 698–715.
Oiarzabal, Agustín M., and Pedro J. Oiarzabal. 2005. *La identidad Vasca en el mundo: Narrativas sobre Identidad más allá de Fronteras.* Bilbao: Erroteta.
Oiarzabal, Pedro J. 2006. The Basque diaspora Webscape: Online discourses of Basque diaspora identity, nationhood, and homeland. Ph.D. diss., Center for Basque Studies, University of Nevada.
———. Forthcoming. The Basque diaspora Webscape. Reno: University of Nevada Press.
Raento, Pauliina, and Cameron J. Watson. 2000. Guernika, Guernica, Guernica? Contested meanings of a Basque place. *Political Geography* 19: 707–36.
Smith, Anthony D. 1992. *National identity.* Reprint. Reno: University of Nevada Press.

Contributors

JOSÉ LUIS BENÍTEZ (Ph.D. in communications from Ohio University) is a professor in the Department of Communication and Journalism and director of the Master's in Communication Program at the Central American University José Simeón Cañas in El Salvador. He coordinates the "Migration and Human Development" research project, which is sponsored by the United Nations Development Program, El Salvador, and the European Union. He has presented numerous papers at various academic conferences in Latin America and the United States on topics such as the Salvadoran media, freedom of expression, radio, Internet, migration, and transnational communication processes.

VICTORIA BERNAL (Ph.D. in anthropology from Northwestern University in Evanston, Illinois) is an associate professor of anthropology at the University of California–Irvine. Her work has addressed issues of gender, nationalism, war, transnationalism, development, civil society, cyberspace, and Islam. She has conducted research in Eritrea, Tanzania, and Sudan. She published *Cultivating Workers: Peasants and Capitalism in a Sudanese Village*. Her articles have appeared in edited collections and in *American Ethnologist, American Anthropologist, Comparative Studies in Society and History, Cultural Anthropology, African Studies Review, Political and Legal Anthropology Review,* and *Global Networks,* among others.

JENNIFER M. BRINKERHOFF (Ph.D. in public administration from the University of Southern California) is an associate professor of public administration, international affairs, and international business at George Washington University. She is the director and cofounder of George Washington's Diaspora Research Program. She consults for multilateral development banks, bilateral assistance agencies, NGOs, and foundations. Combining her research with this work, she published *Partnership for International Development: Rhetoric or Results?* and *Digital Diasporas: Identity and Transnational Engagement.* She is also the editor of *Diasporas and International Development: Exploring the Potential* and completed an edited volume and commissioned research for the Asia Development Bank, culminating in *Converting Migration Drains into Gains: Harnessing the Resources of Overseas Professionals.*

JAVIER BUSTAMANTE (Ph.D. in philosophy and sciences of education from Complutense University in Madrid) is a professor of moral philosophy in the Department of Ethics and Sociology at Complutense University; Director of the Ibero-American Center for Science, Technology, and Society; and vice president of FIAP. He has been a Visiting Professor in several universities. He currently coordinates an international project for developing "Ibero-American Digital Citizenship" with the participation of several Spanish, Brazilian, and Bolivian universities. He was awarded the Fundesco Prize in 1993 for his book *Sociedad Informatizada, ¿Sociedad Deshumanizada?* (Computerized Society, Dehumanized Society?).

BRENDA CHAN (Ph.D. in communication studies from Nanyang Technological University in Singapore) is an assistant professor in the Wee Kim Wee School of Communication and Information at Nanyang Technological University, where she teaches courses on cultural studies and qualitative research methods. Her research interests are in the area of media and mobility, particularly how media influence, and intervene in, the processes of migration and tourism. Her doctoral research examined how the Internet was used by Chinese migrants in the construction of their cultural identities, particularly national and ethnic identities.

XABIER CID is a visiting lecturer on Galician Studies at the University of Stirling in Scotland. He holds a B.A. in Hispanic Studies with emphasis on Galician and Portuguese languages from the University of Santiago de Compostela in Spain. He worked as a journalist and as an Internet consultant for more than a decade. He has also delivered papers at international conferences, such as "Galician Blog System: Can Bytes Pull Down the Concept of National Literature?" at Oxford University.

CYBERGOLEM: ANDONI ALONSO (Ph.D. in philosophy from the University of the Basque Country in Spain) is a professor at the University of Extremadura, Spain. IÑAKI ARZOZ is a freelance cultural writer and artist. Together they form the Cybergolem Collective Author. Their work has focused on the effects of communication technologies on society and communities. Among some of their books on this issue are *Basque Cyberculture: From Digital Euskadi to Cyber Euskalerria* and *La Quinta Columna Digital,* which was the winner of the Epson Prize for the best book on technoethics. They have also developed diverse activist projects, including Artamugarriak (http://www.artamugarriak.org/) and La Quinta Columna Digital (http://www.quintacolumna.org).

RON EGLASH (Ph.D. in history of consciousness from the University of California) is an associate professor of science and technology studies at Rensselaer Polytechnic Institute. A Fulbright postdoctoral fellowship enabled his field research on African ethnomathematics, which was published as *African Fractals: Modern Computing and Indigenous Design.*

RADHIKA GAJJALA (Ph.D. in rhetoric and communication from the University of Pittsburgh) is an associate professor and graduate coordinator in the School of Communication Studies at Bowling Green State University in Ohio. She published *Cyberselves: Feminist Ethnographies of South Asian Women* in 2004. She is currently working on a single-author work, "Technocultural Agency: Production of Identity at the Interface," and on an edited volume titled "Digital Embodiment in 3D Worlds."

HEATHER A. HORST (Ph.D. in anthropology from University College London) is a postdoctoral scholar at the Institute for the Study of Social Change at the University of California–Berkeley. She is the author of *The Cell Phone: An Anthropology of Communication* and *Jamaican-Americans.* She has also published in *Identities: Global Studies in Culture and Power, Global Networks,* and *Jamaica Journal* on the relationship between transnationalism, place, and belonging.

MICHEL S. LAGUERRE (Ph.D. in social anthropology from the University of Illinois at Urbana-Champaign) is a professor and director of the Berkeley Center for Globalization and Information Technology at the University of California–Berkeley. He has published several books, among them *The Digital City: The American Metropolis and Information Technology* and *Global Neighborhoods: Jewish Quarters in Paris, London, and Berlin.* He is completing a new volume titled "Jerusalem, Rome, and Mecca: Network Governance of Global Religions in the Digital Age."

TOLU ODUMOSU is a postdoctoral research fellow in the Program on Science, Technology, and Society of the Kennedy School of Government at Harvard University. His dissertation studied mobile phones and the telecommunications industry in Nigeria.

IOLANDA OGANDO (Ph.D. in Galician Studies from University of Santiago de Compostela in Spain) is a lecturer in the Galician and Portuguese Department at the University of Extremadura, Spain. Her work has focused on theater and visual representations of national and historical identity. She published *Teatro histórico: Construcción dramática e construcción nacional,* an account of the historical theater written in Galician. In recent years, her research has dealt with the relationship between literature and self-identity building in nineteenth-century Portugal.

PEDRO J. OIARZABAL (Ph.D. in Basque Studies and political science from the University of Nevada, Reno) is a research scholar at the University of Deusto (Bilbao). He has published *La identidad Vasca en el mundo* (The Basque identity in the world) and *A Candle in the Night: Basque Studies at the University of Nevada, 1967–2007.* He is currently working on a book on the Basque diaspora webscape.

DWAINE PLAZA (Ph.D. in sociology from York University in Canada) is an associate professor in the Sociology Department at Oregon State University. He has written extensively on the topic of Caribbean migration within the international diaspora. His most recent book is *Returning to the Source: The Final Stage of the Caribbean Migration Circuit.* He has also published numerous book chapters in edited collections as well as journal articles in *Identity: An International Journal of Theory and Research, Canadian Journal of Latin American and Caribbean Studies, International Journal of Intercultural Relations,* and *Journal of Eastern Caribbean Studies,* among others.

KHALIL RINNAWI (Ph.D. in political sociology from the Free University of Berlin, Germany) is a lecturer in the Department of Behavioral Science at the College of Management in Tel-Aviv, Israel. He is the author of *Instant Nationalism: McArabism, al-Jazeera, and Transnational Media in the Arab World,* which discusses the role of Arab transnational media and the emergence of a new Pan-Arabism. He is the author of two other books and numerous scholarly journal articles.

ADELA ROS (Ph.D. in sociology from the University of California–San Diego) is the director of the Migration and Information Society Research Program at the Internet Interdisciplinary Institute, where she conducts research on communication and information technologies and flows in migratory contexts. She was a professor at the Universitat Oberta de Catalunya, coordinator of the Human Movements and Immigration World Conference, and immigration director at the Regional Autonomous Government of Catalonia. She specializes in ethnicity and immigration in Catalonia.

GINA SÁNCHEZ GIBAU (Ph.D. in anthropology from the University of Texas at Austin) is an associate professor in the Department of Anthropology at Indiana University–Purdue University, Indianapolis. Her research interests include race and ethnicity, identity, migration, and gender studies. She conducted fieldwork on identity formation among Cape Verdeans in Boston and in the islands. She has presented papers at numerous conferences in the United States and abroad and has published articles in *Cimboa: Journal of Cape Verdean Letters, Arts, and Studies, Transforming Anthropology, Identities,* and *Western Journal of Black Studies.*

YITZHAK SHICHOR (Ph.D. in international relations from the London School of Economics and Political Science) is a professor of political science and East Asian Studies at the University of Haifa and professor emeritus at the Hebrew University of Jerusalem, where he is a senior research fellow at the Harry S. Truman Research Institute for the Advancement of Peace. A former Michael William Lipson Chair in Chinese Studies and dean of students at the Hebrew University and head of the Tel-Hai Academic College, his research and publications cover aspects of China's military modernization and defense conversion, Middle East policy and labor export, and international energy policy; East Asian democratization processes; Sino-Uyghur relations and the Uyghur diaspora; and the Eastern Turkestan Independence Movement.

YU ZHOU (Ph.D. in geography from the University of Minnesota) is a professor of geography at Vassar College. She has done research in the areas of ethnic business, gender, ethnic communities, and transnational business networks. Her current research is on globalization and the high-tech industry in China, especially in Beijing's Zhongguancun region, the so-called Chinese Silicon Valley.

Index

Page numbers in italics refer to tables

ABC News, 94, 101
Activism online, 24, 89, 244, 251
Advanced Research Projects Agency, 9
Afewerki, Isaias, 127, 130
Africa, 87, 111, 112, 152, 168; and slave trade, 100, 126
African Americans, 86, 89, 90, 91, 137; and gender of names, *92,* 92, 93; heritage of, 85, 86, 92; and religion, 89
African Caribbeans. *See* Caribbean diaspora
African diaspora, 3, 4, 78, 85, 93, 94, 137, 138, 172; and ancient Egypt, 87, 88; heritage, 91; and Indian clothing, 87; and Indian hairstyle, 91. *See also* Nigerian diaspora
Africanpath.com, 116–17
AfroFuturists, 86, 88
Alexa.com, 48, 96, 100, 103, 107, 177, 328
Alexandre Bóveda Association of Fuerteventura (Canary Islands), 328
al-Jazeera (Lebanon), 266, 269, 275, 276, 277, 278, 279, 284, 288
Allied Committee of the Peoples of Eastern Turkestan, 298, 301. *See also* Uyghur diaspora
"All-India Radio," 210
al-Manar (Lebanon), 266, 275, 276, 277, 278, 279, 281
al-Mustaqbal (TV), 276
"Alphaworld," 217
Alptekin, Erkin, 311
Alptekin, Isa Yusuf, 296, 297, 311
al-Qaeda: in Afghanistan, 269; and Madrid bombings, 8
Analogical communication, 66, 72

ANN (TV), 269, 270, 275, 277, 278, 279
Arab diaspora, 265–66, 268; in Berlin, 265, 270, 271, 272; and gender, 270, 274, 275, 276, 282; and identity, 285–86; Islamic heritage, 275, 280, 282; and language maintenance, 281–82, 286; and movies/*telenovelas,* 269, 275, 276, 277; and Satellite TV, 272–73, 274, 279–80; and TV and Islam, 280–81, 286; and type of Satellite TV viewing by, 275–76; in U.K., 267, 273, 274, 275, 277, 278, 282, 284. *See also* Lebanese diaspora; Palestinian diaspora
Arab-Israeli Conflict, 269, 283
Arab transnational media, 265, 266; and Satellite TV, 266, 267, 270. *See also* ICTs; Trasnationalism
Arana, Sabino, 346
Armenian diaspora, 3, 4
Arpanet, 67
ART TV, 266
AsiaInfo.com, 248
Asmarino.com, 133, 172
ASOSAL, 198. *See also* Salvadoran diaspora
AutumnLeaves (virtual community), 232–37
Auzolan (Basque-community-based collaborative work), 68, 76
Awate.com, 127, 130

Baidu.com, 242–43, 256
Baker, Josephine, 92
Baldwin, James, 92
Bank of Eritrea, 131
Basque Autonomous Community (Euskadi), 339, 340

357

Basque Collectivity of Concordia (Argentina), 346. *See also* Basque diaspora

Basque Country, 9, 66, 75, 76, 77, 321, 323, 329, 339, 342, 343, 345, 347

Basque diaspora, 69, 71, 75, 76, 77, 79, 338–39; and associationism, 339; and cyberculture, 66, 67; and digitalization of, 76; and digital nationalism, 4; heritage, 342; identity, 66, 67, 75, 341, 342, 346; Internet presence of, 339; and migration to American West, 69; and migration to Argentina, 1, 76, 339, 346; and nationalism, 340; and nationalist iconography, 342; and political refugees and exiles, 344; quantification of, 340

Basque diaspora online, 75; as activist commons, 76, 78, 338, 339; and domains, country of origin of, 343–44; and homeland/hostlands representations, 343; identity, 343, 344, 346; and nationalism, 339, 340, 341, 346; and nationalist iconography, 342; and symbols, 341, 342

Basque diaspora Webscape, 338

Basque digital diaspora. *See* Basque diaspora online

Basque government, 76, 340, 342

Basque institutional diaspora. *See* Basque diaspora

Basque institutional diaspora online. *See* Basque diaspora online

Batzart! 77, 79

B-Boys, 86, 102

bbs. *See* Bulletin board systems (bbs)

BiDil, 90

bin Laden, Osama, 269

Biotechnology: and genetics corporations, 88; industry, 88. *See also* DNA

Bizkaia, 1, 79, 104, 312, 344

Black Power, 86

Blog, 69, 71, 79, 94, 110, 147, 149, 152, 196, 197, 199, 204, 218, 219, 224, 328, 329, 333, 335, 339, 352

Blogosphere, 73, 74, 76

Bohos, 86

Bollywood: music, 209, 222; and Second Life, 218–19. *See also* Indian diaspora; South Asian diaspora

Branco (white), 112

Brazil: and African slave trade, 171–73; ethnic and cultural diversity of, 171; and European immigration to, 173; and internal migration, 174; and Japanese immigration to, 174–75

Brazilian diaspora(s), 170–71, 175; and destinations of, 175; and quantification of, 175; and remittances of, 175

Brazilian digital diaspora: online social networks of, 176, 179; and Orkut, 180, 185–86

British Arab diaspora, 278. *See also* Arab diaspora

British, Empire, 3; and colonies, in America, 172; colonies, in Southeast Asia, 210, 231

British Caribbeans, 152, 164. *See also* Caribbean diaspora

British North America, 172

Buğra, Mehmet Emin, 296

Bulletin board systems (bbs), 227, 229, 232, 230, 233, 236. *See also* ICTs

Buppies, 86

Buyukokkten, Orkut, 176

Cable, communication through, 54, 142

Caboverdeonline.com, 113

Cape Verdean diaspora, 125, 128–31; and African heritage, 115, 117; and African identity of, 117, 118; and colonial (online) discourses of, 110, 111–13, 118; and European identity of, 113; heritage, 110; and identity, 110–20; and nationalism, 113; and online chats, 114; and Mauritius identity, 116; and Portuguese identity, 111, 112, 116–19; postcolonial (online) discourses of, 110, 115, 118. *See also* African diaspora

Caribana (Caribbean festival), 142

Caribbean Creole, 151, 153, 154, *158*, 161, 164, 165. *See also* Creole

Caribbean diaspora, 3, 88, 136, 139–42, 165–68, 172, 354; African origin of, 152, 161; and colonialism, 153; heritage, 139; Indian origin of, 152, 161; Jamaican heritage, 139; settlement of, in Canada, 153, 155, 156, 160; West Indian heritage, 139. *See also* Jamaican diaspora
Caribbean Vibes Radio, 139
Carifesta (Caribbean festival), 142
Casa de la Cultura de El Salvador (Los Angeles), 197. *See also* Salvadoran diaspora
Casal, Manuel, 329, 330
Castro de, Rosalía, 318, 320, 330
Catalonia, migration to, 19, 20, 24, 25, 321, 328; and ICTs, 28–30
Cellular (cell) phone, 8, 73, 74; and diasporas, 69, 192, 193, 203; and immigrants' usage of, 25, 30–32, 35–36, 201, 202; industry, 145; and Latinos' usage of, 202. *See also* Chinese returnee entrepreneurs; ICTs; Jamaican diaspora; Salvadoran diaspora
Central Asia–Caucasus Institute's School of Advances International Studies (Johns Hopkins University), 301. *See also* Uyghur diaspora
Centro Deportivo, 195. *See also* Salvadoran diaspora
Chat (online): and diasporas, 69, 73, 74; and migrants, 20, 29, 34. *See also* ICTs
Chen, Steve, 250
China-Japan war, 237. *See also* War
Chinese Cultural Revolution, 226, 228, 244
Chinese cyberattacks against Uyghur pro-independence movement, 291, 301, 302, 304, 305. *See also* Uyghurs China relations
Chinese diaspora, 3, 4; and associationism, 226, 227; and Chinese language and media, 226–27; and foreign education, 226, 231, 234, 242, 246; history of, 225–26; and the Internet, 227–29; and "old emigrants," 226
Chinese migrants in Singapore, 227, 228, 230–36; colonial past, 231; and PRC Chinese, 232; and discrimination, 235–36; and nationalism, 235
Chinese migrants in the U.S.: brain drain and students, 245–46; history of, 243, 244; professionals, 243; and socioeconomic profile of, 244, 245
Chinese News Digest, 228
Chinese returnee entrepreneurs, 243, 247–51; and business difficulties, 256–57; and business opportunities, 254; and domestic capital, 259–61; enterprises, 253–54; and locations of settlement of, 251; mobile phone industry, 254, 255, 257; and R&D strategies, 257–59. *See also* Chinese diaspora
Chinese Singaporeans, 232, 235. *See also* Chinese migrants in Singapore
Churchward, Jack, 300
Chuza! 328
Cinema, 7, 317
Cipotes.net, 196
Citizens Against Communist Chinese Propaganda, 300
Classical diasporas, 3. *See also* Diaspora
"Cloud computing," 8
ClubNino.com, 113
CMC, 299. *See also* ICTs; Telecommunication technologies, and diasporas
CNN, 100–102, 104, 137, 267
Cold War, 244, 292, 298. *See also* War
Common Voice, 298, 301
Commons: and activist, 71–76, 78; definition of, 68, 69, 71–78; and digital, 69, 72, 75, 76; and diasporas, 69, 72, 74, 75; and digital diasporas, 69, 71, 72
Community online. *See* Virtual community
Computer-Mediated Communication. *See* CMC
Conase.se, 198. *See also* Salvadoran diaspora
Connectivity: and ICTs participation, 50, 51, 61, 62, 78, 88, 122, 148, 194, 201, 227, 230, 234, 268, 274; and Internet, 200, 201, 204, 214; and migrants, 25, 50; wireless, 206. *See also* ICTs; Digital divide

Convergence: and cellular phone, 202, 204; and culture, 147; and technologies, 74. *See also* ICTs

Creole, 111; and Crioulu, 118; and Kriolu, 115, 120

Cuban diaspora, 42

Cultural diaspora, 3. *See also* Diaspora

Cybercafés, 30, 34, 35, 133, 200, 206. *See also* Internet cafés

Cyberculture, 66, 67, 70, 71, 73, 74, 77, 181, 195, 199; and commons, 68, 69; and cybercultural mediascape, 180; definition of, 194. *See also* Commons

Cyber–Euskal Herria (Cyber–Basque Country), 75, 76, 77

Cybergolem, 65–67

Cybernation (free e-nation), 72, 73, 78

Cyberspace: and citizenship, 129, 179; and commons, 68, 69; definition of, 2; and deterritorialization, 8, 110, 124, 130, 180, 181, 193; and diasporic identity, 110, 111, 195, 344; and digital diasporas, 11, 39, 41, 42, 70, 78, 74, 122, 179; and ethnicity, 70, 72; fragmentation of, 132, 228; and freedom, 128; and political participation, 126, 133, 134, 237; and race, 51, 88. *See also* ICTs

Cyborg identity, 10, 70, 71, 72; and community, 72; and nation, 73. *See also* Ethnicity; Identity

Cyborg diaspora, 71, 211. *See also* Digital diaspora

Cyborg nation. *See* Cybernation (free e-nation)

Dehai (dehai.org), 124–34; Dehainers, 130

Dekagasi (Brazilian of Japanese origin returnees), 175

Democracy: and "Bollystan," 219; and digital democracy, 207; and Eastern Turkestan, 302; and e-democracy, 76, 78; and Eritreans, 134; participative, 68, 179; and "racial democracy," 112, 171; and Spain, 323, 324; and the West, 298

Development: aid projects, 164; ICT industry, 247, 257, 261; educational, 322; international, 44–46; and Internet, 10, 124, 237, *306;* local projects, 198; manufacturing industry, 247; personal, 246; religious, 269; social, 269; satellite technology, 279; socioeconomic, 191, 207

Dalai Lama, 311

Diaspora: and communication media, 6, 9; concept of, 2–4, 7, 9, 170, 192, 213, 339; and digitalization of, 78; and deterritorialization, 190, 309; and displacement, 4, 123, 138, 192, 228; economic remittances, 34, 39, 43, 45, 145, 196, 204; and gender, 192; and ICTs, 6, 9, 11; identity, 2, 7, 9, 40, 42, 70, 72, 73, 85, 87, 88, 93, 292, 293, 295; and imagined communities, 9; and socioeconomic development, 39, 40; security, 39, 42, 46, 52. *See also* Commons; ICTs; Digital diaspora

Diasporic. *See* Diaspora; Digital Diaspora

Digital diaspora: concept of, 10, 11, 75, 78, 211–13, 219, 223, 243, 304; and deterritorialization, 180–81; and digital citizenship, 179; and empowerment, 27, 51, 52, 55, 56, 57, 75; and ICTs, 11; identity, 11, 71; and virtual communities, 11. *See also* Commons; ICTs; Diaspora

Digital divide, 35, 49, 51, 53, 74, 78, 190, 191, 193, 194, 200, 201, 202, 204–7; and African Americans, 52; and characteristics of, in El Salvador, 200–201, 204, 206; and diasporas, 193–94; and digitalization, 55, 60; and education, 33, 128, 194, 200, 204, 206; and gender, 51, 53, 128, 194; and religion, 128. *See also* Connectivity; Digital exclusion; Digital (diasporic) marginality

Digital exclusion, 55. *See also* Digital divide

Digital (diasporic) marginality, 52, 55. *See also* Digital Divide

Discrimination: cultural, 54; and immigrants, 123, 228, 250, 302; and racism, 117, 137

Displacement. *See* Diaspora; Gentrification
DjaBraba, 113
DNA: and African diaspora identity, 86–89, 94, 106, 107; ancestry tracing, 90, 93; and "biopolitics," 91
Doğu Türkistan'in Sesi (Voice of Eastern Turkistan), 297
Dot-com, 213, 254; dot-commers, 52, 57, 58
Durango (Bizkaia, Spain), 344

Eastern Turkestan Information Bulletin, 297, 301
Eastern Turkestan Information Center (ETIC), 297, 301, 302
Eastern Turkestan National Center, 297
Eastern Turkestan National Freedom Center, 302
Eastern Turkestan Republic, 293, 299
EFPL (Eritrean People's Liberation Front), 123, 124, 125, 127, 130, 134. *See also* Eritrean diaspora
El Diario de Hoy, 197
Ellis Island, 2
E-mail: and diasporas, 69, 73, 74; and governments' usage of, 44; and immigrants' usage of, 20, 28, 29, 30, 34, 62, 116, 141, 143, 152, 197, 291
E-nation. *See* Cybernation
Eritrean diaspora: and citizenship, 129; and ICTs, 122, 124–27, 134; in Europe, 128, 129; and Internet, 128, 130, 132, 133, 134; and nationalism, 123, 124, 130, 134; and political associationism, 123, 134; and satellite television, 127, 128; and transnationalism, 134
Eritrean Ethiopian war, and diaspora, 131–36. *See also* Eritrean diaspora; War
Eri-TV, 125, 127
Estrada, Joseph, 8
ETA (Basque Country and Freedom), 76, 345
Ethiopia, 123, 125, 130, 131. *See also* Eritrean Ethiopian war

Ethnic identity. *See* Ethnicity; Identity
Ethnicity: and assimilation, 41; and ICTs, 60, 128; and cyborg identity, 71; construction of, 94, 347; and migration, 192, 230; and second-generation diasporans, 152, 161, 164; virtual performance of, 115. *See also* Identity
Ethnopolis. *See* Panethnopolis
Europe, vernacular culture of, 68
Euskadi. *See* Basque Autonomous Community
Euskal etxeak (community-based social clubs), 339
Euskal Herria. *See* Basque Country
"Euskarians." *See* Second Life (3D digital environment)
Exclusion: and assimilation, 41; social, 52–55, 216

Facebook, 94, 103–6, 140, 176, 199, 219, 327, 332, 333
Facebook en Galego (Facebook in Galician), 332
Fax, 129, 310
Fillos de Galicia (Children of Galicia), 329, 330, 331
5 percenters, 86
Flickr, 146
FMLN (Farabundo Martí National Liberation Front), 191
Foral Community of Navarre, 340
Forcv.com, 113, 117, 118
Foundation for the Research of Turkestan, Azerbaijan, Crimea, Caucasus, and Siberia, 300
"419" (Nigerian legal code), 95, 97, 99, 106; fraudsters (scammers), 97, 98. *See also* Nigerian diaspora; United States (U.S.)
Franco, Francisco, 2, 76, 322, 323; and dictatorship of, 322, 323, 344, 346. *See also* Spanish Civil War
Fueros (Basque consuetudinary laws), 344
Future TV (Lebanon), 266

INDEX 361

Galicia—Fogar de Breogán (Galicia—the Home of Breogán), 332. *See also* Galician diaspora

Galician diaspora: and associationism, 322, 324; destination of, 318–19, 321–25; emigrants' profile and image of, 319–20; emigration history of, 317–18; and Galician nationalism, 325, 326, 328–30, 332; and Galician regionalism, 324, 331, 332; migration to Argentina, 321, 323, 330, 331, 333; and negative stereotypes, 320; and political exiles, 322; and political participation of, in Galicia, 320, 330; and returnees' impact, 319

Galicia Global, 330, 331

Galician Web presence: and Celtic iconography, 326; and Galician identity, graphic representation of, 325, 326, 328–30; and Galician-language usage, 326; history of, 328; and Spanish identity, graphic representation of, 326

Gandhi, Mahatma, 213

Gangtas, 86

Garveyism, 86, 88

Gates, Henry Louis, Jr., 85, 86, 87

Gentrification, 51, 52, 57, 58

Gernika (Bizkaia, Spain), 344, 346; Assembly House of, 344–45; and bombing of, 344, 346; Oak Tree of, 344–45

German diaspora, in Brazil, 173, 186. *See also* Arab diaspora

Go-Jamaica, 137

Golding, Bruce, 137

Google, 176

Google Earth, 181

Greek diaspora, 3, 4

Green Revolution (Iran), 5

Guanacos.com, 195, 196. *See also* Salvadoran diaspora

Gulf war, 279. *See also* War

Habbo Hotel, 74

Hafash.com, 127

Haitian diaspora, 159, *159*, 164. *See also* Caribbean diaspora

Hemings, Sally, 92

Hezbollah, 266, 275, 279

Hi5, 140, 146, 219

High-Level Committee on the Indian Diaspora (HLC), 213. *See also* Indian diaspora

Hina Group, 260

Homies Unidos (United Homies), 199. *See also* Salvadoran diaspora

Hong Kong, 226, 234, 245. *See also* Chinese diaspora; ICTs

Hsinchu Science Park (Taiwan), 259

Hua Yuan Science and Technology Association (HYSTA), 260

Huaxia Wenzhai, 228

Human Genome Diversity Project, 89, 90

Hurricane Dean, 136, 137, 138, 140, 147

Hurricane Gilbert, 136

Hurricane Katrina, 137

Hybrid identity. *See* Hybridity

Hybridity, 152; and digital diasporas, 42, 51; and migrants, 40, 41, 192. *See also* Diaspora; Identity

Hypertext, 65

Ibarretxe, Juan José, 77

Ibo, 100, 101

ICTs: and diasporas, 2, 8, 9, 11, 74, 75, 130; and European diasporas' usage of, 75; and identity, 5, 9, 10, 11, 40, 41, 45–46, 50, 69, 70, 110, 138, 219, 267, 296, 307; industry, 57, 59, 242, 247, 249, 250, 252, 258, 260, 261; and international migration, 5, 7, 8, 19, 21, 22–28, 30–35, 44, 50, 56, 62, 65, 67, 73, 110, 122, 124, 126, 128, 138, 139, 142, 143, 145, 146, 151, 180, 182, 190–94, 195, 198, 199, 202, 204, 205, 217, 229, 243, 291, 292, 296, 300, 306, 311; and penetration of, in Europe, 30; revolution, 22, 23, 24, 61. *See also* CMC, Connectivity; Migration; Telecommunication technologies, and diasporas

Identity: and authenticity, 110; building process of, 86, 94, 110, 139, 141, 179, 180, 190, 209, 217, 285, 346; collective, 39, 42, 195, 205, 216, 253, 292, 293, 295, 296, 306, 307, 309, 310, 340; and deterritorialization, 110; essentialist, 152; ethnic, 3, 72; heterogeneous discourse of, 8; homogeneous discourse of, 7; online/virtual, 10, 63; oppositional, 86, 106; politics of, 100, 102, 110, 111, 113; situational, 111. *See also* Cyborg identity; Ethnicity

Imperial diaspora, 4. *See also* Diaspora

Indian diaspora, 3, 213, 219; high-tech workers, 52, 258; immigrant women, 213; Indian technopolis, 56; telemarketers, 51; and the U.S., 213. *See also* Silicon Valley; South Asian diaspora; Technopolis

Indian digital diaspora, conceptualization of, 213. *See also* Bollywood; Second Life (3D digital environment)

Indo-Caribbeans. *See* Caribbean diaspora

Industrial Revolution, 173, 319

Info-sphere (information environment), 7–8

Information and Communication Technologies. *See* ICTs

Information Society, 5, 9, 20–26, 28, 30, 32, 35, 36, 319; pre-information society, 1. *See also* Knowledge: Society

Informationalism (information processing), 23, 32

Infoseek, 242

International Basque Organization of Human Rights in the United States, 346. *See also* Basque diaspora

International Salvadoran Women Association (Los Angeles), 199. *See also* Salvadoran diaspora

International Taklamakan Human Rights Association (ITHRA), 301

International Uyghur Human Rights and Democracy Foundation, 302

Internet, access to, 6, 41, 44, 68, 143–44, 244, 250, 318. *See also* ICTs; Digital Divide

Internet cafés, 20, 54, 143, 194, 234. *See also* Cybercafés

Internet nationalism. *See* Long-distance nationalism; Transnationalism

Intipuacity.com, 198. *See also* Salvadoran diaspora

Iparralde, 76, 340

Iqra TV, 275, 281

Iraqi crisis and the U.S., 269, 283, 287

Irish diaspora, 3

Islamic Republic of Eastern Turkestan, 293

Italian diaspora, 3; in Brazil, 173, 186; in Latin America, 318

Jamaica Day, 142

Jamaica Gleaner, 137, 139

Jamaica National Building Society, 137

Jamaican Creole, 140, 141, 142. *See also* Caribbean diaspora; Creole; Jamaican diaspora

Jamaican diaspora, 136–50, 160, 162, 164; in Canada, 137–139, 142, 146, 147; and cell phone usage, 24, 145, 147, 157; in Europe, 139, 140; and ICTs, 138, 139–41, 159; and Indian hairstyle, 91; heritage of, 138, 139; and transnationalism, 147. *See also* Caribbean diaspora

Jamaican Labour Party, 137

Jamaica Observer, 139

Japanese diaspora, in Brazil, 173, 174, 175, 186

Jewish diaspora, 2, 3, 4, 278; and the Exodus, 87

Josu Lariz Askatu Campaign, 345–46. *See also* Basque diaspora

Kadeer, Rebiya, 311

Kingston, 136, 143

Kodak, 255

Kodak Gallery, 146, 255

Knowledge: and assimilation, 41; communal, 67; economy, 231; industry, 248; religious, 286; Society, 5; and technology, 7, 8, 31, 34, 35, 56, 60, 125; transfer of, 6, 39; unequal distribution of, 194, 201,

206. See also Digital divide; Information Society
Kpelle, 86

Labor diaspora, 3. See also Diaspora
La Charamusca (Lacharamusca.net), 196. See also Salvadoran diaspora
La Prensa Gráfica, 197
Lebanese Broadcasting Corporation International (LBCI; Lebanon), 266, 269, 276
Lebanese diaspora, 3; in Brazil, 174; refugees in Berlin, 270–71; and Satellite TV, 266, 275, 276. See also Arab diaspora
Lebanon war, 271. See also War
"Leisure-time ethnicity," 341. See also Ethnicity
Lenovo, 255, 260
Li, Robin (Li Hongyan), 242–43
Liberia, 86
Licklider, Joseph C. R., 10, 11
LifeJournal, 219
Linux communities, 217
Locality, as synchronicity, 9
London Galician Centre, 327. See also Galician diaspora
Long-distance nationalism, 233; and Internet, the role of, 234–35, 237–38; and overseas Chinese, 237. See also "McArabism" (virtual nationalism); Nationalism; Transnationalism

Mafia Gallega, 331. See also Galician diaspora
"McArabism" (virtual nationalism): and components of, 268–70, 287–89, *288*; concept of, 268–83. See Long-distance nationalism; Nationalism
Meskerem.com, 127
Middle East Broadcasting Centre (MBC) (Saudi Arabia), 266, 269, 276, 277
Migration: international, causes of, 4, 5; quantification of, 6. See also ICTs; Diaspora, Social network
Mitxelena, Koldo, 77

Mobile communication, 5, 20, 33. See also ICTs
Mobile phone. See Cellular (cell) phone
Monajah TV, 275, 281
MOOS (Multiuser Domains Object Oriented), 217; and Linguamoo, 217; and PMCMOO, 217
MTV (Lebanon), 266, 276
MUDS (Multiuser Domains), 217
Multimedia, 62, 63, 206, 338, 342; industry, 250, 254–56
MySpace, 74, 104, 140, 176, 199, 219

Nairaland.com, 96, 99, *100*, 100, 104, 105, 106
Nanjing Massacre, 228
NASDAQ, 242, 248, 249, 254
Nationalism: and diasporas, 72; and globalization, 8, 9, 69; and identity, 129, 174, 186, 196, 232, 288, 292, 283, 340. See also Long-distance nationalism; "McArabism" (virtual nationalism)
Navarre, Foral Community of, 76, 340
Negritude, 86
Netizens, 50, 305; religious netizens, 52
"New Pan-Arabism." See "McArabism" (virtual nationalism)
Newspapers, 7, 9, 59, 62, 78, 139, *158*, 162, *163*, 192, 193, 197, 198, 210, 227, 269, 272, 296
New transnational media, 265; and diasporas, 267
New TV (Lebanon), 266
New Yorker, 88
New World, 112, 172, 173, 174, 346
Nigeria, 94–106
Nigerian diaspora, 94. See also African diaspora
"Nigerian scams." See "419" (Nigerian legal code)
Nigeriaworld.com, 96, *96*, 97, 100, 105, 106
Ning, 176
Nova Sintra, 113

Office of Disaster Preparedness and Emergency Management, 137
Open Society Institute's Central Eurasia Project in New York, 301
Orbit TV, 266
Orkut, 176–77, 181–82, 219, 222; and demographic data of, 177; and digital diasporas, 181; and *orkontros,* 181, 185; and social network theories, 183–85; and social relationships of, 178; users of, 177. *See also* Brazilian digital diaspora; Social network
"Ourense the Best," 332

Palestinian diaspora, 3; and refugees of, in Berlin, 270–71; in the U.K., 267. *See also* Arab diaspora
Palestinian Intifida, 269, 275, 277, 278, 282, 283, 284, 287
Pan-Africanism. *See* African Diaspora
"Pan-Arabism." *See* "McArabism" (virtual nationalism)
Panethnopolis, 58, 60–62
Paraná Galician Club-Spanish Centre, 326
People's National Party, 137
People's Republic of China (PRC), 225, 226; and *xin yimin* (new emigrants), 226
People's Telecom, 137
PFDJ (People's Front for Democracy and Justice), 127, 130, 133, 134
Picasso, Pablo, and *Guernica,* 341, 345. *See also* Gernika (Bizkaia, Spain)
PNV (Basque Nationalist Party), 342
Portmore, 143
Portugal: and Galician migration, 173, 177; and West African colonies, 111, 112. *See also* Galician diaspora
Portuguese diaspora, 111, 112, 116–19; in Brazil, 173. *See also* Cape Verdean diaspora
Portuguese language, 317
Power 106 FM, 137, 138
Pravasi Bharatiya Divas (Overseas Indian Day), 213

Public Broadcasting System, 85
Public good, 40, 42, 68, 69. *See also* Commons
Public sphere: and Chinese cultural sphere, 230, 237; diasporic, 229, 233; and the Internet, 236–37; transnational, 191, 193, 205, 229–30, 288

Quino, 320, 337

Race, 6, 8, 51, 53, 89, 90, 110–17, 137, 158, 171, 192, 230, 243, 320
Racism, 89, 90, 117, 153
Radio, 6, 7, 8, 9, 78, 137, 139, *158,* 162, *163,* 192, 193, 196, 197, 202, 210, 227, 296, 301, 304. 328; and Internet, 138
Radio America, 197
Radio Free Asia, 304
Radio Free Europe, 301, 304
Radio La Campeona, 197
Radio Liberty, 301, 304
Radio Pipiles, 197
Ramadan, 34, 280–81
Rastafarians, 86, 148, *158*
"Real" community, 214–16. *See also* Virtual community
Religion, 114, 182; and migration, 192; and new technologies, 66, 70; online, 52. *See also* Islam *under* Arab diaspora; Uyghur diaspora
Rodríguez Castelao, Alfonso Daniel, 330
Ross, Brian, 94, 95
Russia, 7, 294, 304, *307*

Salvadoran civil war, 191
Salvadoran diaspora: history of, 191; and online chats, 196, 198, 201, 202, 203; and reasons for immigration, 191–92; settlement of, in Canada, 198, 204; settlement of, in Honduras, 191; settlement of, in U.S., 191–92; teleconferencing of, 203–4; usage of cellular phones of, 201–3; usage of video usage of Internet and ICTs of, 191, 195–97, 204–5

Salvadoran Diaspora in Canada, 198
Salvadoran Honduras war, 191. *See also* War
Salvadoreños en el Mundo (Salvadoreans Around the World), 196
Salvadorenosenlinea.com, 196
San Francisco, 48; Mission District of, 52, 58
SARS epidemic, 234; crisis of, in Singapore, 234–35
Satellite communication, 8, 193, 299, 319. *See also* ICTs
Satellite phone, 202. *See also* Cellular (cell) phone; ICTs
Satellite TV, 2, 127, 227, 265, 266, 267, 269, 270, 272–77, 279–81, 283–89. *See also* Arab diaspora; ICTs
São Lourenço, 113
São Vicente, 113, 120
Scamming, online. *See* "419" (Nigerian legal code)
ScanStone (software), 71
Second Life (3D digital environment), 10, 74, 209, 213, 217–20, 222–23; and Bollywood Island, 218; and crowdsourcing and outsourcing, 222; Euskarian Etxea (Home of the Euskarians), 77; and Indian avatars, 221–22; and Indianness on, 218–19; and Indian-theme dance clubs, 220–21. *See also* Bollywood; Indian diaspora; South Asian diaspora(s)
Senegalese immigrants: and merchants in Cape Verde, 118; in Spain, 24, 30, 31, 34, 35. *See also* African diaspora
September 11, 267, 282, 284. *See also* United States (U.S.)
Shanghai Information Industry Base (Beijing, China), 252
Shutterfly, 146
Sikh diaspora, 3
Silicon Valley, 49, 60, 249, 254, 256, 258, 260, 262; high-tech industry, 57. *See also* Technopolis; United States (U.S.)
Simpson-Miller, Portia, 136, 137
6 de México, 346. *See also* Basque diaspora

Skype, 203
SMS, 20, 26. *See also* Cellular (cell) phone
Social network: definition of, 175–76; and international migration, 5, 42, 60, 176; and *Mathew Effect*, 178; and social relationships of, 122, 182; and theories of, 183–85; and territory, 180–81; and virtual or digital social networks, 8, 103, 113, 140, 146, 147, 171, 176. *See also* Migration: international
Software, free (libre), 68, 74
Sohu.com, 248
South Asian diaspora(s), 211, 213; colonial/postcolonial discourses of, 211–13; conceptualization of, 213; and gender, 213, 219; and nonresident Indian, 212; and Overseas Indian, 211; queer and transgender, 219–20. *See also* Indian diaspora
South Korea, 7, 307
Soviet Union, 294, 296, 298, 300. *See also* Uyghur diaspora
Spain, 2, 6, 8, 9, 19, 30, 31, 76, 173, 186, 191, 192, 317–18, 320–21, 323–24, 326, 328, 332, 334, 344
Spanish Civil War, 2, 76, 322, 344, 346. *See also* Franco, Francisco
Spanish diaspora, 3, 318, 323; in Brazil, 173
Spanish language, 58, 317, 320, 326, 327, 331; and radio stations in the U.S., 197 (*see also* Salvadoran diaspora)
Springdale (virtual community), 232, 233, 235, 237
The Star, 137, 139
The Straits Times, 232
"Symbolic ethnicity," 341. *See also* Ethnicity
Swahili, 87

Taiwan, 226, 234, 245, 246, 247, 252, 259, 261. *See also* Chinese diaspora; ICTs
Taylor, Robert, 10, 11
Technopolis, 52, 58, 59, 60. *See also* Silicon Valley

Telecommunication technologies, and diasporas, 7, 8, 45, 50, 142, 145, 250. *See also* ICTs; CMC
Telegram, 142, 210
Teleliban (Lebanon), 276
Telephone, 20, 24, *31*, 34, 36, 73, 129, 131, 137, 142, 210, 234, 310, 323; call shops, 32; centers, 20, 26; companies, 22; prepaid cards, 302. *See also* Cellular (cell) phone
Telepresence, 55, 73, 74
Tiananmen Square incident, 225, 228
Tibetans, 291, 310
Tigrinya, 125, 133
Time (magazine), 11
Tong, Chok Goh, 232, 235
Top5Jamaica, 147
Trade diaspora, 3. *See also* Diaspora
Transnational, communities, 4, 190, 201, 310; communication, 138, 145, 146, 191, 192, 195; culture, 160, 164; families, 35, 145, 146, 201–3; networking, 157; networks, 134, 199, 201, 265, 339. *See also* ICTs; Social network; Transnationalism
Transnationalism, definition of, 307. *See also* Long-distance nationalism; Uyghur diaspora
Transportation, revolution, 19, 24
Trinidadian diaspora, *159*, 159, 160, 162. *See also* Caribbean diaspora
Tsinghua University, incubator of, 253
Turkistan Newsletter, 300
Tütünçü, Mehmet, 300

Uighur Affairs Survey, 297
Union of East Turkestani Youth, 297
United States (U.S.): and Information Society, 23; Internet boom, 248; migration to, 6, 76, 86, 111–14, 137, 138, 142, 146, 153, 155, 156, 160, 161, 175, 191, 192, 195–200, 202–4, 211–13, 228, 229, 232, 242, 243, 244, 245, 246, 247, 250, 300, 341; political hegemony of, 75, 235; religious fundamentalism, 89; skilled professionals in, 249. *See also* Chinese migrants; Chinese returnee entrepreneurs; "419" (Nigerian legal code)
UTstar.com, 248
Uyghur American Association (UAA), 291, 302
Uyghur diaspora, 292; and CMC and computer-mediated communication (CMC), 299–303; and CMC conventional media, 308; and conventional media, 296–99; development of, 293–94; and digital/virtual transnationalism, 292, 307, 308, 311; dimension of, 294–95; and history of, 292; and impact of CMC on Uyghur nationalism, 303, 306, 307, 309–11; and Islam, 293; nationalism, 292–96, 298, 299, 307, 310, 311
Uyghur Human Rights Coalition, 302
Uyghur Human Rights Project, 302
Uyghur-L, 301
Uyghur Perspectives, 300
Uyghurs China relations, 291, 292, 295, 298

Venetian diaspora, 3
Vernacular culture, 67–73, 80, 222
Victim diasporas, 2, 3. *See also* Diaspora
Vimicro Corporation, 254, 255, 260
Virtual community: concept of, 10, 11, 50, 77, 111, 195, 209, 213–15, 233, 234, 283, 289; versus "real" community, 214–15, 217. *See also* "Real" community
"Virtual life," 10, 70, 308
Virtual nationalism. *See* "McArabism" (virtual nationalism)
Viva Galicia (Long Live Galicia), 332, 333
The Voice, 137

War, and forced migration, 3, 123, 125, 130, 266
Web 2.0, 8, 31, 217, 219. *See also* Social network
Webcam, immigrants' usage of, 19, 28, 29, 73, 202, 203

Web pages, 113, 151, 152, 156, 157, 158, 159, 162, 164, 165, 198, 234, 326. *See also* ICTs; Web sites

Web sites: as "communities of discourse," 94, 96; purposes of, 113, 124, 126; dissemination of information, 153; and cultural participation, 164. *See also* ICTs; Web pages

Wikipedia, 176, 221

Winfrey, Oprah, 85, 93–107; heritage of, 86, 87; Kpelle identity, 86; Zulu identity, 93. *See also* "419" (Nigerian legal code)

Wired, 74

World Uighur Network News, 302

World Uyghur Congress (WUC), 291, 297, 302, 310. *See also* Uyghur diaspora

World War II, 3, 85

Wright, Richard, 92

Xinjiang ("New Dominion" or Eastern Turkestan) 2009 riots, 291; and Uyghurs, 292

Xinjiang Uyghur Autonomous Region of the People's Republic of China, 292–93; and Internet usage in, 305, 306

Yoruba, 101

YouTube, 74, 140, 141, 199, 218–19

Zhongguancun International Incubator (ZGD) (Beijing, China), 252, 253, 256, 259, 260. *See also* Zhongguancun Science Park (Beijing, China)

Zhongguancun Science Park (Beijing, China), 243, 248, 251, 253. *See also* Zhongguancun International Incubator

Zulu (ethnic group), 85, 86, 93, 107. *See also* Winfrey, Oprah